Student's Solutions Manual

to accompany

College Algebra
Second Edition

Student's Solutions Manual

Edgar Reyes
Southeastern Louisiana University

to accompany

College Algebra
Second Edition

Mark Dugopolski

Reading, Massachusetts • Menlo Park, California • New York • Harlow, England
Don Mills, Ontario • Sydney • Mexico City • Madrid • Amsterdam

Reprinted with corrections, April 1999

Reproduced by Addison-Wesley from camera-ready copy supplied by the author.

Copyright © 1999 Addison Wesley Longman.

All rights reserved. No part of this publication may be reproduced, stored in a retrieval system, or transmitted, in any form or by any means, electronic, mechanical, photocopying, recording, or otherwise, without the prior written permission of the publisher. Printed in the United States of America.

ISBN   0-201-38392-6

4 5 6 7 8 9 10 VG01

TABLE OF CONTENTS

Chapter P	Prerequisites	1
Chapter 1	Equations and Inequalities	14
Chapter 2	Functions and Graphs	39
Chapter 3	Polynomial and Rational Functions	72
Chapter 4	Exponential and Logarithmic Functions	117
Chapter 5	Systems of Equations and Inequalities	140
Chapter 6	Matrices and Determinants	176
Chapter 7	The Conic Sections	205
Chapter 8	Sequences, Series, and Probability	229

REAL NUMBERS AND THEIR PROPERTIES

For Thought

1. False, 0 is not an irrational number.
2. True 3. True
4. False, since 0 has no multiplicative inverse.
5. True
6. False, since the statement fails when $a = w = z$.
7. False, since $a - (b - c) = a - b + c$.
8. False, the distance is $|a - b|$.
9. False, a TI-92 can give an exact answer like π.
10. False, the opposite of $a + b$ is $-a - b$.

P.1 Exercises

1. True 3. True
5. True 7. True
9. All
11. $\{-\sqrt{2}, \sqrt{3}, \pi, 5.090090009...\}$
13. $\{0, 1\}$ 15. $x + 7$
17. $5x + 15$ 19. $5(x + 1)$
21. $(-13 + 4) + x$ 23. $\dfrac{1}{0.125} = 8$
25. $\sqrt{3}$ 27. $y^2 - x^2$
29. 7.2 31. $\sqrt{5}$
33. $|13.5 - 8| = 5.5$ 35. $|17 - (-5)| = 22$
37. $\dfrac{4}{5} \cdot 18 = \dfrac{72}{5}$ 39. $4 - 42 = -38$
41. 52 43. $49 \cdot (6 + 4) = 49 \cdot 10 = 490$
45. $\dfrac{26}{5} \cdot \dfrac{2}{1} \cdot 5 = 52$ 47. 1
49. $1.6 - 5 = -3.4$ 51. 143
53. $2 - 3|3 - 24| = 2 - 3 \cdot 21 = -61$
55. $(2 - 4)(4 + 3) = -14$ 57. $\dfrac{4}{4} = 1$
59. $(598 + 432)(0) = 0$

61. $-2x$ 63. $0.85x$
65. $-6xy$ 67. $-24xw$
69. $-3x - 6$ 71. $0.97x - 6$
73. $3 - 4zy - 5 + 6zy = 2zy - 2$
75. $\dfrac{4}{12}x + \dfrac{3}{12}x = \dfrac{7x}{12}$ 77. $\dfrac{3xy}{4}$
79. $3 - 2x$ 81. $\dfrac{6x}{2} - \dfrac{2y}{2} = 3x - y$
83. $\dfrac{-3}{2} + \dfrac{3}{4}y + \dfrac{9}{2} - 2y = 3 - \dfrac{5y}{4}$
85. $\dfrac{-7 - 5}{3 - (-2)} = -\dfrac{12}{5}$ 87. $\dfrac{3 - 3}{5 - (-8)} = 0$
89. $-\dfrac{1}{2}, -\dfrac{5}{12}, -\dfrac{1}{3}, 0, \dfrac{1}{3}, \dfrac{5}{12}, \dfrac{1}{2}$
91. In 1945-85, Americans consumed at least 175 packs per year. The highest cigarette consumption per person was in 1964.
93. $132 - 0.6a + 0.4r$
95. Cars 3 and 4 since $\dfrac{27}{260} > 0.1$ and $\dfrac{35}{342} > 0.1$
97. The product of your number and 9 is one of 18, 27, 36, 45, 54, 63, 72, 81. The sum of the digits is always 9 and the letter corresponding to 4 is D. The state is Delaware and the color of an elephant is gray. We are thinking of a big gray elephant.
99. Let x (in quarts) be the amount of nightshade she pours into 1 quart of snakeroot. After mixing and ending up with 1 quart of each mixture, the amount of nightshade in the snakeroot is $x - x\dfrac{x}{x+1} = \dfrac{x}{x+1}$ while the amount of snakeroot in nightshade $x\dfrac{1}{x+1} = \dfrac{x}{x+1}$. Thus, amount of nightshade in the snakeroot is exactly the same as the amount of snakeroot in nightshade.
101. First, locate a length of $\sqrt{2}$ units as in Exercise 100. In a right triangle whose shorter sides have length 1 unit and $\sqrt{2}$ units, the hypotenuse has a length of $c = \sqrt{1^2 + \sqrt{2}^2} = \sqrt{3}$.

For Thought

1. True, since $\frac{1}{2} + \frac{1}{2} = 1$.

2. True, since $2^{100} = (2^2)^{50}$.

3. True, since $9^{16} = (9^2)^8 = 81^8$.

4. True

5. False, since $\frac{5^{10}}{5^{-12}} = 5^{22}$.

6. False, since $2 \cdot 2 \cdot 2 \cdot 2^{-1} = 4$.

7. True

8. True

9. False, since $10^{-4} = 0.0001$

10. False, since $98.6 \times 10^8 = 9.86 \times 10^9$.

P.2 Exercises

1. 64   3. -16 5. $\frac{1}{8}$

7. $49 - 36 = 13$ 9. $25 + (5)^2 = 50$

11. $4 - 9 = -5$ 13. $(-5)(1) = -5$

15. $(-5)(4 - 6 + 9) = -35$ 17. $(-8) + (-64) = -72$

19. $\frac{1}{3^4} = \frac{1}{81}$ 21. $5^2 = 25$

23. $\frac{1}{6} + \frac{1}{5} = \frac{5}{30} + \frac{6}{30} = \frac{11}{30}$

25. $\frac{6^3}{3^2} = 24$ 27. $2^3 = 8$ 29. $64 \cdot \frac{1}{4} = 16$

31. $5 \cdot 1000 + 3 \cdot 100 + 60 = 5360$

33. $-6x^{11}y^{11}$ 35. $25 \cdot 9 = 225$

37. $-4x^6$ 39. $\frac{(-2)^3(x^2)^3}{27} = \frac{-8x^6}{27}$

41. $3x^4$ 43. $\frac{25}{y^4}$

45. $\frac{1}{6}x^0y^{-3} = \frac{1}{6y^3}$ 47. $\frac{n^2}{2}$

49. $8a^6 + 9a^6 = 17a^6$

51. $-3\frac{1}{9}(x^{-1})^{-2}(y^3)^{-2} = -\frac{x^2}{3y^6}$

53. $\left(\frac{-2x^5}{3y^2}\right)^4 = \frac{16x^{20}}{81y^8}$

55. $(x^{3b-3})(x^{8-2b}) = x^{b+5}$

57. $-125a^{6t}b^{-9t} = -125\frac{a^{6t}}{b^{9t}}$ 59. $\frac{-3y^{6v}}{2x^{5w}}$

61. $a^{-3s+15}b^{6t-27}$ 63. $43,000$

65. -0.0000356 67. 5×10^6

69. -6.72×10^{-5} 71. 0.000000007

73. -2×10^{10} 75. $20 \times 10^{15} = 2 \times 10^{16}$

77. $2 \times 10^{-6+3} = 2 \times 10^{-3}$

79. $40 \times 10^{-30} = 4 \times 10^{-29}$

81. $\frac{(8 \times 10^{18})(5 \times 10^{-6})}{(4 \times 10^{-10})} = \frac{40 \times 10^{12}}{4 \times 10^{-10}} = 1 \times 10^{23}$

83. -9.936×10^{-5}

85. $\approx \frac{(178.45 \times 10^{-18})(0.00666 \times 10^{-28})}{3.14(79.21 \times 10^{-8})} \approx \frac{1.188 \times 10^{-46}}{248.72 \times 10^{-8}} \approx 4.78 \times 10^{-41}$

87. $(2.7 \times 10^9) + (0.000036) \approx 2.7 \times 10^9$

89. BMI $= \frac{703(250)}{78^2} \approx 29$.

91. Since 1 ton = 2000 pounds, the ratio is
    $D = \frac{2200(2000)175^{-3}10^6}{2240} \approx 367$.

93. Each person owes $\frac{1.3 \times 10^{12}}{2.53 \times 10^8} = \$5,138.34$

95. The number of seconds in 10 billion years is $60^2 \cdot 24 \cdot 365 \cdot 10^{10} = 3.1536 \times 10^{17}$. The number of tons of mass transformed by the sun in 10 billion years is 5 million $\times\, 3.1536 \times 10^{17} \approx 1.577 \times 10^{24}$.

97. The radius of the earth in km. is radius of the sun divided by 109.1 which is approximately 6379 km.

99. The number of times the mass of the sun is larger than that of the earth is mass of the sun divided by the mass of the earth which is 3.3×10^5.

101. The energy consumption of the U.S. is $0.24(3.52 \times 10^{17}) \approx 8.45 \times 10^{16}$ Btu.

For Thought

1. False; since $8^{-\frac{1}{3}} = \frac{1}{2}$.

2. True, since $16^{\frac{1}{4}} = 2 = 4^{\frac{1}{2}}$.

3. False, since $\sqrt{\frac{4}{6}} = \frac{2}{\sqrt{3}}$.   4. True

5. False, since $(-1)^{\frac{1}{2}}$ is not a real number.

6. False, since $\sqrt[3]{7^2} = 7^{\frac{2}{3}}$.   7. False, since $9^{\frac{1}{2}} = 3$.

8. True, since $\frac{1}{\sqrt{3}} \cdot \frac{\sqrt{3}}{\sqrt{3}} = \frac{\sqrt{3}}{3}$.

9. True, since $\frac{2^{\frac{1}{2}}}{2^{\frac{1}{3}}} = 2^{\frac{1}{2}-\frac{1}{3}} = 2^{\frac{1}{6}} = \sqrt[6]{2}$.

10. True, since $\sqrt[3]{7^5} = \sqrt[3]{7^3 \cdot 7^2} = 7\sqrt[3]{49}$.

P.3 Exercises

1. -3 3. 8 5. -4

7. 81 9. $(8^{1/3})^{-4} = \frac{1}{16}$

11. $|x|$ 13. a^3 15. a^2

17. $x^{3 \cdot (1/3)} y^{6 \cdot (1/3)} = xy^2$

19. $\sqrt{a^2}\sqrt{b^4} = |a|b^2$ 21. $x^2 y^{1/2}$

23. $6a^{1/2+1} = 6a^{3/2}$ 25. $3a^{1/2-1/3} = 3a^{1/6}$

27. $a^{2+1/3}b^{1/2-1/2} = a^{7/3}$

29. $(a^{2m}b^{3n})b^n = a^{2m}b^{4n}$

31. $\frac{x^2 y}{z^3}$ 33. 30 35. -2

37. -2 39. $\frac{\sqrt[3]{8}}{\sqrt[3]{1000}} = \frac{1}{5}$

41. $\sqrt[4]{(2^4)^3} = \sqrt[4]{2^{12}} = 2^{12 \cdot (1/4)} = 2^3 = 8$

43. $10^{2/3} = \sqrt[3]{10^2}$ 45. $\frac{3}{y^{3/5}} = \frac{3}{\sqrt[5]{y^3}}$ 47. $x^{-1/2}$

49. $(x^3)^{1/5} = x^{3/5}$ 51. $4x$

53. $2y^3$ 55. $\frac{\sqrt{xy}}{10}$

57. $\frac{-2a}{b^{15 \cdot (1/3)}} = -\frac{2a}{b^5}$ 59. $\sqrt{4(7)} = 2\sqrt{7}$

61. $\frac{1}{\sqrt{5}} \cdot \frac{\sqrt{5}}{\sqrt{5}} = \frac{\sqrt{5}}{5}$

63. $\frac{\sqrt{x}}{\sqrt{8}} \cdot \frac{\sqrt{2}}{\sqrt{2}} = \frac{\sqrt{2x}}{\sqrt{16}} = \frac{\sqrt{2x}}{4}$

65. $\sqrt[3]{8 \cdot 5} = 2\sqrt[3]{5}$

67. $\sqrt[3]{-250x^4} = \sqrt[3]{(-125)(2)x^3 x} = -5x\sqrt[3]{2x}$

69. $\sqrt[3]{\frac{1}{2}} \cdot \sqrt[3]{\frac{4}{4}} = \sqrt[3]{\frac{4}{8}} = \frac{\sqrt[3]{4}}{2}$

71. $2\sqrt{2} + 2\sqrt{5} - 2\sqrt{3}$

73. $-10\sqrt{18} = -30\sqrt{2}$ 75. $12(5a) = 60a$

77. $(-5)^2(3) = 75$

79. $\sqrt{\frac{9}{a^3}} = \frac{3}{a\sqrt{a}} \cdot \frac{\sqrt{a}}{\sqrt{a}} = \frac{3\sqrt{a}}{a^2}$

81. $\frac{5}{\sqrt{x}} \cdot \frac{\sqrt{x}}{\sqrt{x}} = \frac{5\sqrt{x}}{x}$

83. $2x\sqrt{5x} + 3x\sqrt{5x} = 5x\sqrt{5x}$

85. $\sqrt[6]{3^2}\sqrt[6]{2^3} = \sqrt[6]{72}$

87. $\sqrt[12]{3^4}\sqrt[12]{x^3} = \sqrt[12]{81x^3}$

89. $\sqrt[6]{(xy)^3}\sqrt[6]{(2xy)^2} = \sqrt[6]{x^3y^3}\sqrt[6]{4x^2y^2} = \sqrt[6]{4x^5y^5}$

91. $(7^{1/2})^{1/3} = \sqrt[6]{7}$

93. Since $E = \sqrt{\frac{2(25)(6000)}{140}} \approx 46.3$, the most economic order quantity is $E = 46$.

95. $S = \frac{16(42,700)}{[(2200)(2000)]^{2/3}} \approx 25.4$

97. The depreciation rate is $r = 1 - \left(\frac{200}{5000}\right)^{1/5} \approx 0.47469$ or 47.5% per year.

99. $D = \sqrt{4^2 + 6^2 + 12^2} = 14$

101. $A = \sqrt{12(6)(5)(1)} = \sqrt{360} \approx 19.0$ ft^2.

103. Each expression is equivalent to $t^{4/5}$ except for $\sqrt[4]{t^5}$, the expression in (b).

105. No, since $\sqrt{9+16} = 5 \neq 7 = \sqrt{9} + \sqrt{16}$.

For Thought

1. False, since polynomials do not have negative exponents.   2. False, since its degree is 4.

3. False, since $(8)^2 \neq 9 + 25$.

4. True, since $50^2 - 1^2 = 2499$.

5. False, since $(x+3)^2 = x^2 + 6x + 9$.

6. False, since the left-hand side is $-x - 6$.

7. False, since $(a+b)^3 = a^3 + 3a^2b + 3ab^2 + b^3$.

8. False, since divisor times quotient plus the remainder is equal to the dividend.

9. False, since the equation is not defined at $x = -2$.

10. True, since $P(17) + Q(17) = 2310 = S(17)$.

P.4 Exercises

1. Degree 3 polynomial with leading coefficient 1

3. It is not a polynomial.

5. Degree 0 polynomial with leading coefficient 79

7. $8x^2 + 3x - 1$

9. $4x^2 - 3x - 9x^2 + 4x - 3 = -5x^2 + x - 3$

11. $4ax^3 - a^2x - 5a^2x^3 + 3a^2x - 3 = (-5a^2 + 4a)x^3 + 2a^2x - 3$

13. $2x - 1$ 15. $3x^2 - 3x - 6$

17. $-18a^5 + 15a^4 - 6a^3$

19. $(3b^2 - 5b + 2)b - (3b^2 - 5b + 2)3 =$
 $3b^3 - 5b^2 + 2b - 9b^2 + 15b - 6 =$
 $3b^3 - 14b^2 + 17b - 6$

21. $2x(4x^2 + 2x + 1) - (4x^2 + 2x + 1) =$
 $8x^3 + 4x^2 + 2x - 4x^2 - 2x - 1 =$
 $8x^3 - 1$

23. $(x-4)z + (x-4)3 = xz - 4z + 3x - 12$

25. $a(a^2 + ab + b^2) - b(a^2 + ab + b^2) =$
 $a^3 + a^2b + ab^2 - a^2b - ab^2 - b^3 = a^3 - b^3$

27. $a^2 - 2a + 9a - 18 = a^2 + 7a - 18$

29. $2y^2 + 18y - 3y - 27 = 2y^2 + 15y - 27$

31. $4x^2 + 18x - 18x - 81 = 4x^2 - 81$

33. $4x^2 + 10x + 10x + 25 = 4x^2 + 20x + 25$

35. $6x^4 + 10x^2 + 12x^2 + 20 = 6x^4 + 22x^2 + 20$

37. $(1+\sqrt{2})(3+\sqrt{2}) = 3 + \sqrt{2} + 3\sqrt{2} + 2 = 5 + 4\sqrt{2}$

39. $20 - 15\sqrt{2} + 4\sqrt{2} - 3\sqrt{2}\sqrt{2} = 14 - 11\sqrt{2}$

41. $6\sqrt{2}\sqrt{2} - 3\sqrt{6} + 2\sqrt{6} - \sqrt{3}\sqrt{3} = 9 - \sqrt{6}$

43. $\sqrt{5}\sqrt{5} + \sqrt{15} + \sqrt{15} + \sqrt{3}\sqrt{3} = 8 + 2\sqrt{15}$

45. $4 + 2\sqrt{x} + 2\sqrt{x} + \sqrt{x}\sqrt{x} = 4 + 4\sqrt{x} + x$

47. $(3x)^2 + 2(3x)(5) + (5)^2 = 9x^2 + 30x + 25$

49. $(x^n)^2 - 3^2 = x^{2n} - 9$

51. $(\sqrt{2})^2 - 5^2 = -23$

53. $(3\sqrt{6})^2 - 2(3\sqrt{6})(1) + 1 = 55 - 6\sqrt{6}$

55. $(3x^3)^2 - 2(3x^3)(4) + 4^2 = 9x^6 - 24x^3 + 16$

57. $\dfrac{\sqrt{10}}{\sqrt{5}-2} \cdot \dfrac{\sqrt{5}+2}{\sqrt{5}+2} = \dfrac{\sqrt{50}+2\sqrt{10}}{5-4} = 5\sqrt{2} + 2\sqrt{10}$

59. $\dfrac{\sqrt{6}}{6+\sqrt{3}} \cdot \dfrac{6-\sqrt{3}}{6-\sqrt{3}} = \dfrac{6\sqrt{6}-3\sqrt{2}}{33} = \dfrac{2\sqrt{6}-\sqrt{2}}{11}$

61. $\dfrac{36x^6}{-4x^3} = -9x^3$

63. $\dfrac{3x^2}{-3x} + \dfrac{6x}{3x} = -x + 2$

65. Factor & simplify: $\dfrac{(x+3)^2}{x+3} = x+3$

67. Factor & simplify: $\dfrac{(a-1)(a^2+a+1)}{a-1} = a^2+a+1$

69. Quotient $x+5$, remainder 13 since

$$\begin{array}{r} x+5 \\ x-2\overline{\smash{\big)}\,x^2+3x+3} \\ \underline{x^2-2x} \\ 5x+3 \\ \underline{5x-10} \\ 13 \end{array}$$

71. Quotient $2x - 6$, remainder 13 since
$$\begin{array}{r} 2x - 6 \\ x + 3 \overline{\smash{)}2x^2 + 0x - 5} \\ \underline{2x^2 + 6x} \\ -6x - 5 \\ \underline{-6x - 18} \\ 13 \end{array}$$

73. Quotient $x^2 + x + 1$, remainder 0 since
$$\begin{array}{r} x^2 + x + 1 \\ x - 3 \overline{\smash{)}x^3 - 2x^2 - 2x - 3} \\ \underline{x^3 - 3x^2} \\ x^2 - 2x \\ \underline{x^2 - 3x} \\ x - 3 \\ \underline{x - 3} \\ 0 \end{array}$$

75. Quotient $ab + b$, remainder 0 since
$$\begin{array}{r} ab + b \\ a + 3 \overline{\smash{)}a^2b + 4ab + 3b} \\ \underline{a^2b + 3ab} \\ ab + 3b \\ \underline{ab + 3b} \\ 0 \end{array}$$

77. $x + 1 + \dfrac{-1}{x - 1}$ since
$$\begin{array}{r} x + 1 \\ x - 1 \overline{\smash{)}x^2 + 0x - 2} \\ \underline{x^2 - x} \\ x - 2 \\ \underline{x - 1} \\ -1 \end{array}$$

79. $\dfrac{2x^2}{x} - \dfrac{3x}{x} + \dfrac{1}{x} = 2x - 3 + \dfrac{1}{x}$

81. $\dfrac{x^2}{x} + \dfrac{x}{x} + \dfrac{1}{x} = x + 1 + \dfrac{1}{x}$

83. $x + \dfrac{1}{x - 2}$ since
$$\begin{array}{r} x \\ x - 2 \overline{\smash{)}x^2 - 2x + 1} \\ \underline{x^2 - 2x} \\ 1 \end{array}$$

85. $P(-2) = 4 + 6 + 2 = 12$

87. $P(10) = 100 - 30 + 2 = 72$

89. $M(-3) = 27 + 45 + 3 + 2 = 77$

91. $M\left(\dfrac{1}{3}\right) = -\left(\dfrac{1}{27}\right) + 5 \cdot \dfrac{1}{9} - \dfrac{1}{3} + 2 =$
$-\dfrac{1}{27} + 5 \cdot \dfrac{3}{27} - \dfrac{9}{27} + \dfrac{54}{27} = \dfrac{59}{27}$

93. $x^2 + 2x - 24$ **95.** $2a^{10} - 3a^5 - 27$

97. $-y - 9$ **99.** $6a - 6$

101. $w^2 + 8w + 16$ **103.** $16x^2 - 81$

105. $3y^5 - 9xy^2$ **107.** $2b - 1$

109. $x^6 - 64$ **111.** $9w^4 - 12w^2n + 4n^2$ **113.** 1

115. The area is $(x + 3)(2x - 1) = 2x^2 + 5x - 3$.

117. Since the volume divided by the height is the area and
$$\begin{array}{r} x^2 + 5x + 6 \\ x + 4 \overline{\smash{)}x^3 + 9x^2 + 26x + 24} \\ \underline{x^3 + 4x^2} \\ 5x^2 + 26x \\ \underline{5x^2 + 20x} \\ 6x + 24 \\ \underline{6x + 24} \\ 0 \end{array}$$
then the area of the bottom is $x^2 + 5x + 6$.

119. The dimensions of the rectangular habitat are $6 - 2x$ and $4 - 2x$. The area is the product of the dimensions which is $4x^2 - 20x + 24$.

121. After 3 years, the higher yielding account would have 45 cents more $(A(.1724) - A(.0502) \approx .45)$.

123. The fine is $F(25) = 500 + 200(25) = \$5,500$.

125. The NHL total revenue is $R = 38(1.85 \times 10^7) + 10^9 + (8.5 \times 10^7) = \$1.788$ billion dollars. The percentage of total revenue coming from the broadcasts is $\dfrac{8.5 \times 10^7}{1.788 \times 10^9}(100) \approx 4.75\%$.

127. Let x be the length of a side of the square. The areas of the square and rectangle are x^2 and $(x + 10)(x - 10) = x^2 - 100$, respectively. He had 100 ft² less area than he thought.

129. Let $x \gg 0$ represent a sufficiently large number. If $x \gg 0$, then the sign of $P(x)$ is the same as the sign of the leading coefficient. If $x < 0$, $|x| \gg 0$, and the degree is even, then the sign of $P(x)$ is the same as the sign of the leading coefficient. If $x < 0$, $|x| \gg 0$, and the degree is odd, then the sign of $P(x)$ is opposite the sign of the leading coefficient.

For Thought

1. False, since it factors as $(x+8)(x-2)$.

2. True   3. True

4. False, since $a^2 - 1 = (a-1)(a+1)$.

5. False, since $a^3 - 1 = (a-1)(a^2+a+1)$.

6. False, since $x^3 - y^3 = (x-y)(x^2+xy+y^2)$.

7. False, it factors as $(2a+3b)(4a^2-6ab+9b^2)$.

8. False, since $(x+2)(x-6) = x^2 - 4x - 12$.

9. True   10. True

P.5 Exercises

1. $6x^2(x-2), -6x^2(-x+2)$

3. $4a(1-2b), -4a(2b-1)$

5. $ax(-x^2+5x-5), -ax(x^2-5x+5)$

7. $1(m-n), -1(n-m)$

9. $x^2(x+2) + 5(x+2) = (x^2+5)(x+2)$

11. $y^2(y-1) - 3(y-1) = (y^2-3)(y-1)$

13. $ady + d - awy - w = d(ay+1) - w(ay+1) = (ay+1)(d-w)$

15. $x^2y^2 - ay^2 - (bx^2 - ab) = y^2(x^2-a) - b(x^2-a) = (x^2-a)(y^2-b)$

17. $(x+2)(x+8)$ 19. $(x-6)(x+2)$

21. $(m-2)(m-10)$ 23. $(t-7)(t+12)$

25. $(2x+1)(x-4)$ 27. $(4x+1)(2x-3)$

29. $(3y+5)(2y-1)$ 31. $(t-u)(t+u)$

33. $(t+1)^2$ 35. $(2w-1)^2$

37. $(y^{2t}-5)(y^{2t}+5)$ 39. $(3zx+4)^2$

41. $(t-u)(t^2+ut+u^2)$ 43. $(a-2)(a^2+2a+4)$

45. $(3y+2)(9y^2-6y+4)$

47. $(3xy^2)^3 - (2z^3)^3 = (3xy^2 - 2z^3)(9x^2y^4 + 6xy^2z^3 + 4z^6)$

49. Replace y^3 by w and y^6 by w^2 and get $w^2 + 10w + 25 = (w+5)^2 = (y^3+5)^2$.

51. Replace $2a^2b^4$ by w and $4a^4b^8$ by w^2 and get $w^2 - 4w - 5 = (w+1)(w-5) = (2a^2b^4+1)(2a^2b^4-5)$.

53. Replace $(2a+1)$ by w and $(2a+1)^2$ by w^2 and get $w^2 + 2w - 24 = (w+6)(w-4) = ((2a+1)+6)((2a+1)-4) = (2a+7)(2a-3)$.

55. Replace (b^2+2) by w and $(b^2+2)^2$ by w^2 and get $w^2 - 5w + 4 = (w-4)(w-1) = ((b^2+2)-4)((b^2+2)-1) = (b^2-2)(b^2+1)$.

57. $-3x^3 + 27x = -3x(x^2-9) = -3x(x-3)(x+3)$

59. $2t(8t^3 + 27w^3) = 2t(2t+3w)(4t^2-6tw+9w^2)$

61. $a^3 + a^2 - 4a - 4 = a^2(a+1) - 4(a+1) = (a+1)(a^2-4) = (a+1)(a+2)(a-2)$

63. $x^4 - 2x^3 - 8x + 16 = x^3(x-2) - 8(x-2) = (x-2)(x^3-8) = (x-2)^2(x^2+2x+4)$

65. $-2x(18x^2 - 9x - 2) = -2x(6x+1)(3x-2)$

67. $a^7 - a^6 - 64a + 64 = a^6(a-1) - 64(a-1) = (a^6-64)(a-1) = (a^3-8)(a^3+8)(a-1) = (a-2)(a^2+2a+4)(a+2)(a^2-2a+4)(a-1)$

69. $-(3x+5)(2x-3)$

71. Replace (a^2+2) by w and $(a^2+2)^2$ by w^2. $w^2 - 4w + 3 = (w-3)(w-1) = ((a^2+2)-3)((a^2+2)-1) = (a^2-1)(a^2+1) = (a+1)(a-1)(a^2+1)$

73. Yes, since $1261 = 13(97)$ and

```
        97
   13 )1261
       117
        91
        91
         0
```

75. Yes, since

$$\begin{array}{r} x^2 + x + 1 \\ x+3 \overline{\smash{\big)}x^3 + 4x^2 + 4x + 3} \\ \underline{x^3 + 3x^2} \\ x^2 + 4x \\ \underline{x^2 + 3x} \\ x + 3 \\ \underline{x + 3} \\ 0 \end{array}$$

and $x^3 + 4x^2 + 4x + 3 = (x+3)(x^2 + x + 1)$.

77. No, since

$$\begin{array}{r} 3x^2 + 8x - 4 \\ x-1 \overline{\smash{\big)}3x^3 + 5x^2 - 12x - 9} \\ \underline{3x^3 - 3x^2} \\ 8x^2 - 12x \\ \underline{8x^2 - 8x} \\ -4x - 9 \\ \underline{-4x + 4} \\ -13 \end{array}$$

and the remainder is not 0.

79.
$$\begin{array}{r} x^2 + 5x + 6 \\ x-1 \overline{\smash{\big)}x^3 + 4x^2 + x - 6} \\ \underline{x^3 - x^2} \\ 5x^2 + x \\ \underline{5x^2 - 5x} \\ 6x - 6 \\ \underline{6x - 6} \\ 0 \end{array}$$

So $x^3 + 4x^2 + x - 6 = (x-1)(x^2 + 5x + 6) = (x-1)(x+3)(x+2)$.

81.
$$\begin{array}{r} x^2 + 2x + 2 \\ x-3 \overline{\smash{\big)}x^3 - x^2 - 4x - 6} \\ \underline{x^3 - 3x^2} \\ 2x^2 - 4x \\ \underline{2x^2 - 6x} \\ 2x - 6 \\ \underline{2x - 6} \\ 0 \end{array}$$

So $x^3 - x^2 - 4x - 6 = (x-3)(x^2 + 2x + 2)$.

83.
$$\begin{array}{r} x^3 + 3x^2 - x - 3 \\ x+2 \overline{\smash{\big)}x^4 + 5x^3 + 5x^2 - 5x - 6} \\ \underline{x^4 + 2x^3} \\ 3x^3 + 5x^2 \\ \underline{3x^3 + 6x^2} \\ -x^2 - 5x \\ \underline{-x^2 - 2x} \\ -3x - 6 \\ \underline{-3x - 6} \\ 0 \end{array}$$

So
$$x^4 + 5x^3 + 5x^2 - 5x - 6 =$$
$$(x+2)(x^3 + 3x^2 - x - 3) =$$
$$(x+2)(x^2(x+3) - (x+3)) =$$
$$(x+2)(x^2 - 1)(x+3) =$$
$$(x+2)(x-1)(x+1)(x+3) =$$

85. $(x^m + 1)(x^m - 1)$

87. $(x^q - 3)(x^q - 2)$

89. $(x^n - 1)(x^{2n} + x^n + 1)$

91. $x^{w+1}(x^w + 1) + 2(x^w + 1) = (x^w + 1)(x^{w+1} + 2)$

93. The area of the bottom is the volume divided by the height.

$$\begin{array}{r} x^2 + x + 1 \\ x-1 \overline{\smash{\big)}x^3 + 0x^2 + 0x - 1} \\ \underline{x^3 - x^2} \\ x^2 + 0x \\ \underline{x^2 - x} \\ x - 1 \\ \underline{x - 1} \\ 0 \end{array}$$

So $x^2 + x + 1$ is the area of the bottom.

95. A formula for the volume is $V = x(6 - 2x)(7 - 2x)$. Among the tabulated values,

x	0.5	1	2
$Vol.$	15	20	12

$x = 1$ in. produces the largest volume.

97. Part (b) is not a perfect square trinomial since $(\sqrt{1000}a - b)^2 \neq 1000a^2 - 200b + b^2$. The others are perfect squares.

99. No, since $1^3 + 1^3 = 2 \neq n^3$ for any integer n.

For Thought

1. False, since 2 is not factor.   2. True

3. False, since the first expression is not defined at $x=0$.

4. False, since $\dfrac{(a-b)(a+b)}{a-b} = a+b$.

5. False, since the LCD is $x(x+1)$.   6. True

7. True   8. True

9. True, since $\dfrac{2(500)+1}{500-3} = \dfrac{1001}{497} \approx 2.014$

10. True, since $\dfrac{5x+1}{x} \approx \dfrac{5x}{x} = 5$ when $|x|$ is large.

P.6 Exercises

1. $\{x | x \neq -2\}$ 3. $\{x | x \neq 4, -2\}$

5. $\{x | x \neq \pm 3\}$

7. All real numbers since $x^2 + 3 \neq 0$ for any real number x.

9. $\dfrac{3(x-3)}{(x-3)(x+2)} = \dfrac{3}{x+2}$

11. $\dfrac{10a - 8b}{12b - 15a} = \dfrac{2(5a - 4b)}{-3(5a - 4b)} = -\dfrac{2}{3}$

13. $\dfrac{a^3 b^6}{a^2 b^3 - a^4 b^2} = \dfrac{a^3 b^6}{a^2 b^2 (b - a^2)} = \dfrac{ab^4}{b - a^2}$

15. $\dfrac{y^2 z}{x^3}$

17. $\dfrac{a^3 - b^3}{a^2 - b^2} = \dfrac{(a-b)(a^2 + ab + b^2)}{(a-b)(a+b)} = \dfrac{a^2 + ab + b^2}{a+b}$

19. $\dfrac{2a}{3b^2} \cdot \dfrac{9b}{14a^2} = \dfrac{1}{b} \cdot \dfrac{3}{7a} = \dfrac{3}{7ab}$

21. $\dfrac{12a}{7} \cdot \dfrac{49}{2a^3} = \dfrac{42}{a^2}$

23. $\dfrac{(a-3)(a+3)}{3(a-2)} \cdot \dfrac{(a-2)(a+2)}{(a-3)(a+2)} = \dfrac{a+3}{3}$

25. $\dfrac{(x-y)(x+y)}{9} \cdot \dfrac{9(2)}{(x+y)^2} = \dfrac{2x - 2y}{x+y}$

27. $\dfrac{(x-y)(x+y)}{-3xy} \cdot \dfrac{3xy(2xy^2)}{-2(x-y)} = x^2 y^2 + xy^3$

29. $\dfrac{16a}{12a^2}$

31. $\dfrac{x-5}{x+3} \cdot \dfrac{x-3}{x-3} = \dfrac{x^2 - 8x + 15}{x^2 - 9}$

33. $\dfrac{x}{x+5} \cdot \dfrac{x+1}{x+1} = \dfrac{x^2 + x}{x^2 + 6x + 5}$

35. $12a^2 b^3$

37. Since $3a + 3b = 3(a+b)$ and $2a + 2b = 2(a+b)$, the LCD is $6(a+b)$.

39. Since $x^2 + 5x + 6 = (x+3)(x+2)$ and $x^2 - x - 6 = (x-3)(x+2)$, the LCD is $(x+3)(x-3)(x+2)$.

41. $\dfrac{3(3)}{2x(3)} + \dfrac{x}{6x} = \dfrac{9+x}{6x}$

43. $\dfrac{(x+3)(x+1)}{(x-1)(x+1)} - \dfrac{(x+4)(x-1)}{(x+1)(x-1)} =$
$\dfrac{x^2 + 4x + 3}{(x-1)(x+1)} - \dfrac{x^2 + 3x - 4}{(x+1)(x-1)} =$
$\dfrac{x+7}{(x+1)(x-1)}$

45. $\dfrac{3a}{a} + \dfrac{1}{a} = \dfrac{3a+1}{a}$

47. $\dfrac{(t-1)(t+1)}{t+1} - \dfrac{1}{t+1} = \dfrac{t^2 - 2}{t+1}$

49. $\dfrac{x}{(x+2)(x+1)} + \dfrac{x-1}{(x+3)(x+2)} =$
$\dfrac{x(x+3)}{(x+2)(x+1)(x+3)} +$
$\dfrac{(x-1)(x+1)}{(x+3)(x+2)(x+1)} =$
$\dfrac{2x^2 + 3x - 1}{(x+1)(x+2)(x+3)}$

51. $\dfrac{1}{x-3} - \dfrac{5}{-2(x-3)} = \dfrac{2}{2(x-3)} - \dfrac{-5}{2(x-3)} = \dfrac{7}{2x-6}$

53.
$$\frac{y^2}{(x-y)(x^2+xy+y^2)} +$$
$$\frac{(x+y)(x-y)}{(x^2+xy+y^2)(x-y)} =$$
$$\frac{y^2}{(x-y)(x^2+xy+y^2)} +$$
$$\frac{x^2-y^2}{(x^2+xy+y^2)(x-y)} =$$
$$\frac{x^2}{x^3-y^3}$$

55.
$$\frac{(x+1)(x-1)}{x(x+1)(x-1)} + \frac{x(x+1)}{(x-1)(x)(x+1)} -$$
$$\frac{x(x-1)}{(x+1)(x)(x-1)} =$$
$$\frac{x^2-1}{x(x+1)(x-1)} + \frac{x^2+x}{(x-1)(x)(x+1)} -$$
$$\frac{x^2-x}{(x+1)(x)(x-1)} =$$
$$\frac{x^2+2x-1}{x(x^2-1)}$$

57.
$$\frac{\left(\frac{4}{a}-\frac{3}{b}\right)(ab^2)}{\left(\frac{1}{ab}+\frac{2}{b^2}\right)(ab^2)} = \frac{4b^2-3ab}{b+2a}$$

59.
$$\frac{\frac{1}{b^2}-\frac{1}{ab^2}}{\frac{3}{a^2}+\frac{1}{a^2b}} = \frac{\left(\frac{1}{b^2}-\frac{1}{ab^2}\right)(a^2b^2)}{\left(\frac{3}{a^2}+\frac{1}{a^2b}\right)(a^2b^2)} = \frac{a^2-a}{3b^2+b}$$

61.
$$\frac{\left(a+\frac{4}{a+4}\right)(a+4)}{\left(a-\frac{4a+4}{a+4}\right)(a+4)} = \frac{a^2+4a+4}{a^2+4a-(4a+4)} =$$
$$\frac{(a+2)^2}{(a+2)(a-2)} = \frac{a+2}{a-2}$$

63.
$$\frac{\left(\frac{t+2}{t-1}-\frac{t-3}{t}\right)((t-1)t)}{\left(\frac{t+4}{t}+\frac{t-2}{t-1}\right)((t-1)t)} =$$
$$\frac{(t^2+2t)-(t-3)(t-1)}{(t+4)(t-1)+(t^2-2t)} =$$
$$\frac{(t^2+2t)-(t^2-4t+3)}{(t^2+3t-4)+(t^2-2t)} = \frac{6t-3}{2t^2+t-4}$$

65. $\dfrac{(x^{-1}+1)x}{(x^{-1}-1)x} = \dfrac{1+x}{1-x}$

67. $\dfrac{(a^2+a^{-1}b^{-3})ab^3}{ab^3} = \dfrac{a^3b^3+1}{ab^3}$

69. $\dfrac{(x^2-y^2)xy}{(x^{-1}-y^{-1})xy} = \dfrac{(x-y)(x+y)xy}{y-x} =$
$-(x+y)xy = -x^2y-xy^2$

71. $\left(\dfrac{1}{m}-\dfrac{1}{n}\right)^{-2} = \left(\dfrac{n-m}{mn}\right)^{-2} =$
$\left(\dfrac{mn}{n-m}\right)^2 = \dfrac{m^2n^2}{n^2-2mn+m^2}$

73. $\dfrac{3}{7}$ 75. $\dfrac{3}{506}$

77. $S(2) = \dfrac{4-5}{4-9} = \dfrac{1}{5}$

79. $\dfrac{1200-5}{1200-9} = \dfrac{1195}{1191}$

81. $\dfrac{9(16)-1}{3(16)-2} \approx 3.1087$

83. $T(-400) = \dfrac{9(160,000)-1}{3(160,000)-2} \approx 3.00001$

85 (a) The average cost decreases as the truck capacity increases.

(b) Tha average costs in dollars are

n	7	12	22
$R(n)$	27.14	24.17	22.27

87. The costs in millions of dollars are

p	50%	75%	99%
$C(p)$	6	18	594

and the domain is $\{p|0 \leq p < 100\}$.

89. The portion of the invoices Gina and Bert can file in one hour are $\dfrac{1}{4}$ and $\dfrac{1}{6}$, respectively. The part they can file together in one hour is $\dfrac{1}{4}+\dfrac{1}{6}=\dfrac{5}{12}$.

91. Let d be the distance between the restaurant and his home. The number of hours dashing home and returning to the restaurant are $\dfrac{d}{250}$ and $\dfrac{d}{300}$, respectively. Then his average speed is

$$\dfrac{2d}{\dfrac{d}{250}+\dfrac{d}{300}} = \dfrac{2}{\dfrac{1}{250}+\dfrac{1}{300}} \approx 272.7 \text{ mph}$$

93. If a rational expression simulates an application, then the domain describes the extent to which the simulation applies.

95. (a) Since $1 + \dfrac{1}{2} = \dfrac{3}{2}$, $1 + \dfrac{1}{3/2} = \dfrac{5}{3}$, and so on..., then the exact answer is $\dfrac{8}{13}$.

   (b) Since $1 - \dfrac{1}{3} = \dfrac{2}{3}$, $1 - \dfrac{1}{2/3} = -\dfrac{1}{2}$, and so on..., then the exact answer is $\dfrac{3}{2}$.

Review Exercises

1. False, since $\sqrt{2}$ is an irrational number.

3. False, since -1 is a negative number.

5. False, since terminating decimal numbers are rational numbers.

7. False, since $\{1, 2, 3, ...\}$ is the set of natural numbers.

9. False, since $\dfrac{1}{3} = 0.3333...$

11. False, since the additive inverse of 0.5 is -0.5.

13. $-3x - 12 + 20x = 17x - 12$

15. $\dfrac{2x}{10} + \dfrac{x}{10} = \dfrac{3x}{10}$

17. $\dfrac{3(x-2)}{3 \cdot 3} = \dfrac{x-2}{3}$ 19. $\dfrac{-6}{8} = -\dfrac{3}{4}$

21. $3 - 5 = -2$ 23. $|-4| = 4$

25. $8 - 18 \div 3 + 5 = 8 - 6 + 5 = 7$

27. $3 \cdot 3 \div 6 + 3^3 = 9 \div 6 + 27 = 1.5 + 27 = 28.5$

29. 625 31. $4 + 40 = 44$

33. $\dfrac{1}{2} + 1 = \dfrac{3}{2}$ 35. $\dfrac{-2^3 \cdot 2^1}{3^1} = \dfrac{-16}{3}$

37. $(8^2)^{-1/3} = 64^{-1/3} = \dfrac{1}{4}$ 39. $5x^2$ 41. 11

43. $\sqrt{4 \cdot 7s^2 \cdot s} = 2s\sqrt{7s}$

45. $\sqrt[3]{-1000(2)} = -10\sqrt[3]{2}$

47. $\sqrt{\dfrac{5}{2a}} \cdot \sqrt{\dfrac{2a}{2a}} = \dfrac{\sqrt{10a}}{2a}$

49. $\sqrt[3]{\dfrac{2}{5}} \sqrt[3]{\dfrac{25}{25}} = \dfrac{\sqrt[3]{50}}{5}$

51. $3n\sqrt{2n} + 5n\sqrt{2n} = 8n\sqrt{2n}$

53. $\dfrac{2\sqrt{3}}{\sqrt{3}-1} \cdot \dfrac{\sqrt{3}+1}{\sqrt{3}+1} = \dfrac{6+2\sqrt{3}}{3-1} = 3 + \sqrt{3}$

55. $\dfrac{\sqrt{6}}{2\sqrt{2}+3\sqrt{2}} = \dfrac{\sqrt{6}}{5\sqrt{2}} = \dfrac{\sqrt{12}}{10} = \dfrac{\sqrt{3}}{5}$

57. $320{,}000{,}000$ 59. -0.000185 61. 5.6×10^{-5}

63. -2.34×10^6 65. $125 \cdot 10^{18} = 1.25 \times 10^{20}$

67. $\dfrac{(8 \times 10^2)^2(10^{-5})^{-3}}{(2 \times 10^6)^3(2 \times 10^{-5})} = \dfrac{(64 \times 10^4)(10^{15})}{(8 \times 10^{18})(2 \times 10^{-5})} = \dfrac{64}{8(2)} \times 10^6 = 4 \times 10^6$

69. $2x^2 + x - 7$

71. $-4x^4 - 3x^3 + x - x^4 + 6x^3 + 2x = -5x^4 + 3x^3 + 3x$

73. $(3a^2 - 2a + 5)a - (3a^2 - 2a + 5)2 =$
    $3a^3 - 2a^2 + 5a - 6a^2 + 4a - 10 =$
    $3a^3 - 8a^2 + 9a - 10$

75. $b^2 - 6by + 9y^2$

77. $3t^2 - 7t - 6$ 79. $-\dfrac{35y^5}{7y^2} = -5y^3$

81. $9 - 2 = 7$

83. $4(5) + 2(2\sqrt{5})\sqrt{3} + 3 = 23 + 4\sqrt{15}$

85. $1 + 2\sqrt{2x-1} + (2x-1) = 2x + 2\sqrt{2x-1}$

87. Quotient $x^2 + 4x - 1$, remainder 1, since

$$\begin{array}{r}
x^2 + 4x - 1 \\
x-2 \overline{\smash{)}x^3 + 2x^2 - 9x + 3} \\
\underline{x^3 - 2x^2} \\
4x^2 - 9x \\
\underline{4x^2 - 8x} \\
-x + 3 \\
\underline{-x + 2} \\
1
\end{array}$$

REVIEW EXERCISES

89. Quotient $3x + 2$, remainder 4, since

$$\begin{array}{r} 3x + 2 \\ 2x - 1 \overline{\smash{)}6x^2 + x + 2} \\ \underline{6x^2 - 3x} \\ 4x + 2 \\ \underline{4x - 2} \\ 4 \end{array}$$

91. $x - 2 + \dfrac{1}{x + 2}$ since

$$\begin{array}{r} x - 2 \\ x + 2 \overline{\smash{)}x^2 + 0x - 3} \\ \underline{x^2 + 2x} \\ -2x - 3 \\ \underline{-2x - 4} \\ 1 \end{array}$$

93. $2 + \dfrac{13}{x - 5}$ since

$$\begin{array}{r} 2 \\ x - 5 \overline{\smash{)}2x + 3} \\ \underline{2x - 10} \\ 13 \end{array}$$

95. $6x(x^2 - 1) = 6x(x - 1)(x + 1)$ **97.** $(3h + 4t)^2$

99. $(t + y)(t^2 - ty + y^2)$

101. $x^2(x + 3) - 9(x + 3) = (x^2 - 9)(x + 3) = (x - 3)(x + 3)(x + 3) = (x + 3)^2(x - 3)$

103. $t^6 - 1 = (t^3 + 1)(t^3 - 1) = (t + 1)(t^2 - t + 1)(t - 1)(t^2 + t + 1)$

105. $(6x + 5)(3x - 4)$

107. $ab(a^2 + 3a - 18) = ab(a + 6)(a - 3)$

109. $x^2(2x + y) - (2x + y) = (x^2 - 1)(2xy + 1) = (x + 1)(x - 1)(2x + y)$

111. $\dfrac{2x + 6}{x + 3} = \dfrac{2(x + 3)}{x + 3} = 2$

113. $\dfrac{(x - 1)(x + 4)}{(x - 2)(x + 4)} - \dfrac{(x + 3)(x - 2)}{(x + 4)(x - 2)} = \dfrac{x^2 + 3x - 4 - (x^2 + x - 6)}{(x - 2)(x + 4)} = \dfrac{2(x + 1)}{(x + 4)(x - 2)}$

115. $\dfrac{(x - 3)(x + 3)}{x + 3} \cdot \dfrac{1}{-2(x - 3)} = -\dfrac{1}{2}$

117. $\dfrac{c^7}{a^6 b^2} \cdot \dfrac{a^2 b^6 c^{10}}{a^4 b^3} = \dfrac{bc^{17}}{a^8}$

119. $\dfrac{1}{(x - 2)(x + 2)} + \dfrac{3(x + 2)}{(x - 2)(x + 2)} = \dfrac{3x + 7}{x^2 - 4}$

121. $\dfrac{5x}{30x^2} - \dfrac{21}{30x^2} = \dfrac{5x - 21}{30x^2}$

123. $\dfrac{(a - 5)(a + 5)}{(a - 5)(a + 1)} \cdot \dfrac{(a - 1)(a + 1)}{2(a + 5)} = \dfrac{a - 1}{2}$

125. $\dfrac{(x - 4)(x + 4)}{(x + 4)(x + 1)} \cdot \dfrac{(x + 1)(x^2 - x + 1)}{-2(x - 4)} = \dfrac{-x^2 + x - 1}{2}$

127.
$\dfrac{a - 2}{(a + 5)(a + 1)} + \dfrac{2a + 1}{(a - 1)(a + 1)} =$
$\dfrac{(a - 2)(a - 1)}{(a + 5)(a + 1)(a - 1)} + \dfrac{(2a + 1)(a + 5)}{(a - 1)(a + 1)(a + 5)} =$
$\dfrac{(a^2 - 3a + 2) + (2a^2 + 11a + 5)}{(a + 5)(a + 1)(a - 1)} =$
$\dfrac{3a^2 + 8a + 7}{(a + 1)(a - 1)(a + 5)}$

129. $\dfrac{\left(\dfrac{5}{2x} - \dfrac{3}{4x}\right)(4x)}{\left(\dfrac{1}{2} - \dfrac{2}{x}\right)(4x)} = \dfrac{10 - 3}{2x - 8} = \dfrac{7}{2x - 8}$

131. $\dfrac{\left(\dfrac{1}{y^2 - 2} - 3\right)(y^2 - 2)}{\left(\dfrac{5}{y^2 - 2} + 4\right)(y^2 - 2)} = \dfrac{1 - 3(y^2 - 2)}{5 + 4(y^2 - 2)} = \dfrac{7 - 3y^2}{4y^2 - 3}$

133. $\dfrac{(a^{-2} b^{-3}) a^2 b^3}{(a^{-1} b^{-1}) a^2 b^3} = \dfrac{b^3}{ab^2} \cdot \dfrac{a^2}{}$

135. $(p^{-1} + pq^{-3}) \dfrac{pq^3}{pq^3} = \dfrac{q^3 + p^2}{pq^3}$

137. $P(2) = (2)^3 - 3(2)^2 + 2 - 9 = 8 - 3(4) - 7 = -11$

139. $P(0) = -9$

141. $R(-1) = \dfrac{3(-1) - 1}{2(-1) - 9} = \dfrac{-4}{-11} = \dfrac{4}{11}$

143. $R(50) = \dfrac{3(50) - 1}{2(50) - 9} = \dfrac{149}{91}$

145. $d = \dfrac{126^2}{32} \approx 496.1$ feet

147. The number of hydrogen atoms in one kilogram of hydrogen is $\dfrac{1000}{1.7 \times 10^{-24}} \approx 5.9 \times 10^{26}$.

149. The distance is $|-2.35 - 8.77| = |-11.12| = 11.12$

151. The part of the lawn Howard and Will can mow in 2 hours are $\dfrac{2}{6}$ and $\dfrac{2}{4}$, respectively. In 2 hours, the portion they will mow together is $\dfrac{2}{6} + \dfrac{2}{4} = \dfrac{5}{6}$.

Chapter P Test

1. All   2. $\{-1.22, -1, 0, 2, 10/3, \}$

3. $\{-\pi, -\sqrt{3}, \sqrt{5}, 6.020020002...\}$ 4. $\{0, 2\}$

5. $|6 - 25| - 6 = |-19| - 6 = 13$ 6. $8^{1/3} = 2$

7. $-\dfrac{1}{(27^{1/3})^2} = -\dfrac{1}{3^2} = -\dfrac{1}{9}$

8. $\dfrac{9 + 12 + 9}{(-5)(6)} = \dfrac{30}{-30} = -1$

9. $6x^5 y^7$ 10. $2x^2 - 2x^2 = 0$

11. $\dfrac{(ab(b+a))^2}{a^2 b^2} = \dfrac{a^2 b^2 (b+a)^2}{a^2 b^2} = a^2 + 2ab + b^2$

12. $\dfrac{-8a^{-3} b^{18}}{2^{-2} a^{-6} b^8} = -32 a^3 b^{10}$

13. $3\sqrt{3} - 2\sqrt{2} + 4\sqrt{2} = 3\sqrt{3} + 2\sqrt{2}$

14. $\dfrac{2\sqrt{2}}{\sqrt{6} - \sqrt{2}} \cdot \dfrac{\sqrt{6} + \sqrt{2}}{\sqrt{6} + \sqrt{2}} = \dfrac{2\sqrt{12} + 4}{4} = \dfrac{2(2\sqrt{3}) + 4}{4} = \sqrt{3} + 1$

15. $\dfrac{1}{x\sqrt[3]{4x}} = \dfrac{\sqrt[3]{2x^2}}{x\sqrt[3]{4x}\sqrt[3]{2x^2}} = \dfrac{\sqrt[3]{2x^2}}{2x^2}$

16. $\sqrt{4(3)x^2 xy^8 y \cdot 1} = 2xy^4 \sqrt{3xy}$

17. $3x^3 + 3x^2 - 12x$

18. $-x^2 + 3x - 4 - 4x^2 + 6x - 9 = -5x^2 + 9x - 13$

19. $x(x^2 - 2x - 1) + 3(x^2 - 2x - 1) = x^3 - 2x^2 - x + 3x^2 - 6x - 3 = x^3 + x^2 - 7x - 3$

20. $\dfrac{(2h - 1)(4h^2 + 2h + 1)}{2h - 1} = 4h^2 + 2h + 1$

21. $x^2 - 6xy - 27y^2$

22.
$$\begin{array}{r}
x^2 + 2x + 2 \\
x - 2 \overline{\smash{)}\, x^3 + 0x^2 - 2x - 4} \\
\underline{x^3 - 2x^2} \\
2x^2 - 2x \\
\underline{2x^2 - 4x} \\
2x - 4 \\
\underline{2x - 4} \\
0
\end{array}$$

So $\dfrac{x^3 - 2x - 4}{x - 2} = x^2 + 2x + 2$.

23. $9x^2 - 48x + 64$ 24. $4t^8 - 1$

25. $\dfrac{x(x^2 - 5x + 6)}{2x(x - 3)} \cdot \dfrac{4(x^3 + 8)}{2x(x^2 - 4)} = \dfrac{x(x-3)(x-2)}{2x(x-3)} \cdot \dfrac{4(x+2)(x^2 - 2x + 4)}{2x(x-2)(x+2)} = \dfrac{x^2 - 2x + 4}{x}$

26. $\dfrac{(x+5)(x+4)}{(x-3)(x-1)(x+4)} + \dfrac{(x-1)^2}{(x+4)(x-3)(x-1)} = \dfrac{(x^2 + 9x + 20) + (x^2 - 2x + 1)}{(x+4)(x-3)(x-1)} = \dfrac{2x^2 + 7x + 21}{(x+4)(x-1)(x-3)}$

27. $\dfrac{a - 1}{(2a - 3)(2a + 3)} + \dfrac{a - 2}{2a - 3} = \dfrac{a - 1}{(2a - 3)(2a + 3)} + \dfrac{(a - 2)(2a + 3)}{(2a - 3)(2a + 3)} = \dfrac{(a - 1) + (2a^2 - a - 6)}{(2a - 3)(2a + 3)} = \dfrac{2a^2 - 7}{4a^2 - 9}$

28. $\dfrac{\dfrac{1}{2a^2 b} - 2a}{\dfrac{1}{4ab^3} + \dfrac{1}{3b}} \cdot \dfrac{12a^2 b^3}{12a^2 b^3} = \dfrac{6b^2 - 24a^3 b^3}{3a + 4a^2 b^2}$

29. $a(x^2 - 11x + 18) = a(x-9)(x-2)$

30. $m(m^4 - 1) = m(m^2 - 1)(m^2 + 1) = m(m-1)(m+1)(m^2+1)$

31. $(3x - 1)(x + 5)$

32. $bx(x-3) + w(x-3) = (bx + w)(x-3)$

33. The amount spent by each person is
$\dfrac{1.1 \times 10^9}{2.9 \times 10^6} \approx \$379.31$

34. The number of miles in a light year is
$(92.95582 \times 10^6) \cdot (6.3240 \times 10^4) \approx 5.9 \times 10^{12}$.

35. The altitude is $A(2) = -64 + 240 = 176$ feet.

For Thought

1. True, since $5(1) = 6 - 1$.
2. True, since $x = 3$ is the solution to both equations.
3. False, $\sqrt{-2}$ is not a real number.   4. True
5. False, $x = 0$ is the solution.   6. True
7. False, since $|x| = -8$ has no solution.
8. False, $\dfrac{x}{x-5}$ is undefined at $x = 5$.
9. False, since we should multiply by $-\dfrac{3}{2}$.
10. False, $0 \cdot x + 1 = 0$ has no solution.

1.1 Exercises

1. No, since $2(3) - 4 = 2 \neq 9$.
3. Yes, since $(-4)^2 = 16$.
5. Since $3x = 5$, the solution set is $\left\{\dfrac{5}{3}\right\}$.
7. Since $-3x = 6$, the solution set is $\{-2\}$.
9. Since $14x = 7$, the solution set is $\left\{\dfrac{1}{2}\right\}$.
11. Since $7 + 3x = 4x - 4$, the solution set is $\{11\}$.
13. Since $x = -\dfrac{4}{3} \cdot 18$, the solution set is $\{-24\}$.
15. Multiplying by 6,
$$3x - 30 = -72 - 4x$$
$$7x = -42$$
and the solution set is $\{-6\}$.
17. Since $2x = 2\sqrt{2} - 4$, the solution set is $\{\sqrt{2} - 2\}$.

19.
$$\sqrt{3}w = \sqrt{6} + 5$$
$$w = \dfrac{\sqrt{6}+5}{\sqrt{3}} \cdot \dfrac{\sqrt{3}}{\sqrt{3}}$$
$$w = \dfrac{\sqrt{18} + 5\sqrt{3}}{3}$$
The solution set is $\left\{\dfrac{3\sqrt{2} + 5\sqrt{3}}{3}\right\}$.

21.
$$-2 = 2\pi x$$
$$x = -\dfrac{2}{2\pi}$$
The solution set is $\left\{-\dfrac{1}{\pi}\right\}$.

23.
$$\sqrt[3]{2}x = 6$$
$$x = \dfrac{6}{\sqrt[3]{2}} \cdot \dfrac{\sqrt[3]{4}}{\sqrt[3]{4}}$$
$$x = \dfrac{6\sqrt[3]{4}}{2}$$
The solution set is $\{3\sqrt[3]{4}\}$.

25. Since $3(x-6) = 3x - 18$ is true by the distributive law, this is an identity and the solution set is R.

27. Since
$$3x - 18 = 3x + 18$$
$$-18 = 18$$
which is inconsistent, there is no solution.

29. Identity and the solution set is $\{x | x \neq 0\}$
31. Identity and the solution set is $\{w | w \neq 1\}$
33. Multiplying by $6x$,
$$6 - 2 = 3 + 1$$
$$4 = 4.$$
An identity with solution set $\{x | x \neq 0\}$.

35. Multiplying by $3(z-3)$,
$$3(z+2) = -5(z-3)$$
$$3z + 6 = -5z + 15$$
$$8z = 9.$$
This is conditional and the solution set is $\left\{\dfrac{9}{8}\right\}$.

37. Multiplying by $x(x-3)$,
$$(x-3) + x = 9$$
$$2x = 12$$
This is conditional and the solution set is $\{6\}$.

39. Multiplying by $(y-3)$,
$$4(y-3) + 6 = 2y$$
$$4y - 6 = 2y$$
$$y = 3$$
and since division by zero is not allowed, the equation is inconsistent and there is no solution.

41. Multiplying by $t+3$,
$$t + 4t + 12 = 2$$
$$5t = -10.$$
This is conditional and the solution set is $\{-2\}$.

43. Since $-4.19 - 0.21x$, $-19.952 \approx x$ and the solution set is $\{-19.952\}$.

45.
$$\sqrt{2}x = \sqrt{6} + \sqrt{3}$$
$$x = \frac{\sqrt{6} + \sqrt{3}}{\sqrt{2}}$$
$$x \approx \frac{2.4494 + 1.7320}{1.4142}$$
$$x \approx 2.957$$
The solution set is $\{2.957\}$.

47.
$$2a = -1 - \sqrt{17}$$
$$a = \frac{-1 - \sqrt{17}}{2}$$
$$a \approx \frac{-1 - 4.1231}{2}$$
$$a \approx -2.562$$

49.
$$0.001 = 3(y - 0.333)$$
$$\frac{0.001}{3} = y - 0.333$$
$$0.00033 + 0.333 \approx y$$
$$0.333 \approx y$$
The solution set is $\{0.333\}$.

51. Factor an x,
$$x\left(\frac{1}{0.376} + \frac{1}{0.135}\right) = 2$$
$$x(2.6596 + 7.4074) \approx 2$$
$$10.067x \approx 2$$
$$x \approx 0.199$$
and the solution set is $\{0.199\}$.

53.
$$x^2 + 6.5x + 3.25^2 = x^2 - 8.2x + 4.1^2$$
$$14.7x = 4.1^2 - 3.25^2$$
$$14.7x = 16.81 - 10.5625$$
$$14.7x = 6.2475$$
$$x = 0.425$$

55.
$$(2.3 \times 10^6)x = 1.63 \times 10^4 - 8.9 \times 10^5$$
$$x = \frac{1.63 \times 10^4 - 8.9 \times 10^5}{2.3 \times 10^6}$$
$$x \approx -0.380$$
The solution set is $\{-0.380\}$.

57. $x = \pm 9$

59. Since $2x - 3 = 7$ or $2x - 3 = -7$, then $2x = 10$ or $2x = -4$. The solution set is $\{5, -2\}$.

61. Dividing $2|x+5| = 10$ by 2 we obtain $|x+5| = 5$.
Then $x + 5 = 5$ or $x + 5 = -5$.
The solution set is $\{0, -10\}$.

63. Since $3x - 2 = 0$, the solution set is $\left\{\frac{2}{3}\right\}$.

65. Solving for $|x|$, we find $|x| = -\frac{1}{2}$.
The equation has no solution

67. Since $0.95x = 190$, the solution is $\{200\}$.

69.
$$0.1x - 0.05x + 1 = 1.2$$
$$0.05x = 0.2$$
The solution is $\{4\}$.

71. Since $x^2 + 4x + 4 = x^2 + 4$, $4x = 0$ and the solution is $\{0\}$.

73. Multiplying by 4,
$$2x + 4 = x - 6$$
$$x = -10$$
and the solution is $\{-10\}$.

75. Multiplying by 30,
$$15(y - 3) + 6y = 90 - 5(y + 1)$$
$$15y - 45 + 6y = 90 - 5y - 5$$
$$26y = 130$$
and the solution is $\{5\}$.

77. Since $7|x + 6| = 14$, $|x + 6| = 2$.
Then $x + 6 = 2$ or $x + 6 = -2$.
The solution set is $\{-4, -8\}$.

79. Multiplying by $(x - 2)(x + 2)$,
$$3(x + 2) + 4(x - 2) = 7x - 2$$
$$3x + 6 + 4x - 8 = 7x - 2$$
$$7x - 2 = 7x - 2$$
An identity with solution set $\{x | x \neq 2, x \neq -2\}$.

81. The equation is equivalent to
$$\frac{4}{x + 3} + \frac{3}{x - 2} = \frac{7x + 1}{(x + 3)(x - 2)}.$$
Multiplying by $(x + 3)(x - 2)$,
$$4(x - 2) + 3(x + 3) = 7x + 1$$
$$4x - 8 + 3x + 9 = 7x + 1$$
$$7x + 1 = 7x + 1$$
This is an identity and the solution set is $\{x \mid x \neq 2 \text{ and } x \neq -3\}$.

83. Multiplying by $(x - 3)(x - 4)$,
$$(x - 4)(x - 2) = (x - 3)^2$$
$$x^2 - 6x + 8 = x^2 - 6x + 9$$
$$8 = 9$$
This is inconsistent and there is no solution.

85. In the year 2003, when $t = 13$, telecom investment is predicted to be
$I = 7.5(13) + 115 = \$212.5$ billion.
By solving $35 = 7.5t$, we find $t = \dfrac{35}{7.5} \approx 4.7$.
In 1995 ($\approx$ 1990 + 4.7), global investments reached \$150 billion.

87. From $B = 21{,}000 - 0.15B$, we obtain $1.15B = 21{,}000$. Then $B = \dfrac{21{,}000}{1.15} \approx \$18{,}260.87$

89. Rewrite the left-hand side as a sum.
$$10{,}000 + \frac{500{,}000{,}000}{x} = 12{,}000$$
$$\frac{500{,}000{,}000}{x} = 2{,}000$$
$$500{,}000{,}000 = 2000x$$
$$250{,}000 = x$$
So $250{,}000$ vehicles must be sold.
to be $x \approx \dfrac{55}{.520195} \approx \$105.7$ thousands.

For Thought

1. False, $P(1 + rt) = S$ implies $P = \dfrac{S}{1 + rt}$.

2. False, since the perimeter is twice the sum of the length and width. 3. False, since $n + 1$ and $n + 3$ are even integers if n is odd. 4. True

5. True, since $x + (-3 - x) = -3$. 6. False

7. False, if the house sells for x dollars then
$$x - 0.09x = 100{,}000$$
$$0.91x = 100{,}000$$
and the house sells for $x = \$109{,}890.11$

8. True 9. False, a correct one is $4(x - 2) + 5 = 3x$.

10. False, since 9 and $x + 9$ differ by x.

1.2 Exercises

1. $r = \dfrac{I}{Pt}$

3. Since $F - 32 = \dfrac{9}{5}C$, $C = \dfrac{5}{9}(F - 32)$.

5. $r = \dfrac{C}{2\pi}$

7. Since $By = C - Ax$, $y = \dfrac{C - Ax}{B}$.

9. Multiplying by $RR_1R_2R_3$,
$$R_1R_2R_3 = RR_2R_3 + RR_1R_3 + RR_1R_2$$
$$R_1R_2R_3 - RR_2R_3 - RR_1R_2 = RR_1R_3$$
$$R_2(R_1R_3 - RR_3 - RR_1) = RR_1R_3$$
and so $R_2 = \dfrac{RR_1R_3}{R_1R_3 - RR_3 - RR_1}$.

11. Since $a_n - a_1 = (n - 1)d$, $d = \dfrac{a_n - a_1}{n - 1}$.

13. Multiplying by 2.37, one obtains
$$2.4(2.37) = L + 2D - F\sqrt{S}$$
$$5.688 - L + F\sqrt{S} = 2D$$
and $D = \dfrac{5.688 - L + F\sqrt{S}}{2}$.

15. By using the formula $I = Prt$, we find
$$51.30 = 950r \cdot 1$$
$$0.054 = r$$
The simple interest rate is 5.4%.

17. Since $D = RT$,
$$5570 = 2228 \cdot T$$
$$2.5 = T$$
and the surveillance takes 2.5 hours.

19. Since $V = \pi r^2 h$, the height is $h = \dfrac{126\pi}{6^2 \pi} = 3.5$ in.

21. Using linear regression we get $p = -6.9A + 40.3$.

    If $A = 4$, then $p = -(6.9) \cdot 4 + 40.3 = -27.6 + 40.3 = 12.7\%$.

23.
$$3.458x = 2.347 + 4.782$$
$$x = \frac{2.347 + 4.782}{3.458}$$
$$x \approx 2.0616$$
The solution is $\{2.0616\}$.

25.
$$\frac{3.33}{x} = 9.876 - \frac{2.391}{3.4}$$
$$3.33 = x\left(9.876 - \frac{2.391}{3.4}\right)$$
$$\frac{3.33}{9.876 - \frac{2.391}{3.4}} = x$$
$$0.3630 \approx x$$
The solution set is $\{0.3630\}$.

27.
$$x - 3.45 = 7.53(x + 4.98)$$
$$x - 3.45 = 7.53x + 7.53(4.98)$$
$$-3.45 - 7.53(4.98) = 6.53x$$
$$\frac{-3.45 - 7.53(4.98)}{6.53} = x$$
$$-6.2710 \approx x$$
The solution set is $\{-6.2710\}$.

29. If x is the cost of the car before taxes, then
$$1.08x = 28,728$$
$$x = \$26,600.$$

31. Let S be the saddle height and let L be the inside measurement. Then
$$S = 1.09L$$
$$37 = 1.09L$$
$$\frac{37}{1.09} = L$$
$$33.9 \approx L$$
The inside leg measurement is 33.9 inches.

33. If her game-show winnings is x (in dollars), then
$$0.14\frac{x}{3} + 0.12\frac{x}{6} = 4000$$
$$6\left(0.14\frac{x}{3} + 0.12\frac{x}{6}\right) = 24000$$
$$0.28x + 0.12x = 24000$$
$$0.40x = 24000$$
$$x = \$60,000.$$

35. If x is the length of the shorter piece in feet then the length of the longer side is $2x + 2$. Then
$$x + x + (2x + 2) = 30$$
$$4x = 28$$
$$x = 7$$
and so the lengths of the short piece and longer piece are 7 feet and 16 feet, respectively.

37. If x is the length of the side of the larger square lot then $2x$ is the amount of fencing needed to divide the square lot into four smaller lots. The solution to $4x + 2x = 480$ is $x = 80$. The side of the larger square lot is 80 feet and its area is 6400 ft^2.

39. As Ricky is getting his car onto the track Bobby will complete the remaining 8 laps in $\frac{8}{90}$ of an hour. If Ricky is to finish at the same time as Bobby then Ricky's average speed s over 10 laps must satisfy $\frac{10}{s} = \frac{8}{90}$. $\left(Note: time = \frac{distance}{speed}\right)$. This is equivalent to $900 = 8s$ whose solution is $s = 112.5$, which is Ricky's average speed to be even with Bobby at the end of 10th lap.

41. Let d be the halfway distance between San Antonio and El Paso; and let s be the speed in the last half of the trip. Junior took $\frac{d}{80}$ hours to get to the halfway point and the last half took $\frac{d}{s}$ hours to drive. Since the total distance is $2d$ and $distance = rate \times time$,

$$2d = 60\left(\frac{d}{80} + \frac{d}{s}\right)$$
$$160sd = 60\left(sd + 80d\right)$$
$$160sd = 60sd + 4800d$$
$$100sd = 4800d$$
$$100d(s - 48) = 0$$

Since $d \neq 0$, the speed for the last half of the trip is $s = 48$ mph.

43. If x is the part of the start-up capital invested at 5% and $x + 10,000$ is the part invested at 6% then
$$0.05x + 0.06(x + 10,000) = 5880$$
$$0.11x + 600 = 5880$$
$$0.11x = 5280$$
$$x = 48,000.$$
Norma invested $48,000 at 5% and $58,000 at 6% for a total start-up capital of $106,000.

45. Let x and $1500 - x$ be the number of employees from the Northside and Southside, respectively. Then
$$(0.05)x + 0.80(1500 - x) = 750$$
$$0.05x + 1200 - 0.80x = 750$$
$$450 = 0.75x$$
$$600 = x.$$

There were 600 and 900 employees at the Northside and Southside, respectively.

47. Let x be the number of hours it takes both combines working together to harvest an entire wheat crop.

	rate
old	1/72
new	1/48
combined	1/x

Then $\frac{1}{72} + \frac{1}{48} = \frac{1}{x}$. Multiply both sides by $144x$ and get $2x + 3x = 144$. The solution is $x = 28.8$ hrs. which is the time it takes both combines to harvest the entire wheat crop.

49. Let t be the number of hours since 8 : 00 a.m.

	rate	time	work completed
Batman	1/8	t-2	(t-2)/8
Robin	1/12	t	t/12

$$\frac{t-2}{8} + \frac{t}{12} = 1$$
$$24\left(\frac{t-2}{8} + \frac{t}{12}\right) = 24$$
$$3(t-2) + 2t = 24$$
$$5t - 6 = 24$$
$$t = 6$$

Robin had been crime-fighting for 6 hours and all the crime was cleaned up at 2 : 00 p.m.

51. Since $2y = -3x - 6$, $y = -\frac{3}{2}x - 3$.

53. Multiplying by 6,
$$3x - 2y = 18$$
$$3x - 18 = 2y$$
$$y = \frac{3}{2}x - 9.$$

55. Multiplying by 4,
$$4y - 12 = 3x - 3$$
$$4y = 3x + 9$$
$$y = \frac{3}{4}x + \frac{9}{4}.$$

57. Multiplying by $x - x_1$,
$$y - y_1 = mx - mx_1$$
$$y = mx - mx_1 + y_1.$$

59. Since $y(1 + x) = 3$, $y = \dfrac{3}{1 + x}$.

61. Since there are 5280 feet to a mile and the circumference of a circle is $C = 2\pi r$, the radius r of the race track is $r = \dfrac{5280}{2\pi}$.

Since the length of a side of the square plot is twice the radius, the area of the plot is
$$\left(2 \cdot \dfrac{5280}{2\pi}\right)^2 \approx 2,824,677.3 \text{ ft}^2.$$

Dividing this number by 43,560 results to 64.85 acres which is the acreage of the square lot.

63. The area of a trapezoid is $A = \dfrac{1}{2}h(b_1 + b_2)$. Since
$$90,000 = \dfrac{1}{2}h(500 + 300)$$
$$90,000 = 400h$$
$$225 = h$$
the streets are 225 feet apart.

65. Since the volume of a circular cylinder is $V = \pi r^2 h$,
$$\dfrac{22,000}{7.5} = \pi 15^2 \cdot h. \text{ Solving for } h,$$
$h = 4.15$ ft., which is the depth of the pool.

67. If r is the radius of the semicircular turns and since the circumference of a circle is given by $C = 2\pi r$,
$$514 = 2\pi r + 200. \text{ So } r = \dfrac{157}{\pi} \approx 49.9747 \text{ m.}$$
The dimension of the rectangular lot is 99.9494 m by 199.9494 m; its area is 19,984.82 m², which is equivalent to 1.998 hectares.

69. If $n, n+1$, and $n+2$ are the three integers then
$$n + (n+1) + (n+2) = 105$$
$$3n + 3 = 105$$
$$3n = 102$$
$$n = 34.$$
The three integers are 34, 35, and 36.

71. If x is Lorinda's taxable income, then
$$13,152 + 0.31(x - 58,150) = 18,731.10$$
$$0.31x - 4,874.50 = 18,731.10$$
$$0.31x = \$23,605.60$$
Her taxable income is $76,147.10

73. Let x be the amount of water that must be added. The volume of the resulting solution is $4 + x$ liters and the amount of pure baneberry in it is $0.05(4)$ liters. Since the resulting solution is a 3% extract,
$$0.03(4 + x) = 0.05(4)$$
$$0.12 + 0.03x = 0.20$$
$$x = \dfrac{0.08}{0.03}$$
$$x = \dfrac{8}{3}$$
The amount of water to be added is $\dfrac{8}{3}$ liters.

75. Let x and $2x$ be the number of years of experience of Eric and Kim, respectively. In four more years, Eric will have $x + 4$ years of experience and Kim $2x + 4$ years. Their combined experience after 4 more years is 50 i.e. $(x+4) + (2x+4) = 50$. The solution is $x = 14$. Kim has now 28 years of experience.

77. The costs of x pounds of dried apples is $(1.20)4x$ and the cost of $(20 - x)$ pounds of dried apricots is $4(1.80)(20-x)$. Since the mixture costs $1.68 per quarter-pound,
$$4(1.68)(20) = (1.20)4x + 4(1.80)(20 - x)$$
$$134.4 = 4.80x + 144 - 7.20x$$
$$2.40x = 9.6$$
$$x = 4$$
The mix needs 4 lbs of dried apples and 16 lbs of dried apricots.

79. Let x and $8 - x$ be the number of dimes and nickels, respectively. Since 55 cents is the value of the coins, $55 = 10x + 5(8-x)$, and the solution is $x = 3$. Dana has 3 dimes and 5 nickels.

81. By solving
$$13.5n + 190 = 7.5n + 225$$
$$6n = 35$$
$$n \approx 5.8$$
we find that in the year 1992 (=1986+6), restaurant revenue will exceed supermarket revenue. By solving the equation
$$13.5n + 190 = 2(7.5n + 225)$$
$$13.5n + 190 = 15n + 450$$
$$-1.5n = 260$$
we see that its solution is a negative number, thus, according to this model the restaurant revenue will never be twice the supermarket revenue in any year after 1986.

For Thought

1. True, since $i \cdot (-i) = 1$.

2. True, since $\overline{0+i} = 0 - i = -i$.

3. False, the set of real numbers is a subset of the complex numbers.

4. True, $(\sqrt{3} - i\sqrt{2})(\sqrt{3} + i\sqrt{2}) = 3 - 2 = 1$.

5. False, since $(2+5i)(2+5i) = 4 + 20i + 25i^2 = 4 + 20i - 25 = -21 + 20i$.

6. False, $5 - \sqrt{-9} = 5 - 3i$.

7. True, since $P(3i) = (3i)^2 + 9 = -9 + 9 = 0$.

8. True, since $(3i)^2 + 9 = -9 + 9 = 0$.

9. True, since $i^4 = i^2 \cdot i^2 = (-1)(-1) = 1$.

10. False, $i^{18} = (i^4)^4 i^2 = (1)^4(-1) = -1$.

1.3 Exercises

1. $0 + 6i$, imaginary 3. $\frac{1}{3} + \frac{1}{3}i$, imaginary

5. $\sqrt{7} + 0i$, real 7. $\frac{\pi}{2} + 0i$, real

9. $7 + 2i$ 11. $1 - i - 3 - 2i = -2 - 3i$

13. $-18i + 12i^2 = -12 - 18i$

15. $8 + 12i - 12i - 18i^2 = 26 + 0i$

17. $(5-2i)(5+2i) = 25 - 4i^2 = 25 - 4(-1) = 29$

19. $(\sqrt{3} - i)(\sqrt{3} + i) = 3 - i^2 = 3 - (-1) = 4$

21. $9 + 24i + 16i^2 = -7 + 24i$

23. $5 - 4i\sqrt{5} + 4i^2 = 1 - 4i\sqrt{5}$

25. $(i^4)^4 \cdot i = (1)^4 \cdot i = i$

27. $(i^4)^{24} i^2 = 1^{24}(-1) = -1$

29. $(i^4)^{-1} = 1^{-1} = 1$

31. Since $i^4 = 1$ then $i^{-1} = i^{-1}i^4 = i^3 = -i$.

33. $(3-9i)(3+9i) = 9 - 81i^2 = 90$

35. $\left(\frac{1}{2} + 2i\right)\left(\frac{1}{2} - 2i\right) = \frac{1}{4} - 4i^2 = \frac{1}{4} + 4 = \frac{17}{4}$

37. $i(-i) = -i^2 = 1$

39. $(3 - i\sqrt{3})(3 + i\sqrt{3}) = 9 - 3i^2 = 9 - 3(-1) = 12$

41. $\frac{1}{2-i} \cdot \frac{2+i}{2+i} = \frac{2+i}{5} = \frac{2}{5} + i\frac{1}{5}$

43. $\frac{-3i}{1-i} \cdot \frac{1+i}{1+i} = \frac{-3i+3}{2} = \frac{3}{2} - i\frac{3}{2}$

45. $\frac{-2+6i}{2} = \frac{-2}{2} + \frac{6i}{2} = -1 + 3i$

47. $\frac{-3+3i}{i} \cdot \frac{-i}{-i} = \frac{3i - 3i^2}{1} = 3i - 3(-1) = 3 + 3i$

49. $\frac{1-i}{3+2i} \cdot \frac{3-2i}{3-2i} = \frac{3 - 5i - 2}{13} = \frac{1}{13} - i\frac{5}{13}$

51. $2i - 3i = -i$ 53. $-4 + 2i$

55. $\left(i\sqrt{6}\right)^2 = -6$

57. $(i\sqrt{2})(i\sqrt{50}) = i^2\sqrt{2} \cdot 5\sqrt{2} = (-1)(2)(5) = -10$

59. $\dfrac{-2}{2} + \dfrac{i\sqrt{20}}{2} = -1 + i\dfrac{2\sqrt{5}}{2} = -1 + i\sqrt{5}$

61. $-3 + \sqrt{9-20} = -3 + i\sqrt{11}$

63. $2i\sqrt{2}\left(i\sqrt{2} + 2\sqrt{2}\right) = 4i^2 + 8i = -4 + 8i$

65. $\dfrac{-2 + \sqrt{-16}}{2} = \dfrac{-2 + 4i}{2} = -1 + 2i$

67. $\dfrac{-4 + \sqrt{16-24}}{4} = \dfrac{-4 + 2\sqrt{2}i}{4} = -1 + \dfrac{\sqrt{2}}{2}i$

69. $\dfrac{-6 - \sqrt{-32}}{2} = \dfrac{-6 - 4i\sqrt{2}}{2} = -3 - 2i\sqrt{2}$

71. $\dfrac{-6 - \sqrt{36+48}}{-4} = \dfrac{-6 - 2\sqrt{21}}{-4} = \dfrac{3}{2} + \dfrac{\sqrt{21}}{2}$

73. $(-2+i)^2 + 4(-2+i) + 5 =$
$(4 - 4i + i^2) - 8 + 4i + 5 = 0$

75. $(-2-i)^2 + 4(-2-i) + 5 =$
$(4 + 4i + i^2) - 8 - 4i + 5 = 0$

77. $(1+i)^2 + 4(1+i) + 5 =$
$(1 + 2i + i^2) + 4 + 4i + 5 = 9 + 6i$

79. $(3 + i\sqrt{5})^2 - 6(3 + i\sqrt{5}) + 14 =$
$(9 + 6i\sqrt{5} + 5i^2) - 18 - 6i\sqrt{5} + 14 = 0$

81. $W(-1) = (-1)^2 - 6(-1) + 14 = 1 + 6 + 14 = 21$

83. Yes, since $(2i)^2 + 4 = -4 + 4 = 0$

85. No, since $(1-i)^2 + 2(1-i) - 2 =$
$(1 - 2i + i^2) + 2 - 2i - 2 = -4i$

87. Yes, since $(3 - 2i)^2 - 6(3 - 2i) + 13 =$
$(9 - 12i + 4i^2) - 18 + 12i + 13 = 0$

89. Yes, since $3\left(\dfrac{i\sqrt{3}}{3}\right)^2 + 1 = 3 \cdot \dfrac{-3}{9} + 1 = 0.$

91. $x^2 - i^2 = x^2 + 1$

93. $(x - (1+i))(x - (1-i)) =$
$x^2 - (1-i)x - (1+i)x + (1+i)(1-i) =$
$x^2 - x + ix - x - ix + (1 - i^2) = x^2 - 2x + 2$

95. $(x - (2+3i))(x - (2-3i)) =$
$x^2 - (2-3i)x - (2+3i)x + (2+3i)(2-3i) =$
$x^2 - 2x + 3ix - 2x - 3ix + (4 - 9i^2) = x^2 - 4x + 13$

97. If r is the remainder when n is divided by 4 then $i^n = i^r$. The possible values of r are $0, 1, 2, 3$ and for i^r they are $0, i, -1, -i$, respectively.

99. Note, $w + \overline{w} = (a + bi) + (a - bi) = 2a$ is a real number and $w - \overline{w} = (a + bi) - (a - bi) = 2bi$ is an imaginary number.

When a complex number is added to its complex conjugate the sum is twice the real part of the complex number. When the complex conjugate of a complex number is subtracted from the complex number, the difference is an imaginary number.

101. Its reciprocal is $\dfrac{1}{a+bi} = \dfrac{a-bi}{(a+bi)(a-bi)}$
$= \dfrac{a-bi}{a^2+b^2} = \dfrac{a}{a^2+b^2} - \dfrac{b}{a^2+b^2}i.$

For Thought

1. False, since $x = 1$ is a solution of the first equation and not of the second equation.   2. False, since $x^2 + 1 = 0$ cannot be factored with real coefficients.

3. False, $\left(x + \dfrac{2}{3}\right)^2 = x^2 + \dfrac{4}{3}x + \dfrac{9}{4}.$

4. False, the solutions to $(x-3)(2x+5) = 0$ are $x = 3$ and $x = -\dfrac{5}{2}.$

5. False, $x^2 = 0$ has only $x = 0$ as its solution.

6. True, since this is a restatement of the quadratic formula except that $a = m, b = -n$, and $c = p$.

7. False, the quadratic formula can be used to solve any quadratic equation.

8. False, $x^2 + 1 = 0$ has only imaginary zeros.

9. True, since $(12)^2 - 4(4)(9) = 144 - 144 = 0.$

10. False, since the imaginary zeros always occur in conjugate pairs.

1.4 Exercises

1. Since $x^2 = 5$, the solution set is $\{\pm\sqrt{5}\}$.

3. Since $x^2 = -\dfrac{2}{3}$, $x = \pm i\dfrac{\sqrt{2}}{\sqrt{3}} = \pm i\dfrac{\sqrt{2}}{\sqrt{3}}\dfrac{\sqrt{3}}{\sqrt{3}} = \pm i\dfrac{\sqrt{6}}{3}$. The solution set is $\left\{\pm i\dfrac{\sqrt{6}}{3}\right\}$.

5. Since $x - 3 = \pm 3$ then $x = 3 \pm 3$. The solution set is $\{0, 6\}$.

7. Since $x - \dfrac{1}{2} = \pm\dfrac{5}{2}$, $x = \dfrac{1}{2} \pm \dfrac{5}{2}$.
The solution set is $\{-2, 3\}$.

9. Since $x + 2 = \pm 2i$, solution set is $\{-2 \pm 2i\}$.

11. Since $x - \dfrac{2}{3} = \pm i\dfrac{2}{3}$, solution set is $\left\{\dfrac{2}{3} \pm i\dfrac{2}{3}\right\}$.

13. Since $(x - 5)(x + 4) = 0$, the solution set is $\{5, -4\}$.

15. Since $a^2 + 3a + 2 = (a + 2)(a + 1) = 0$, the solution set is $\{-2, -1\}$.

17. Since $(2x + 1)(x - 3) = 0$, the solution set is $\left\{-\dfrac{1}{2}, 3\right\}$.

19. Since $(2x - 1)(3x - 2) = 0$, the solution set is $\left\{\dfrac{1}{2}, \dfrac{2}{3}\right\}$.

21. Since $y^2 + y - 12 = 30$, $y^2 + y - 42 = 0$, $(y + 7)(y - 6) = 0$. Solution set is $\{-7, 6\}$.

23. Since $2z^2 + 5z - 3 = 15$, $2z^2 + 5z - 18 = 0$, $(z - 2)(2z + 9) = 0$. Solution set is $\left\{2, -\dfrac{9}{2}\right\}$.

25. $x^2 - 12x + \left(\dfrac{12}{2}\right)^2 = x^2 - 12x + 6^2 = x^2 - 12x + 36$

27. $r^2 + 3r + \left(\dfrac{3}{2}\right)^2 = r^2 + 3r + \dfrac{9}{4}$

29. $w^2 + \dfrac{1}{2}w + \left(\dfrac{1}{4}\right)^2 = w^2 + \dfrac{1}{2}w + \dfrac{1}{16}$

31.
$$x^2 + 6x = -1$$
$$x^2 + 6x + 9 = -1 + 9$$
$$(x + 3)^2 = 8$$
$$x + 3 = \pm 2\sqrt{2}$$
The solution set is $\{-3 \pm 2\sqrt{2}\}$.

33.
$$n^2 - 2n = 1$$
$$n^2 - 2n + 1 = 1 + 1$$
$$(n - 1)^2 = 2$$
$$n - 1 = \pm\sqrt{2}$$
The solution set is $\{1 \pm \sqrt{2}\}$.

35.
$$h^2 + 3h = 1$$
$$h^2 + 3h + \dfrac{9}{4} = 1 + \dfrac{9}{4}$$
$$(h + \dfrac{3}{2})^2 = \dfrac{13}{4}$$
$$h + \dfrac{3}{2} = \pm\dfrac{\sqrt{13}}{2}$$
The solution set is $\left\{\dfrac{-3 \pm \sqrt{13}}{2}\right\}$.

37.
$$x^2 + 2x = -5$$
$$x^2 + 2x + 1 = -5 + 1$$
$$(x + 1)^2 = -4$$
$$x + 1 = \pm 2i$$
The solution set is $\{-1 \pm 2i\}$.

39.
$$x^2 + \dfrac{5}{2}x = 6$$
$$x^2 + \dfrac{5}{2}x + \dfrac{25}{16} = 6 + \dfrac{25}{16}$$
$$\left(x + \dfrac{5}{4}\right)^2 = \dfrac{121}{16}$$
$$x = -\dfrac{5}{4} \pm \dfrac{11}{4}$$
The solution set is $\left\{\dfrac{3}{2}, -4\right\}$.

41.
$$x^2 + \frac{2}{3}x = -\frac{1}{3}$$
$$x^2 + \frac{2}{3}x + \frac{1}{9} = -\frac{3}{9} + \frac{1}{9}$$
$$\left(x + \frac{1}{3}\right)^2 = -\frac{2}{9}$$
$$x + \frac{1}{3} = \pm i\frac{\sqrt{2}}{3}$$
$$x = -\frac{1}{3} \pm i\frac{\sqrt{2}}{3}$$
The solution set is $\left\{-\frac{1}{3} \pm i\frac{\sqrt{2}}{3}\right\}$.

43. Since $a = 1, b = 3, c = -4$ and
$$x = \frac{-3 \pm \sqrt{3^2 - 4(1)(-4)}}{2(1)} = \frac{-3 \pm \sqrt{25}}{2} = \frac{-3 \pm 5}{2},$$
the solution set is $\{1, -4\}$.

45. Since $a = 2, b = -5, c = -3$ and
$$x = \frac{5 \pm \sqrt{(-5)^2 - 4(2)(-3)}}{2(2)} = \frac{5 \pm \sqrt{49}}{4} = \frac{5 \pm 7}{4},$$
the solution set is $\left\{-\frac{1}{2}, 3\right\}$.

47. Since $a = 9, b = 6, c = 1$ and
$$x = \frac{-6 \pm \sqrt{6^2 - 4(9)(1)}}{2(9)} = \frac{-6 \pm 0}{18},$$
the solution set is $\left\{-\frac{1}{3}\right\}$.

49. Since $a = 2, b = 0, c = -3$ and
$$x = \frac{0 \pm \sqrt{0^2 - 4(2)(-3)}}{2(2)} = \frac{\pm\sqrt{24}}{4} = \frac{\pm 2\sqrt{6}}{4},$$
the solution set is $\left\{\pm\frac{\sqrt{6}}{2}\right\}$.

51. In $x^2 - 4x + 5 =$, $a = 1, b = -4, c = 5$. Then
$$x = \frac{4 \pm \sqrt{(-4)^2 - 4(1)(5)}}{2(1)} = \frac{4 \pm \sqrt{-4}}{2} = \frac{4 \pm 2i}{2},$$
the solution set is $\{2 \pm i\}$.

53. Since $9x^2 + 6x - 1 = 0$ then $a = 9, b = 6, c = -1$ and
$$x = \frac{-6 \pm \sqrt{6^2 - 4(9)(-1)}}{2(9)} = \frac{-6 \pm \sqrt{72}}{18} = \frac{-6 \pm 6\sqrt{2}}{18},$$
the solution set is $\left\{\frac{-1 \pm \sqrt{2}}{3}\right\}$.

55. Since $a = 3.2, b = 7.6$, and $c = -9$,
$$x = \frac{-7.6 \pm \sqrt{(7.6)^2 - 4(3.2)(-9)}}{2(3.2)} \approx$$
$$\frac{-7.6 \pm \sqrt{172.96}}{6.4} \approx \frac{-7.6 \pm 13.151}{6.4}.$$
The solution set is $\{0.87, -3.24\}$.

57. Since $a = 3.25, b = -4.6, c = 20$ and
$$x = \frac{4.6 \pm \sqrt{(-4.6)^2 - 4(3.25)(20)}}{2(3.25)} \approx$$
$$\frac{4.6 \pm \sqrt{-238.84}}{6.5} \approx \frac{4.6 \pm 15.454i}{6.5} \approx$$
$0.71 \pm 2.38i$, solution set is $\{0.71 \pm 2.38i\}$.

59. The discriminant is $(-30)^2 - 4(9)(25) = 900 - 900 = 0$. Only one solution and it is real.

61. The discriminant is $(-6)^2 - 4(5)(2) = 36 - 40 = -4$. There are two distinct imaginary solutions.

63. The discriminant is $12^2 - 4(7)(-1) = 144 + 28 = 172$. There are two distinct real solutions.

65. Set the right-hand side to 0,
$$x^2 - \frac{4}{3}x + \frac{5}{9} = 0$$
$$9x^2 - 12x + 5 = 0.$$
By the quadratic formula,
$$x = \frac{12 \pm \sqrt{(-12)^2 - 4(9)(5)}}{2(9)} = \frac{12 \pm 6i}{18}$$
and the solution set is $\left\{\frac{2 \pm i}{3}\right\}$.

67. Since $x^2 = -\sqrt{2}$, $x = \pm\sqrt{-\sqrt{2}} = \pm i\sqrt[4]{2}$.
The solution set is $\{\pm i\sqrt[4]{2}\}$.

69. By the quadratic formula,
$$x = \frac{-\sqrt{6} \pm \sqrt{(-\sqrt{6})^2 - 4(12)(-1)}}{2(12)} =$$
$$\frac{-\sqrt{6} \pm \sqrt{54}}{24} = \frac{-\sqrt{6} \pm 3\sqrt{6}}{24} = \frac{2\sqrt{6}}{24}, \frac{-4\sqrt{6}}{24}.$$
Then $x = \frac{\sqrt{6}}{12}, \frac{-\sqrt{6}}{6}$.

71. Since $x^2 + 6x - 72 = (x+12)(x-6) = 0$, the solution set is $\{-12, 6\}$.

73. Multiply by x to get $x^2 = x + 1$.
So $x^2 - x - 1 = 0$ and by the quadratic formula,
$$x = \frac{1 \pm \sqrt{1 - 4(1)(-1)}}{2} = \frac{1 \pm \sqrt{5}}{2}.$$
The solution set is $\left\{\dfrac{1 \pm \sqrt{5}}{2}\right\}$.

75. Multiplying by $(3-x)(x+7)$,
$$(x-12)(x+7) = (x+4)(3-x)$$
$$x^2 - 5x - 84 = -x^2 - x + 12$$
$$2x^2 - 4x - 96 = 0$$
$$2(x-8)(x+6) = 0$$
the solution set is $\{8, -6\}$.

77. Since $r^2 = \dfrac{A}{\pi}$, $r = \pm\sqrt{\dfrac{A}{\pi}}$.

79. Apply the quadratic formula to $x^2 + (2k)x + 3 = 0$ where $a = 1, b = 2k$, and $c = 3$.
$$x = \frac{-2k \pm \sqrt{(2k)^2 - 4(1)(3)}}{2(1)}$$
$$x = \frac{-2k \pm \sqrt{4k^2 - 12}}{2}$$
$$x = \frac{-2k \pm \sqrt{4(k^2 - 3)}}{2}$$
$$x = \frac{-2k \pm 2\sqrt{k^2 - 3}}{2}$$
$$x = -k \pm \sqrt{k^2 - 3}.$$

81. Apply the quadratic formula to $2y^2 + (4x)y - x^2 = 0$ where $a = 2, b = 4x$, and $c = -x^2$. Then
$$y = \frac{-4x \pm \sqrt{(4x)^2 - 4(2)(-x^2)}}{2(2)} =$$
$$\frac{-4x \pm \sqrt{16x^2 + 8x^2}}{4} = \frac{-4x \pm \sqrt{24x^2}}{4}$$
$$\frac{-4x \pm 2|x|\sqrt{6}}{4} = \frac{-2x \pm x\sqrt{6}}{2}$$
and note that we used $|x| = \sqrt{x^2}$.

83. From the revenue function,
$$x(40 - 0.001x) = 175,000$$
$$40x - 0.001x^2 = 175,000$$
By applying the quadratic formula to $0.001x^2 - 40x + 175,000 = 0$,
$$x = \frac{40 \pm \sqrt{(40)^2 - 4(0.001)(175,000)}}{0.002}$$
$$x = \frac{40 \pm \sqrt{900}}{0.002} = \frac{40 \pm 30}{0.002}$$
$$x = 5,000 \text{ or } 35,000$$
So $5,000$ or $35,000$ units must be produced weekly.

85. The height S (in feet) of the ball from the ground t seconds after it was tossed is given by $S = -16t^2 + 40t + 4$. When the height is 4 feet,
$$-16t^2 + 40t + 4 = 4$$
$$-16t^2 + 40t = 0$$
$$-8t(2t - 5) = 0$$
$$t = 0, \frac{5}{2}.$$
The ball returns to a height of 4 ft in 2.5 secs.

87. Let d be the diagonal distance across the field from one goal to the other. By the Pythagorean Theorem, $d = \sqrt{300^2 + 160^2} = 340$ ft.

89. Let w and $2w + 2$ be the length and width. From the given area of the court, we obtain
$$(2w + 2)w = 312$$
$$2w^2 + 2w - 312 = 0$$
$$w^2 + w - 156 = 0$$
$$(w - 12)(w + 13) = 0$$
Then $w = 12$ yds. and the length is 26 yds. The distance between two opposite corners (by the Pythagorean Theorem) is $\sqrt{12 + 26^2} \approx 28.6$ yds.

91. Substituting the values of d and A, we find that the displacement is
$$\frac{1}{2^{12}}d^2(18.8)^3 - 822^3 = 0$$
$$\frac{1}{2^{12}}d^2(18.8)^3 = 822^3$$
$$d = \sqrt{\frac{822^3(2^{12})}{18.8^3}}$$
$$d \approx 18,503.4 \text{ lbs}$$

93. Let x be the width of the border. The overall dimensions are $8 + 2x$ and $6 + 2x$. Then

$$(8 + 2x)(6 + 2x) = 100$$
$$4x^2 + 28x - 52 = 0$$
$$x^2 + 7x - 13 = 0.$$

By the quadratic formula, $x = \dfrac{-7 \pm \sqrt{101}}{2} \approx$ $1.53, -8.52$. The border should be 1.53 feet wide. The negative value of x is not possible.

95. Let x be the normal speed of the tortoise in ft/hr.

	distance	rate	time
hwy	24	x+2	24/(x+2)
off hwy	24	x	24/x

Since 24 minutes is 2/5 of an hour,

$$\frac{2}{5} + \frac{24}{x+2} = \frac{24}{x}$$
$$2x(x+2) + 24(5)x = 24(5)(x+2)$$
$$2x^2 + 4x + 120x = 120x + 240$$
$$x^2 + 2x - 120 = 0$$
$$(x + 12)(x - 10) = 0$$
$$x = -12, 10.$$

The normal speed of the tortoise is 10 ft/hr.

97. Let x and $x - 2$ be the number of days it takes to design a direct mail package using traditional methods and a computer, respectively.

	rate
together	2/7
computer	1/(x-2)
traditional	1/x

$$\frac{1}{x-2} + \frac{1}{x} = \frac{2}{7}$$
$$7x + (7x - 14) = 2(x^2 - 2x)$$
$$0 = 2x^2 - 18x + 14$$
$$0 = x^2 - 9x + 7$$
$$x = \frac{9 \pm \sqrt{81 - 28}}{2}$$
$$x = \frac{9 \pm \sqrt{53}}{2}$$
$$x \approx 8.1, 0.9$$

Curt using traditional methods can do the job in 8.1 days. Note, $x \approx 0.9$ days has to be excluded since $x - 2$ is negative for $x \approx 0.9$.

99. Let x and $x - 10$ be the number of pounds of white meat in a Party Size bucket and a Big Family Size bucket, respectively. From the ratios,

$$\frac{8}{x} = \frac{3}{x - 10} + 0.10$$
$$8(x - 10) = 3x + 0.10x(x - 10)$$
$$0 = 0.10x^2 - 6x + 80$$
$$0 = x^2 - 60x + 800$$
$$0 = (x - 40)(x - 20)$$
$$x = 40, 20$$

A Party Size bucket weighs 20 or 40 lbs.

101. Using $v_1^2 = v_0^2 + 2gS$ with $S = 1.07$ and $v_1 = 0$, we find that
$$v_0^2 + 2(-9.8)(1.07) = 0$$
$$v_0^2 - 20.972 = 0$$
$$v_0 = \pm\sqrt{20.972} \approx \pm 4.58$$

His initial upward velocity is 4.58 m/sec.

Using $S = \dfrac{1}{2}gt^2 + v_0t$ with $S = 0$ and $v_o = 4.58$, we find that his time t in the air satisfies
$$\frac{1}{2}(-9.8)t^2 + 4.58t = 0$$
$$t(4.58 - 4.9t) = 0$$
$$t \approx 0, 0.93$$

Jordan is in the air for 0.93 seconds.

For Thought

1. True 2. False, since $-2x < -6$ is equivalent to $\dfrac{-2x}{-2} > \dfrac{-6}{-2}$. 3. False, since there is a number between any two distinct real numbers.

4. True, since $|-6-6| = |-12| = 12 > -1$.

5. False, $(-\infty, -3) \cap (-\infty, -2) = (-\infty, -3)$.

6. False, $(5, \infty) \cap (-\infty, -3) = \phi$.

7. False, no real number satisfies $|x-2| < 0$.

8. False, it is equivalent to $|x| > 3$.

9. False, $|x| + 2 < 5$ is equivalent to $-3 < x < 3$.

10. True

1.5 Exercises

1. $x < 12$ 3. $[-8, \infty)$

5. $(-\infty, \pi/2)$ 7. $x \geq -7$

9. Since $3x > 15$ implies $x > 5$, the solution set is $(5, \infty)$ and the graph is <——(===>

11. Since $10 \leq 5x$ implies $2 \leq x$, the solution set is $[2, \infty)$ and the graph is <——[===>

13. Multiply 6 to both sides of the inequality.
$$3x - 24 < 2x + 30$$
$$x < 54$$
The solution is the interval $(-\infty, 54)$ and the graph is <===)——>

15. Multiplying the inequality by 2,
$$7 - 3x \geq -6$$
$$13 \geq 3x$$
$$13/3 \geq x$$
The solution is the interval $(-\infty, 13/3]$ and the graph is <===]——>

17. Multiply the left-hand side.
$$-6x + 4 \geq 4 - x$$
$$0 \geq 5x$$
$$0 \geq x.$$
The solution is the interval $(-\infty, 0]$ and the graph is <===]——>

19. Multiply by 3 on both sides of the inequality.
$$152 + x \geq 240$$
$$x \geq 88$$
The solution is the interval $[88, \infty)$ and the graph is <——[===>

21. $(-3, \infty)$ 23. ϕ

25. $(-\infty, 5]$ 27. $(-3, \infty)$ 29. $[3, 7]$

31. $[5, 7)$ 33. No, since $4 - 2(-3) = 10$.

35. Yes, since if $x = -5$ the inequality becomes $12 > 6$ or $-7 > 9$ which is a true statement.

37. Yes, since if $x = 8$ the inequality becomes $2 \leq 6$, which is a true statement.

39. Solve each simple inequality and find the intersection of their solution sets.
$$x > 3 \text{ and } 0.5x < 3$$
$$x > 3 \text{ and } x < 6$$
The intersection of these values of x is the interval $(3, 6)$ and whose graph is <——(===)——>

41. Solve each simple inequality and find the intersection of their solution sets.
$$2x - 5 > -4 \text{ and } 2x + 1 > 0$$
$$x > \frac{1}{2} \text{ and } x > -\frac{1}{2}$$
The intersection of these values of x is the interval $(1/2, \infty)$ and the graph is <——(===>

43. The union of $x > 2$ or $-3 > x$ is $(2, \infty) \cup (-\infty, -3)$ and the graph is <===)——(===>

45. Solve each simple inequality and find the union of their solution sets.
$$-6 < 2x \text{ or } 3x > -3$$
$$-3 < x \text{ or } x > -1$$
The union of these values of x is $(-3, \infty)$ and the graph is <——(===>

47. Solve each simple inequality and find the union of their solution sets.

$$x + 1 > 6 \text{ or } x < 7$$
$$x > 5 \text{ or } x < 7$$

The union of these values of x is $(-\infty, \infty)$ and the graph is <===>

49. Solve each simple inequality and find the intersection of their solution sets.

$$2 - 3x < 8 \text{ and } x - 8 \leq -12$$
$$-6 < 3x \text{ and } x \leq -4$$
$$-2 < x \text{ and } x \leq -4$$

The intersection is empty and there is no solution.

51. Solve an equivalent compound inequality.

$$-2 < 3x - 1 < 2$$
$$-1 < 3x < 3$$
$$-\frac{1}{3} < x < 1$$

The solution set is the interval $(-1/3, 1)$ and the graph is <—(==)—>

53. Solve an equivalent compound inequality.

$$-1 \leq 5 - 4x \leq 1$$
$$-6 \leq -4x \leq -4$$
$$\frac{3}{2} \geq x \geq 1$$

The solution set is the interval $[1, 3/2]$ and the graph is <—[==]—>

55. Solve an equivalent compound inequality.

$$2x - 1 > 3 \text{ or } 2x - 1 < -3$$
$$2x > 4 \text{ or } 2x < -2$$
$$x > 2 \text{ or } x < -1$$

The solution set is the interval $(-\infty, -1) \cup (2, \infty)$ and the graph is <=)—(=>

57. No solution since an absolute value is never negative.

59. Since an absolute value is always nonnegative, $|2x - 8| \leq 0$ is equivalent to $2x - 8 = 0$. The solution set is $\{4\}$ and the graph is <—•—>

61. No solution since an absolute value is never negative.

63. Solve an equivalent compound inequality.

$$|x - 2| > 3$$
$$x - 2 > 3 \text{ or } x - 2 < -3$$
$$x > 5 \text{ or } x < -1$$

The union is $(-\infty, -1) \cup (5, \infty)$ and the graph is <=)—(=>

65. Solve an equivalent compound inequality.

$$\frac{x - 3}{2} > 1 \text{ or } \frac{x - 3}{2} < -1$$
$$x - 3 > 2 \text{ or } x - 3 < -2$$
$$x > 5 \text{ or } x < 1$$

The solution set is the interval $(-\infty, 1) \cup (5, \infty)$ and the graph is <=)—(=>

67. Since 4 is the midpoint of 3 and 5, the inequality is $|x - 4| > 1$.

69. Since 6 is the midpoint of 4 and 8, the inequality is $|x - 6| < 2$.

71. Since an absolute value is nonnegative, the inequality is $|x - 4| > 0$.

73. Since 2 is the midpoint of -3 and 7, the inequality is $|x - 2| < 5$.

75. $|x| \geq 9$

77. Since 7 is the midpoint, the inequality is $|x - 7| \leq 4$.

79. Since 5 is the midpoint, the inequality is $|x - 5| > 2$.

81. Since $x - 2 \geq 0$, the solution set is $[2, \infty)$.

83. Since $2 - x > 0$ is equivalent to $2 > x$, the solution set is $(-\infty, 2)$.

85. Since $|x| \geq 3$ is equivalent to $x \geq 3$ or $x \leq -3$, the solution set is $(-\infty, -3] \cup [3, \infty)$.

87. If x is the price of a car excluding sales tax then it must satisfy $0 \leq 1.1x + 300 \leq 8000$. This is equivalent to $0 \leq x \leq \frac{7700}{1.1} = 7000$. The price range of Yolanda's car is the interval $[\$0, \$7000]$.

89. Let x be Lucky's score on the final exam.

$$79 < \frac{65 + x}{2} < 90$$
$$158 < 65 + x < 180$$
$$93 < x < 115.$$

The final exam score must lie in $(93, 115)$.

91. Let x be Ingrid's final exam score. Since $\dfrac{2x+65}{3}$ is her weighted average,

$$79 < \dfrac{2x+65}{3} < 90$$
$$237 < 2x+65 < 270$$
$$172 < 2x < 205$$
$$86 < x < 102.5$$

Ingrid's final exam score must lie in $(86, 102.5)$.

93. If h is the height of the box, then

$$40 + 2(30) + 2h \leq 130$$
$$100 + 2h \leq 130$$
$$2h \leq 30.$$

The range of the height is $(0, 15]$.

95. By substituting $N = 50$ and $w = 27$ into $r = \dfrac{Nw}{n}$ we find $r = \dfrac{1350}{n}$. Moreover if $n = 14$, then $r = \dfrac{1350}{14} = 96.4$. Similarly, we can compute the other gear ratios.

n	14	17	20	24	29
r	96.4	79.4	67.5	56.25	46.6

The bicycle has a gear ratio for each of the four types.

97. If x is the price of a BMW 750 iL, then $|x - 66{,}070| > 25{,}000$. An equivalent inequality is

$$x - 66070 > 25000 \text{ or } x - 66070 < -25000$$
$$x > 91{,}070 \text{ or } x < 41{,}070$$

The price of a BMW 750 iL is either under $41,070 or over $91,070.

99. If x is the actual temperature, then

$$\left|\dfrac{x-35}{35}\right| < .01$$
$$-.35 < x - 35 < .35$$
$$34.65 < x < 35.35$$

The actual temperature must lie in the interval $(34.65, 35.35)$.

101. If c is the actual circumference, then $c = \pi d$ and

$$|\pi d - 7.2| \leq 0.1$$
$$-0.1 \leq \pi d - 7.2 \leq 0.1$$
$$7.1 \leq \pi d \leq 7.3$$
$$2.26 \leq d \leq 2.32$$

The actual diameter must lie in the interval $[2.26 \text{ cm.}, 2.32 \text{ cm.}]$.

103. The inequality $|a - b| < 2$ holds for the former Soviet Union, Iran, Venezuela, United Kingdom, United Arab Emirates, and Kuwait.

The inequality $|a - b| > 0.5$ holds for Saudi Arabia, Norway, the former Soviet Union, Iran, Venezuela, United Arab Emirates, and Kuwait.

For Thought

1. False, the sign graph of $(x-3)(x+3) > 0$ is

```
- - - - - - - - 0 + + + +
- - - - 0 + + + + + + + +
<---------------------->
        -3       3
```

and the solution set is $(-\infty, -3) \cup (3, \infty)$.

2. False, since the inequality sign will be reversed if x takes on a negative value.   **3.** False, since there are other possibilities for the signs of the factors.   **4.** True, since we can always find all its real roots and draw its sign graph.   **5.** True

6. True   **7.** False, the inequality fails at $a = 3$.

8. True   **9.** True   **10.** True

1.6 Exercises

1. The solution of $x > \dfrac{3}{2}$ is the interval $(1.5, \infty)$ and the graph is $<\!\!\text{———}(\!\!=\!\!=\!\!=\!\!)\!\!>$ at 1.5.

3. The solution of $w < -1$ is the interval $(-\infty, -1)$ and the graph is $<\!\!=\!\!=\!\!=\!\!)\!\!\text{———}\!\!>$ at -1.

5. The solution of $x > 5$ is the interval $(5, \infty)$ and the graph is $<\!\!\text{———}(\!\!=\!\!=\!\!=\!\!)\!\!>$ at 5.

7. The solution of $z < -3$ is the interval $(-\infty, -3)$ and the graph is $<\!\!=\!\!=\!\!=\!\!)\!\!\text{———}\!\!>$ at -3.

9. The sign graph of $(2x-3)(x+1) < 0$ is

```
- - - - - - - - 0 + + + +
- - - - 0 + + + + + + + +
<---------------------->
        -1       3/2
```

The solution set is the interval $(-1, 3/2)$ and the graph is $<\!\!\text{———}(\!\!=\!\!=\!\!)\!\!\text{——}\!\!>$ at -1, $3/2$.

1.6 MORE INEQUALITIES

11. The sign graph of $(x+3)(x-5) > 0$ is

```
- - - - - - - - 0 + + + +
- - - - 0 + + + + + + +
<———————————————————>
     -3           5
```

The solution set is $(-\infty, -3) \cup (5, \infty)$ and the graph is $\Longleftarrow)\!\!\overset{-3}{\rule{1em}{0.4pt}}\!\!\overset{5}{(}\!\!\Longrightarrow$

13. The sign graph of $(w+2)(w-6) \geq 0$ is

```
- - - - - - - - 0 + + + +
- - - - 0 + + + + + + +
<———————————————————>
     -2           6
```

The solution set is $(-\infty, -2] \cup [6, \infty)$ and the graph is $\Longleftarrow]\!\!\overset{-2}{\rule{1em}{0.4pt}}\!\!\overset{6}{[}\!\!\Longrightarrow$

15. The sign graph of $(t-4)(t+4) \leq 0$ is

```
- - - - - - - - 0 + + + +
- - - - 0 + + + + + + +
<———————————————————>
     -4           4
```

The solution set is $[-4, 4]$ and the graph is $<\!\!\rule{1em}{0.4pt}[\!\!\overset{-4}{\rule{0.5em}{0.4pt}}\overset{4}{=\!=}]\!\!\rule{1em}{0.4pt}\!\!>$

17. The sign graph of $(a+3)^2 \leq 0$ is

```
+ + + + 0 + + + + + + +
<———————————————————>
        -3
```

The solution set is $\{-3\}$ and the graph is

$<\!\!\rule{1em}{0.4pt}\overset{-3}{\bullet}\rule{1em}{0.4pt}\!\!>$

19. The sign graph of $(2x-3)^2 > 0$ is

```
+ + + + 0 + + + + + + +
<———————————————————>
        3/2
```

The solution set is $(-\infty, 3/2) \cup (3/2, \infty)$ and the graph is $<\!\!\Longrightarrow\!\!\overset{3/2}{)(}\!\!\Longleftarrow\!\!>$

21. $P(3) = 9 - 9 + 1 = 1$

23. $R(-1) = \dfrac{-2}{2} = -1$

25. $P(2) = (1)(-3)(-4) = 12$

27. $R(-4) = \dfrac{(-7)(9)}{(-2)(-5)} = -63/10 = -6.3$

29. The sign graph of $\dfrac{x-4}{x+2} \leq 0$ is

```
- - - - - - - - 0 + + + +
- - - - 0 + + + + + + +
<———————————————————>
     -2           4
```

The solution set is $(-2, 4]$ and the graph is $<\!\!\rule{1em}{0.4pt}(\!\!\overset{-2}{\rule{0.5em}{0.4pt}}\overset{4}{=\!=}]\!\!\rule{1em}{0.4pt}\!\!>$

31. The sign graph of $\dfrac{x-7}{x+7} < 0$ is

```
- - - - - - - - 0 + + + +
- - - - 0 + + + + + + +
<———————————————————>
     -7           7
```

The solution set is the interval $(-7, 7)$ and the graph is $<\!\!\rule{1em}{0.4pt}(\!\!\overset{-7}{\rule{0.5em}{0.4pt}}\overset{7}{=\!=})\!\!\rule{1em}{0.4pt}\!\!>$

33. The sign graph of $\dfrac{1}{y} \leq 0$ is

```
- - - - 0 + + + + + + + +
<———————————————————>
        0
```

The solution set is $(-\infty, 0)$ and the graph is $<\!\!\overset{0}{\Longrightarrow}\!\!)\!\!\rule{1em}{0.4pt}\!\!>$

35. The sign graph of $\dfrac{(w-3)(w+2)}{w-6} \geq 0$ is

```
- - - - - - - - - - - - - - 0 + + +
- - - - - - - - - - 0 + + + + + +
- - - - - 0 + + + + + + + + +
<———————————————————>
     -2         3         6
```

The solution set is $(6, \infty) \cup [-2, 3]$ and the graph is $<\!\!\rule{1em}{0.4pt}[\!\!\overset{-2}{\rule{0.5em}{0.4pt}}\overset{3}{=\!=}]\!\!\rule{1em}{0.4pt}\overset{6}{(}\!\!\Longleftarrow\!\!>$

37. The sign graph of $\dfrac{3-t}{(2t+1)(t-5)} > 0$ is

```
- - - - - - - - - - - - - - - 0 + + +
+ + + + + + + + + + 0 - - - - - -
- - - - - 0 + + + + + + + + +
<─────────┼─────────┼─────────┼─────────>
        -1/2        3         5
```

The solution set is $(-\infty, -1/2) \cup (3, 5)$ and the graph is $<\!\!=\!)\underset{-1/2\ \ 3\ \ 5}{-\!(\!=\!)\!-}\!>$

39. The sign graph of $\dfrac{q+8}{q+3} > 0$ is

```
- - - - - - - - - - 0 + + + +
- - - - - 0 + + + + + + + +
<─────────┼─────────┼─────────>
         -8        -3
```

The solution set is $(-\infty, -8) \cup (-3, \infty)$ and the graph is $<\!\!=\!)\underset{-8\ \ \ -3}{-\!\!(\!=\!\!\Rightarrow}$

41. The sign graph of $\dfrac{-5}{(x+2)(x-3)} > 0$ is

```
- - - - - - - - - - - - - - - - - - - - - - - -
- - - - - - - - - - - - - - - 0 + + + + + +
- - - - - - - 0 + + + + + + + + + +
<─────────┼─────────┼─────────>
         -2         3
```

The solution set is $(-2, 3)$ and the graph is $<\!\!-\underset{-2\ \ 3}{(\!=\!)}\!-\!>$

43. The sign graph of $\dfrac{(x-4)(x+2)}{5-x} > 0$ is

```
+ + + + + + + + + + + + + + 0 - - -
- - - - - - - - - - - - - 0 + + + + + +
- - - - - 0 + + + + + + + + +
<─────────┼─────────┼─────────┼─────────>
         -2        4         5
```

The solution set is $(-\infty, -2) \cup (4, 5)$ and the graph is $<\!\!=\!)\underset{-2\ \ \ \ 4\ \ 5}{\ \ \ -\!(\!=\!)\ \ \ }\!\!-\!\!>$

45. Let $R(x) = \dfrac{(x-3)(x+1)}{x-5} < 0$.

Since $R(-2) < 0$, $R(0) > 0$, $R(4) < 0$, and $R(7) > 0$, from the graph

```
        test       test       test       test
         pt         pt         pt         pt
          0          0          U          0
<─────┼────┼────┼────┼────┼────┼────>
      -2   -1   0    3    4    5    7
```

the solution set is $[-1, 3] \cup (5, \infty)$ and the graph is $<\!-\underset{-1\ 3\ \ 5}{[\!=\!]\!-\!(\!=}\!>$

47. Let $R(x) = \dfrac{x^2 - 25}{9 - x^2}$. Since $R(-6) < 0$, $R(-4) > 0$, $R(0) < 0$, $R(4) > 0$, and $R(6) < 0$, from the graph

```
    test      test      test      test      test
     pt        pt        pt        pt        pt
      0         U         U         0
<───┼───┼───┼───┼───┼───┼───┼───>
   -6  -5  -4  -3   0   3   4   5   6
```

the solution set is $(-\infty, -5] \cup (-3, 3) \cup [5, \infty)$ and the graph is $<\!\!=\!]\underset{-5\ \ -3\ \ 3\ \ 5}{-\!(\!=\!)\!-\![\!=}\!>$

49. Let $R(x) = (x-6)(x+4) > 0$.
Since $R(-5) > 0$, $R(0) < 0$, and $R(7) > 0$, from the graph

```
       test       test       test
        pt         pt         pt
         0          0
<─────┼────┼────┼────┼────>
      -5   -4   0    6    7
```

the solution set is $(-\infty, -4) \cup (6, \infty)$ and its graph is $<\!\!=\!)\underset{-4\ \ \ 6}{-\!\!(\!\!=}\!>$

51. Let $R(x) = (2x-3)(2x+3) < 0$.
Since $R(-2) > 0$, $R(0) < 0$, and $R(2) > 0$, from the graph

```
       test       test       test
        pt         pt         pt
         0          0
<─────┼────┼────┼────┼────>
      -2  -3/2  0   3/2   2
```

the solution set is $(-3/2, 3/2)$ and its graph is $<\!\!-\underset{-3/2\ \ 3/2}{(\!=\!)}\!-\!>$

1.6 MORE INEQUALITIES

53. Let $R(y) = y^2 - 10y + 23 > 0$.
Since $R(4) > 0, R(5) < 0$, and $R(6) > 0$, from the graph

```
        test       test       test
        pt    0    pt    0    pt
  <----------------------------------->
      4   5-√2   5   5+√2   6
```
the solution set is $(-\infty, 5 - \sqrt{2}) \cup (5 + \sqrt{2}, \infty)$
and the graph is $\Longleftarrow\ )\ \overset{5-\sqrt{2}}{}\ \overset{5+\sqrt{2}}{(}\ \Longrightarrow$

55. Since $R(t) = t^2 - 6t + 10$ has no real roots then the solution set is $(-\infty, \infty)$ and the graph is $\Longleftarrow\Longrightarrow$

57. Since the sign graph of $\dfrac{w - 1}{w^2} > 0$ is

```
                  - - - - -  0 + + + +
     + + + + 0 + + + + + + +
  <----------------------------------->
            0           1
```
then the solution set is the interval $(1, \infty)$ and the graph is $\Longleftarrow\overset{1}{(}\Longrightarrow$

59. Since the sign graph of $\dfrac{w^2 - 4w + 5}{w - 3} > 0$ is

```
   - - - -  0 + + + + + + + +
  <----------------------------------->
           3
```
the solution is the interval $(3, \infty)$ and the graph is $\Longleftarrow\overset{3}{(}\Longrightarrow$

61. The sign graph is

```
   - - - - - - - - - - - -  0 + + +
   - - - - - - - - - -  0 + + + + + +
   - - - - - 0 + + + | | | | | | |
  <----------------------------------->
        -2/3       1       5
```
The solution set is $(-\infty, -2/3) \cup (1, 5)$ and the graph is $\Longleftarrow\overset{-2/3}{)}\ \overset{1}{-}\ \overset{5}{(}\ =\ )\ \longrightarrow$

63. Let $R(y) = (y - 1)(y - 3)(y - 5)(y - 7) > 0$.
Since $R(0) > 0, R(2) < 0, R(4) > 0, R(6) < 0$ and $R(8) > 0$, from the graph

```
     test      test      test      test      test
     pt   0    pt   0    pt   0    pt   0    pt
  <----------------------------------->
     0   1   2   3   4   5   6   7   8
```
the solution set is $(-\infty, 1) \cup (3, 5) \cup (7, \infty)$
and the graph is $\Longleftarrow\overset{1}{)}\ \overset{3}{-}\ \overset{5}{(=)}\ \overset{7}{-}\ (\Longrightarrow$

65. Since the sign graph of $y(y - 3)(y + 3) > 0$ is

```
   - - - - - - - - - - - -  0 + + +
   - - - - - - - - - -  0 + + + + + +
   - - - - - 0 + + + + + + + + +
  <----------------------------------->
       -3           0           3
```
the solution set is the interval $[-3, 0] \cup [3, \infty)$ and the graph is
$\Longleftarrow\ [\ \overset{-3}{=}\ ]\ \overset{0}{-}\ [\ \overset{3}{=}\ \Longrightarrow$

67. Since the sign graph of $5t^2 - 3t - 20 > 0$ is

```
   - - - - - - - - - - -  0 + + + + + +
   - - - - - - -  0 + + + + + + + + + +
  <----------------------------------->
          -1.7            2.3
```
the solution set is
$\left(-\infty, \dfrac{3 - \sqrt{409}}{10}\right) \cup \left(\dfrac{3 + \sqrt{409}}{10}, \infty\right)$
and the graph is $\Longleftarrow\ )\ \overset{-1.7}{}\ \overset{2.3}{-}\ (\Longrightarrow$

69. Since the sign graph of $\dfrac{(z - 6)(z - 2)}{z - 3} \geq 0$ is

```
   - - - - - - - - - - - - -  0 + + +
   - - - - - - - - - -  0 + + + + + +
   - - - - - 0 + + + + + + + + +
  <----------------------------------->
          2           3           6
```
the solution set is $[2, 3) \cup [6, \infty)$ and the graph is $\Longleftarrow\ [\ \overset{2}{=}\ )\ \overset{3}{-}\ [\ \overset{6}{=}\ \Longrightarrow$

71. Since the sign graph of $\dfrac{(-10d-2)}{(d+2)(d-4)} > 0$ is

```
- - - - - - - - - - - - - - - 0 + + +
+ + + + + + + + + 0 - - - - - -
- - - - - 0 + + + + + + + + +
<-----------|-----|---------|----->
           -2   -1/5        4
```

the solution set is $(-\infty, -2) \cup (-1/5, 4)$ and the graph is

$$<\!\!=\!\!\overset{-2}{)}\!\!-\!\!\overset{-1/5}{(}\!\!=\!\!\overset{4}{)}\!\!-\!\!>$$

73. Let $R(u) = \dfrac{(u-2)(u-1)}{(u-3)(-4)} \leq 0$. Since $R(0) > 0$, $R(1.5) < 0$, $R(2.5) > 0$, $R(3.5) < 0$ and $R(5) > 0$ then from the graph below

```
test    test    test    test    test
 pt  0   pt  0   pt  U   pt  U   pt
<---|---|---|---|---|---|---|---|--->
    0   1  1.5  2  2.5  3  3.5  4   5
```

the solution set is $[1, 2] \cup (3, 4)$ and the graph is $<\!\!-\!\!\overset{1}{[}\!\!=\!\!\overset{2}{]}\!\!-\!\!\overset{3}{(}\!\!=\!\!\overset{4}{)}\!\!-\!\!>$

75. Set the right-hand-side to 0.
Let $R(t) = \dfrac{t^2 - 5t + 2}{(t-1)(t-2)} \geq 0$.
Since $R(0) > 0$, $R(0.6) < 0$, $R(1.5) > 0$, $R(3) < 0$ and $R(6) > 0$, from the graph below

```
test    test    test    test    test
 pt  0   pt  0   pt  U   pt  U   pt
<---|---|---|---|---|---|---|---|--->
    0  0.4 0.6  1  1.5  2   3  4.6  6
```

the solution set is
$(-\infty, (5-\sqrt{17})/2] \cup (1, 2) \cup [(5+\sqrt{17})/2, \infty)$
and the graph is $<\!\!=\!\!\overset{0.4}{]}\!\!-\!\!\overset{1}{(}\!\!=\!\!\overset{2}{)}\!\!-\!\!\overset{4.6}{[}\!\!=\!\!>$

77. The solution set of $2w - 1 \geq 0$ is $[1/2, \infty)$.

79. The solution set of $9 - w^2 \geq 0$ is $[-3, 3]$.

81. Let $R(x) = w^2 - w - 6 > 0$. Since $R(-3) > 0$, $R(0) < 0$, and $R(4) > 0$, from the graph

```
        test        test        test
         pt    U     pt    U     pt
<---|----|----|------|------|----|--->
   -3   -2    0      3      4
```

the solution set is $(-\infty, -2) \cup (3, \infty)$.

83. The sign graph of $\dfrac{w-1}{w+2} \geq 0$ is

```
- - - - - - - - - 0 + + + +
- - - - 0 + + + + + + + +
<-----------|---------|----->
           -2         1
```

The solution set is $(-\infty, -2) \cup [1, \infty)$.

85. If x is the length of the north side then $x(50 - x) \leq 400$. This is equivalent to $x^2 - 50x + 400 \geq 0$ whose sign graph is

```
- - - - - - - - - 0 + + + +
- - - - 0 + + + + + + + +
<-----------|---------|----->
           10        40
```

Since $x < 50$, the possible lengths would lie in $(0, 10] \cup [40, 50)$.

87. Let t be the number of seconds since the flare was fired. The inequality $-16t^2 + 80t + 6 > 100$ is equivalent to $16t^2 - 80t + 94 < 0$ whose sign graph is

```
- - - - - - - - - 0 + + + +
- - - - 0 + + + + + + + +
<-----------|---------|----->
          1.9        3.1
```

The flare was over 100 ft. between 1.9 and 3.1 seconds, which is 1.2 seconds long.

89. If x is the first number then $x(12-x) \geq 10$. This is equivalent to $x^2 - 12x + 10 \leq 0$ whose sign graph is

```
- - - - - - - - 0 + + + +
- - - - 0 + + + + + + + +
←――――――――――――――――→
      0.9        11.1
```

The possible values for the first number are in the interval $\left[6 - \sqrt{26}, 6 + \sqrt{26}\right]$.

91. Since r must satisfy $\sqrt{\dfrac{43}{38}} \leq 1 + r$, $r > 0.0638$ and the annual inflation rate is bigger than 6.38%

93. The solutions to $v_o^2 + 2(-9.8)(1.17) = 0$ are ± 4.788. So Brown's initial velocity is 4.788 m/sec.

95. If $a < 0.5$, then $3.89 \times 10^{-10} h^2 - 3.48 \times 10^{-5} h + 0.5 < 0$. By using the quadratic formula, one finds that the zeros of the left side of the inequality are $h_1 \approx 17{,}982.5$ and $h_2 \approx 71{,}477$. We conclude from the given sign graph of the inequality

```
- - - - - - - - 0 + + + +
- - - - 0 + + + + + + + +
←――――――――――――――――→
       h_1       h_2
```

that the atmospheric pressure is less than 0.5 when the altitude lies in the range $(17{,}982.5, 71{,}477)$. Of course it is also less than 0.5 above 71,477 ft. So the answer is $(17{,}982.5, \infty)$.

97. The inequality $\dfrac{400{,}000p}{100 - p} < 1{,}200{,}000$ can be written as $\dfrac{1{,}600{,}000p - 120{,}000{,}000}{100 - p} < 0$. By using a sign graph, the solution set is $(-\infty, 75] \cup (100, \infty)$. The percentage of coliform that can be removed is less than 75%.

Chapter 1 Review Exercises

1. Since $3x = 2$, the solution set is $\{2/3\}$.

3. The solution set of $x^2 = \pm\dfrac{2}{3}$ is $\{\pm\sqrt{6}/3\}$.

5. Multiply by 60 to get $30y - 20 = 15y + 12$, or $15y = 32$. The solution set is $\{32/15\}$.

7. Multiply by $x(x-1)$ to get $2x - 2 = 3x$. The solution set is $\{-2\}$.

9. Multiply by $(x+1)(x-3)$ and get $-2x - 3 = x - 2$. So $-1 = 3x$. The solution set is $\{-1/3\}$.

11. Multiply by $(x+4)(x-4)$ to get $-2x - 8 = 2x - 8$. So $0 = 4x$. The solution set is $\{0\}$.

13. Multiply by $2x(x-1)$,
$$2x - 2 + 2x = 3x^2 - 3x$$
$$0 = 3x^2 - 7x + 2$$
$$0 = (3x - 1)(x - 2)$$
The solution set is $\{1/3, 2\}$.

15. Rearrange and complete the square,
$$b^2 - 6b = -10$$
$$(b - 3)^2 = -10 + 9$$
$$b = 3 \pm i$$
The solution set is $\{3 \pm i\}$.

17. Rearrange and complete the square,
$$s^2 - 4s = -1$$
$$(s - 2)^2 = -1 + 4 = 3$$
$$s = 2 \pm \sqrt{3}$$
The solution set is $\{2 \pm \sqrt{3}\}$.

19. Solve an equivalent statement
$$3q - 4 = 2 \quad \text{or} \quad 3q - 4 = -2$$
$$3q = 6 \quad \text{or} \quad 3q = 2.$$
The solution set is $\{2/3, 2\}$.

21. We obtain
$$|2h - 3| = 0$$
$$2h - 3 = 0$$
$$h = \dfrac{3}{2}.$$
The solution set is $\left\{\dfrac{3}{2}\right\}$.

23. The solution set of $x > 3$ is the interval $(3, \infty)$ and the graph is <——(===>

25. The solution set of $8 > 2x$ is the interval $(-\infty, 4)$ and the graph is <===)——>

27. Since $-\dfrac{7}{3} > \dfrac{1}{2}x$, the solution set is $(-\infty, -14/3)$ and the graph is <===)——>

29. After multiplying the inequality by 2 we have
$$-4 < x - 3 \leq 10$$
$$-1 < x \leq 13.$$
The solution set is the interval $(-1, 13]$ and the graph is <——(===]——>

31. The solution set of $\dfrac{1}{2} < x$ and $x < 1$ is the interval $(1/2, 1)$ and the graph is <——(===)——>

33. The solution set of $x > -4$ or $x > -1$ is the interval $(-4, \infty)$ and the graph is <——(===>

35. Solve an equivalent statement
$$x - 3 > 2 \quad \text{or} \quad x - 3 < -2$$
$$x > 5 \quad \text{or} \quad x < 1.$$
The solution set is $(-\infty, 1) \cup (5, \infty)$ and the graph is <==)——(==>

37. Since an absolute value is nonnegative, $2x - 7 = 0$. The solution set is $\{7/2\}$ and the graph is <——•——>

39. Since absolute values are nonnegative, the solution set is $(-\infty, \infty)$ and the graph is <=====>

41. Set the right side to zero then factor. The sign graph of $(4x - 1)(2x - 1) < 0$ is

```
- - - - - - - 0 + + + +
- - - - 0 + + + + + + + +
←―――――――――――――――――――→
        1/4     1/2
```

The solution set is the interval $(1/4, 1/2)$.

43. The sign graph of $(3 - x)(x + 5) < 0$ is

```
+ + + + + + + 0 - - - -
- - - - 0 + + + + + + +
←―――――――――――――――――――→
      -5          3
```

The solution set is the interval $[-5, 3]$.

45. The sign graph of $\dfrac{x - 3}{5 - x} \geq 0$ is

```
+ + + + + + + 0 - - - -
- - - - 0 + + + + + + +
←―――――――――――――――――――→
        3         5
```

The solution set is the interval $[3, 5)$.

47. Let $R(x) = \dfrac{x + 10}{x + 2} - 5 = \dfrac{-4x}{x + 2} < 0$. Since $R(-3) < 0, R(-1) > 0$, and $R(1) < 0$ then from the graph

```
test        test        test
 pt          pt          pt
       U           0
←―――――――――――――――――――→
 -3    -2    -1    0    1
```

the solution is $(-\infty, -2) \cup (0, \infty)$.

49. Let $R(x) = \dfrac{12 - 7x}{x^2} + 1 = \dfrac{(x - 3)(x - 4)}{x^2} > 0$. Since $R(-1) > 0, R(1) > 0, R(3.5) < 0$, and $R(5) > 0$,

```
test      test      test      test
 pt        pt        pt        pt
      U         0         0
←―――――――――――――――――――――――→
 -1   0    1    3   3.5   4    5
```

the solution is $(-\infty, 0) \cup (0, 3) \cup (4, \infty)$.

51. Let $R(x) = x^2 - 2x - 4 \leq 0$. Since $R(-2) > 0$, $R(1) < 0$, and $R(4) > 0$, from the graph

```
test          test          test
 pt            pt            pt
         0             0
←―――――――――――――――――――→
 -2   1-√5    1    1+√5    4
```

the solution is $[1 - \sqrt{5}, 1 + \sqrt{5}]$.

REVIEW EXERCISES

53. Let $R(x) = \dfrac{(x-1)(x-2)}{(x-3)(x-4)}$.

    Since $R(0) > 0, R(1.5) < 0, R(2.5) > 0$, $R(3.5) < 0, R(5) > 0$ and the graph below

    the solution is $(-\infty, 1] \cup [2, 3) \cup (4, \infty)$.

    The graph is

55. No solution since $\dfrac{x+3}{x+3} = 1$ for $x \neq -3$.

57. Since $(x-1)^2 \geq 0$, the solution set is $(-\infty, 1) \cup (1, \infty)$ and the graph is

59. Since $2x - 6 = 3y$ then $y = \dfrac{2}{3}x - 2$.

61. Since $y(x-3) = 1$ then $y = \dfrac{1}{x-3}$.

63. Since $by = -ax + c$ then $y = -\dfrac{a}{b}x + \dfrac{c}{b}$.

65. Multiply by $2xy$ and get $2x = 2y + xy$. Factor as $2x = y(2+x)$ and so $y = \dfrac{2x}{x+2}$.

67. $-1 - i$

69. $16 - 40i + 25i^2 = 16 - 25 - 40i = -9 - 40i$

71. $2 + 6i - 6i - 18i^2 = 20$

73. $\dfrac{2}{i} - \dfrac{3i}{i} = \dfrac{2(-i)}{i(-i)} - 3 = \dfrac{-2i}{1} - 3 = -2i - 3$

75. $\dfrac{1-i}{2+i} \cdot \dfrac{2-i}{2-i} = \dfrac{1-3i}{5} = \dfrac{1}{5} - i\dfrac{3}{5}$

77. $\dfrac{1+i}{2-3i} \cdot \dfrac{2+3i}{2+3i} = \dfrac{-1+5i}{4+9} = -\dfrac{1}{13} + i\dfrac{5}{13}$

79. $\dfrac{6 + 2i\sqrt{2}}{2} = 3 + i\sqrt{2}$

81. $\dfrac{-6 + \sqrt{-20}}{-8} = \dfrac{-6 + 2i\sqrt{5}}{-8} = \dfrac{3}{4} - i\dfrac{\sqrt{5}}{4}$

83. $(i^4)^8 \cdot i^2 + (i^4)^4 \cdot i^3 = 1 \cdot (-1) + 1 \cdot (-i) = -1 - i$

85. Since $3x - 4 \geq 0$, solution set is $[4/3, \infty)$.

87. Since the sign graph of $x^2 - 25 \geq 0$ is

    the solution is $(-\infty, -5] \cup [5, \infty)$.

89. The discriminant of $x^2 - 4x + 2$ is $(-4)^2 - 4(2) = 8$. There are two distinct real roots.

91. The discriminant is $(-20)^2 - 4(4)(25) = 0$. Only one root and it is real.

93. Let x be the length of one side of the square. Since dimensions of the base are $8 - 2x$ and $11 - 2x$ then
$$(11 - 2x)(8 - 2x) = 50$$
$$-4x^2 + 38x - 38 = 0$$
$$2x^2 - 19x + 19 = 0$$
$$x = \dfrac{19 \pm \sqrt{209}}{4} \approx 8.36, 1.14$$
But $x = 8.36$ is too big and so $x = 1.14$ inch.

95. Let x be the number of hours it takes Lisa or Taro to drive to the restaurant. Since the sum of the driving distances is 300, x must satisfy
$$300 = 50x + 60x. \text{ So } x = \dfrac{300}{110} \approx 2.7272$$
and Lisa drove $50(2.7272) \approx 136.4$ miles.

97. Let x and $8000 - x$ be the number of fish in Homer Lake and Mirror lake, respectively. Then
$$0.2x + 0.3(8000 - x) = 0.28(8000)$$
$$-0.1x + 2400 = 2240$$
$$1600 = x$$
There were originally 1600 fish in Homer Lake.

99. Let x be the distance she hiked in the northern direction. Then she hiked $32 - x$ miles in the eastern direction. By the Pythagorean Theorem,
$$x^2 + (32 - x)^2 = (4\sqrt{34})^2$$
$$2x^2 - 64x + 480 = 0$$
$$2 \cdot (x - 20)(x - 12) = 0$$
$$x = 20, 12$$
Since the eastern direction was the shorter leg of the journey, the northern direction was 20 miles.

101. Let x and $x+50$ be the cost of a haircut at Joe's and Renee's, respectively. Since 5 haircuts at Joe's is less than one haircut at Renee's, $5x < x + 50$.
So $x < \$12.50$

103. Let x and $x + 2$ be the length and width of a picture frame in inches. Since there are between 32 and 50 inches of molding, x must satisfy
$$32 < 2x + 2(x+2) < 50$$
$$32 < 4x + 4 < 50$$
$$28 < 4x < 46$$
$$7 < x < 11.5$$
The possible widths are between 7 and 11.5 inches.

105. If the average gas mileage is increased from 29.5 mpg to 31.5 mpg then the number of gallons of gas
saved is $\dfrac{10^{12}}{29.5} - \dfrac{10^{12}}{31.5} \approx 2.15 \times 10^9$.
Suppose the mileage is increased to x from 29.5 mpg. Then x must satisfy
$$\dfrac{10^{12}}{29.5} - \dfrac{10^{12}}{x} = \dfrac{10^{12}}{27.5} - \dfrac{10^{12}}{29.5}$$
$$\dfrac{1}{29.5} - \dfrac{1}{x} = \dfrac{1}{27.5} - \dfrac{1}{29.5}$$
$$-\dfrac{1}{x} \approx -0.031433$$
$$x \approx 31.8$$
The mileage must be increased to 31.8 mpg.

Chapter 1 Test

1. Multiply by 6 and get $3x - 2x = 1$.
The solution set is $\{1\}$.

2. Since $x^2 = \dfrac{2}{3}$, $x = \pm\dfrac{\sqrt{2}}{\sqrt{3}} = \pm\dfrac{\sqrt{6}}{3}$.
The solution set is $\left\{\pm\dfrac{\sqrt{6}}{3}\right\}$.

3. By completing the square, we obtain
$$x^2 - 6x = -1$$
$$(x-3)^2 = -1 + 9$$
$$(x-3)^2 = 8$$
$$x - 3 = \pm\sqrt{8}.$$
The solution set is $\{3 \pm 2\sqrt{2}\}$.

4. Factor $x^2 - 9x + 14 = 0$ and get $(x-2)(x-7) = 0$.
The solution set is $\{2, 7\}$.

5. After cross-multiplying, we get
$$(x-1)(x-6) = (x+3)(x+2)$$
$$x^2 - 7x + 6 = x^2 + 5x + 6$$
$$-7x = 5x$$
$$0 = 12x$$
The solution set is $\{0\}$.

6. By completing the square, we obtain
$$x^2 - 2x = -5$$
$$(x-1)^2 = -5 + 1$$
$$(x-1)^2 = -4$$
$$x - 1 = \pm 2i.$$
The solution set is $\{1 \pm 2i\}$.

7. Since $-4 > 2x$ then the solution set is $(-\infty, -2)$ and the graph is $<\!\!=\!\!=\!\!)\!\!-\!\!-\!\!>$ with -2 marked.

8. The solution set to $x > 6$ and $x > 5$ is the interval $(6, \infty)$ and the graph is $<\!\!-\!\!-\!\!(\!\!=\!\!=\!\!>$ with 6 marked.

9. Solve an equivalent statement
$$-3 \le 2x - 1 \le 3$$
$$-2 \le 2x \le 4$$
$$-1 \le x \le 2$$
The solution set is the interval $[-1, 2]$ and the graph is $<\!\!-\!\![\!\!=\!\!=\!\!]\!\!-\!\!>$ with -1 and 2 marked.

10. Rewrite $|x - 3| > 2$ without any absolute values

$$x - 3 > 2 \text{ or } x - 3 < -2$$
$$x > 5 \text{ or } x < 1$$

The solution set is $(-\infty, 1) \cup (5, \infty)$ and the graph is <———1———5———>

11. $16 - 24i + 9i^2 = 7 - 24i$

12. $\dfrac{(2-i)(3-i)}{(3+i)(3-i)} = \dfrac{6 - 5i + i^2}{10} = \dfrac{5 - 5i}{10} = \dfrac{1}{2} - \dfrac{1}{2}i$

13. $i^4 i^2 - (i^4)^8 i^3 = 1(-1) - (1)(-i) = -1 + i$

14. $2i\sqrt{2}(i\sqrt{2} + \sqrt{6}) = -4 + 2i\sqrt{12} = -4 + 4i\sqrt{3}$

15. The sign graph of $(x - 4)(x + 2) < 0$ is

```
- - - - - - - - - 0 + + + +
- - - - 0 + + + + + + + +
<————————————————————>
      -2         4
```

The solution set is the interval $(-2, 4)$.

16. Set the right-hand side to zero and get $\dfrac{2x - 1}{x - 3} > 0$. The sign chart is

```
- - - - - - - - - 0 + + + +
- - - - 0 + + + + + + + +
<————————————————————>
      1/2        3
```

The solution set is $(-\infty, 1/2) \cup (3, \infty)$.

17. The sign chart of $\dfrac{x + 3}{(4 - x)(x + 1)} \geq 0$ is

```
+ + + + + + + + + + + + + + 0 - - -
- - - - - - - - - - - 0 + + + + + +
- - - - - 0 + + | | | + + + +
<————————————————————>
    -3       -1        4
```

The solution set is $(-\infty, -3] \cup (-1, 4)$.

18. $25 - 4(9) = -11$

19. $(7 - 2i)(7 + 2i) = 49 - 4i^2 = 53$

20. Since $1 = 2y + 3xy = y(2 + 3x)$, $y = \dfrac{1}{3x + 2}$.

21. $[5, \infty)$

22. If x is the original length of one side of the square then

$$(x + 20)(x + 10) = 999$$
$$x^2 + 30x + 200 = 999$$
$$x^2 + 30x - 799 = 0$$
$$\dfrac{-30 \pm \sqrt{900 + 4(799)}}{2} = x$$
$$\dfrac{-30 \pm 64}{2} = x$$
$$17, -47 = x$$

So $x = 17$ and the original area is $17^2 = 289$ ft^2.

23. Let x be the number of gallons of the 20% solution. From the concentrations,

$$0.3(10 + x) = 0.5(10) + 0.2x$$
$$3 + 0.3x = 5 + 0.2x$$
$$0.1x = 2$$
$$x = 20.$$

Then 20 gallons of the 20% solution are needed.

Tying It All Together

1. $7x$ 2. $30x^2$ 3. $\dfrac{2}{2x} + \dfrac{1}{2x} = \dfrac{3}{2x}$

4. $x^2 + 6x + 9$ 5. $6x^2 + x - 2$

6. $\dfrac{x^2 + 2xh + h^2 - x^2}{h} = \dfrac{2xh + h^2}{h} = 2x + h$

7. $\dfrac{x + 1}{(x - 1)(x + 1)} + \dfrac{x - 1}{(x + 1)(x - 1)} = \dfrac{2x}{x^2 - 1}$

8. $x^2 + 3x + \dfrac{9}{4}$

9. An identity and the solution set is $(-\infty, \infty)$.

10. Since $30x^2 - 11x = x(30x - 11) = 0$, the solution set is $\{0, 11/30\}$.

11. Since $\dfrac{3}{2x} = \dfrac{3}{2x}$ then the solution set is $(-\infty, 0) \cup (0, \infty)$.

12. Subtract x^2 from $x^2 + 6x + 9 = x^2 + 9$ and get $6x = 0$. The solution set is $\{0\}$.

13. Since $(2x - 1)(3x + 2) = 0$, the solution set is $\left\{\dfrac{1}{2}, -\dfrac{2}{3}\right\}$.

14. Multiply the equation by $8(x+1)(x-1)$

$$8x + 8 + 8x - 8 = 5(x^2 - 1)$$
$$0 = 5x^2 - 16x - 5$$
$$x = \frac{16 \pm \sqrt{356}}{10}$$
$$x = \frac{16 \pm 2\sqrt{89}}{10}$$

The solution set is $\left\{\dfrac{8 \pm \sqrt{89}}{5}\right\}$.

15. Since $7x - 7x^2 = 7x(1-x) = 0$, the solution set is $\{0, 1\}$.   **16.** Since $7x - 7 = 7(x-1) = 0$, the solution set is $\{1\}$.

17. 0   **18.** $-1-3+2 = -2$ **19.** $-\dfrac{1}{8} - 3\dfrac{2}{8} + \dfrac{16}{8} = \dfrac{9}{8}$

20. $-\dfrac{1}{27} - 3\dfrac{3}{27} + \dfrac{54}{27} = \dfrac{44}{27}$ **21.** $-1 - 3 - 4 = -8$

22. $-4 + 6 - 4 = -2$ **23.** -4

24. $-0.25 + 1.5 - 4 = -2.75$

For Thought

1. False, the point $(2,-3)$ is in Quadrant IV.

2. False, the point $(4,0)$ does not belong to any quadrant.   3.   False, since the distance is $\sqrt{(a-c)^2+(b-d)^2}$.   4.   False, since $Ax+By=C$ is a linear equation.

5. True, since the x-intercept can be obtained by replacing y by 0.

6. False, since $\sqrt{7^2+9^2}=\sqrt{130}\approx 11.4$

7. True   8.   True   9.   True

10. False, since $\dfrac{2+3}{3}$ is not between 2 and 3.

2.1 Exercises

1. $(4,1)$, Quadrant I   3. $(1,0)$, x-axis

5. $(5,-1)$, Quadrant IV   7. $(-4,-2)$, Quadrant III

9. $(-2,4)$, Quadrant II

11.

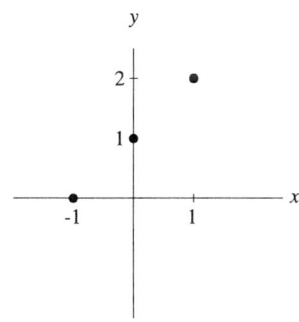

13. $y=x$ goes through $(0,0),(2,2)$.

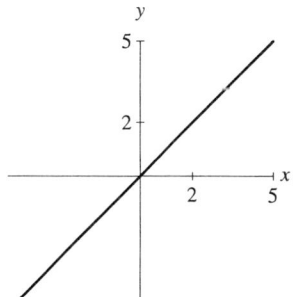

15.

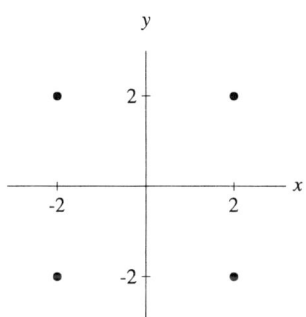

17. $y=-2x+3$ goes through $(0,3),(3/2,0)$

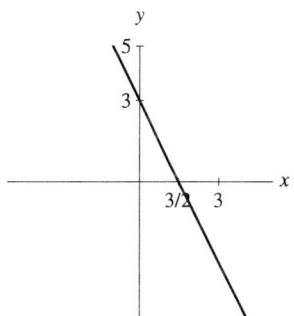

19.

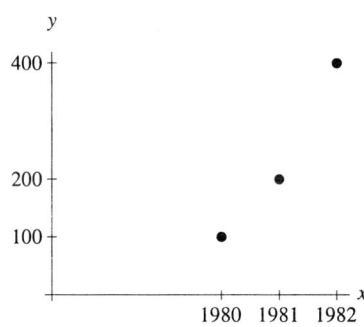

21. $x = 3$ goes through $(3,0), (3,2)$.

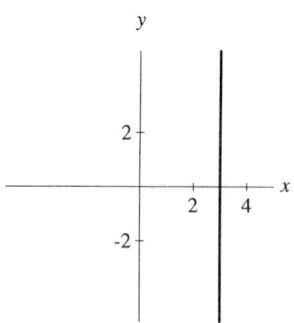

23. $y = 3x - 4$ goes through $(2,2), (0,-4)$.

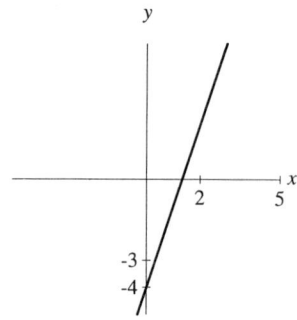

25. $3x - y = 6$ goes through $(0,-6), (2,0)$

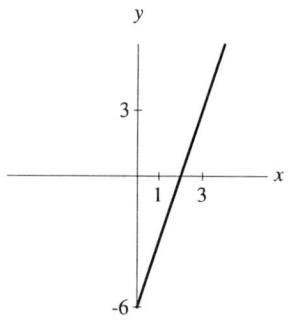

27. $x + y = 80$ goes through $(0,80), (80,0)$

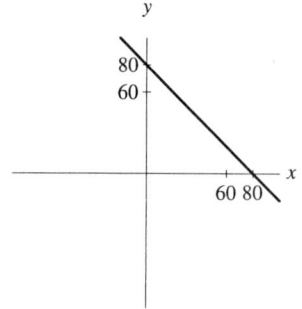

29. $x = 3y - 90$ goes through $(-90, 0), (0, 30)$

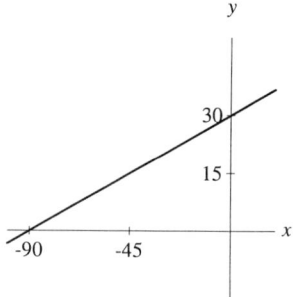

31. $\frac{1}{2}x - \frac{1}{3}y = 600$ goes through $(1200, 0), (0, -1800)$

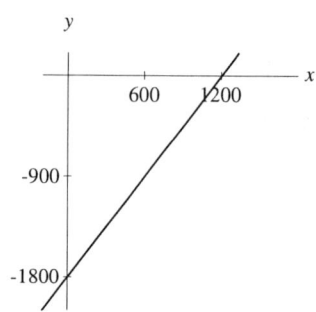

33. Intercepts are $(0,5), (4,0)$

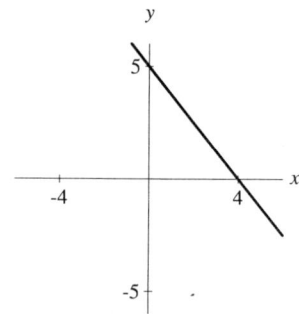

35. Intercepts are $(2/3, 0), (0, -2)$

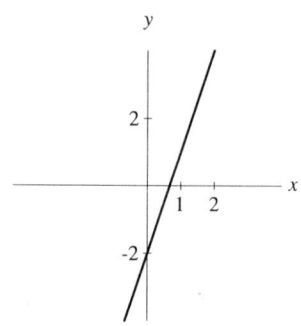

37. Intercepts are $(0.005, 0), (0, 0.0025)$

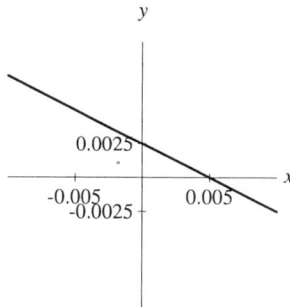

39. Intercepts are $(-20, 0), (0, 15)$

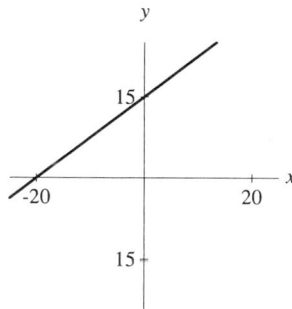

41. Intercepts are $(40, 0), (0, 60)$

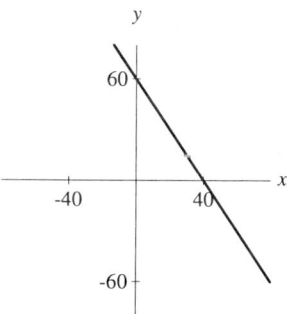

43. Intercepts are $(5000, 0), (0, 2500)$

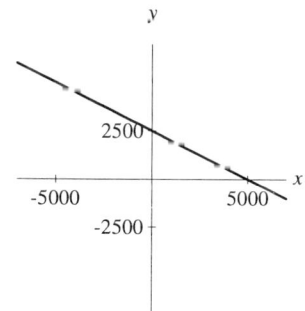

45. The solution is $-\dfrac{3.4}{12} \approx -2.83$

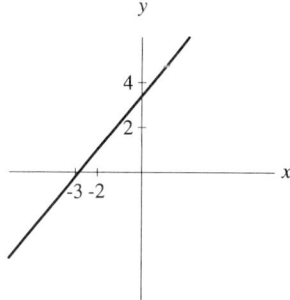

47. The solution is $\dfrac{687}{1.23} \approx 558.54$

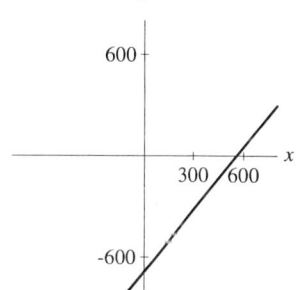

49. Solution is $\dfrac{3497}{0.03} \approx 116,566.67$

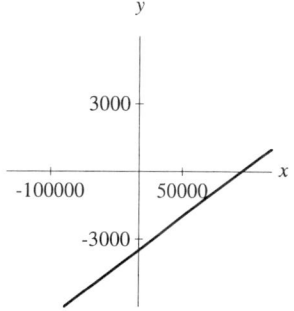

51. Note,

$$4.3 - 3.1(2.3x) + 3.1(9.9) = 0$$
$$4.3 - 7.13x + 30.69 = 0$$
$$34.99 - 7.13x = 0$$
$$x = \frac{3499}{713}$$
$$x \approx 4.91$$

The solution is $x \approx 4.91$.

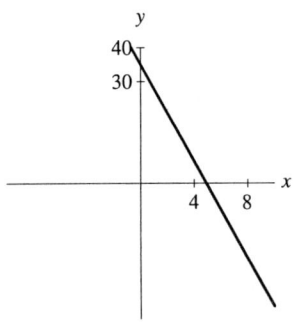

53. $x = 5$

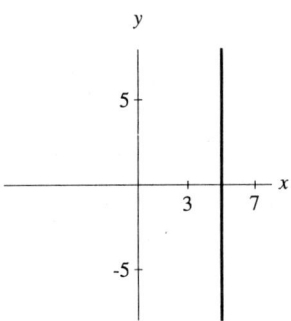

55. $y = 4$

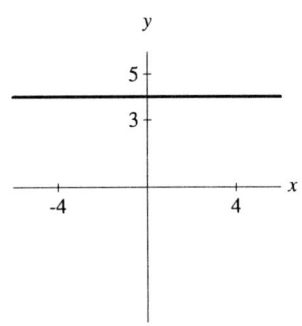

57. $x = -4$

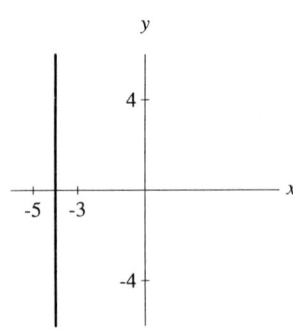

59. $y = 1$

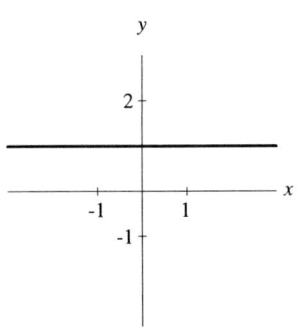

2.1 THE CARTESIAN COORDINATE SYSTEM

61. (a) By using the linear regression key of a TI-82 graphing calculator on the data set

y	E
1990	2688
1991	2902
1992	3144
1993	3331
1994	3510

we obtain $E = 207.3y - 409,826.6$ and its graph is

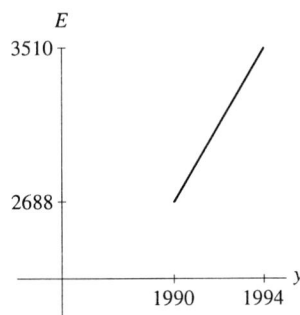

(b) In $y = 1995$, the per capita expenditure is $E = 207.3(1995) - 409,826.6 \approx \$3,737$

(d) If $y = 2001$, then $E \approx \$4,981$. On a monthly average, this is $\dfrac{4,981}{12} \approx \$415$; thus, a self-employed person could buy a good health insurance policy for $500 per month in the year 2001.

63. distance is $\sqrt{(4-1)^2 + (7-3)^2} = \sqrt{9+16} = 5$, midpoint is $(2.5, 5)$

65. distance is $\sqrt{(-1-1)^2 + (-2-0)^2} = \sqrt{4+4} = 2\sqrt{2}$, midpoint is $(0, -1)$

67. distance is $\sqrt{(-1+3\sqrt{3}-(-1))^2 + (4-1)^2} = \sqrt{27+9} = 6$, midpoint is $\left(\dfrac{-2+3\sqrt{3}}{2}, \dfrac{5}{2}\right)$

69. distance is $\sqrt{(1.2+3.8)^2 + (4.4+2.2)^2} = \sqrt{25+49} = \sqrt{74}$, midpoint is $(-1.3, 1.3)$

71. Distance is $\sqrt{(a-b)^2 + 0} = |a-b|$, midpoint is $\left(\dfrac{a+b}{2}, 0\right)$

73. Distance is $\dfrac{\sqrt{\pi^2+4}}{2}$, midpoint is $\left(\dfrac{3\pi}{4}, \dfrac{1}{2}\right)$

75. No, it is not a right triangle since the sides have lengths $\sqrt{41}, \sqrt{85}$, and $\sqrt{34}$.

77. Yes, it is an isosceles triangle since the lengths of the sides are $5, 5$, and $\sqrt{10}$.

79. Since $4 + (k+2)^2 = 36$, $(k+2)^2 = 32$ and $k+2 = \pm 4\sqrt{2}$ by the square root property. So $k = -2 \pm 4\sqrt{2}$.

81. Solving $2 = \dfrac{a-1}{2}$ and $5 = \dfrac{b+3}{2}$ one gets $(a, b) = (5, 7)$.

83. It is a parallelogram since opposite sides have the same lengths. The common lengths are $\sqrt{8}, \sqrt{17}$.

85. The midpoint is $\left(\dfrac{1970+1996}{2}, \dfrac{64.2+33.3}{2}\right) = (1983, 48.75)$, i.e., in 1983, 48.75% of the women in the 20-24 age group have married.

87. If $D = 22,800$ lbs., then a graph of $C = \dfrac{4B}{\sqrt[3]{22,800}}$ is given below.

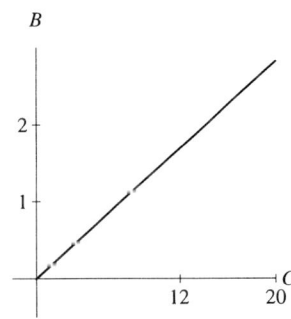

For Island Packet 40, $C = \dfrac{4(12 + \frac{11}{12})}{\sqrt[3]{22,800}} \approx 1.822$

89. Since $a = \dfrac{a+a}{2} < \dfrac{a+b}{2} < \dfrac{b+b}{2} = b$, then $a < \dfrac{a+b}{2} < b$

91. Note, $AB = \sqrt{(1+4)^2+(1+5)^2} = \sqrt{25+36} = \sqrt{61}$. Similarly, $BC = \sqrt{61}$ and $AC = \sqrt{244} = 2\sqrt{61}$. Since $AB + BC = \sqrt{61} + \sqrt{61} = 2\sqrt{61} = AC$, then A, B, and C are collinear.

93. The distance between $(10, 0)$ and $(0, 0)$ is 10.

    If two points have integer coordinates, then the distance between them is of the form $s^2 + t^2$ where $s, t \in \{0, 1, 2^2, 3^2, 4^2, ...\} = \{0, 1, 4, 9, 16, ...\}$.

    Since $1 + 9 = 10$, the distance between $(1, 3)$ and the origin is $\sqrt{10}$.

    Note, there exists no pair s and t in $\{0, 1, 4, 9, 16, ...\}$ satisfying $s^2 + t^2 = 19$. Thus, one cannot find two points with integer coordinates whose distance between them is $\sqrt{19}$.

For Thought

1. False, since $\{(1,2),(1,3)\}$ is not a function.

2. False, since $f(5)$ is not defined.   3. True

4. False, since a student's exam grade is a function of the student's preparation. If two classmates had the same IQ and only one prepared then the one who prepared will most likely achieve a higher grade.

5. False, since $(x+h)^2 = x^2 + 2xh + h^2$

6. False, since the domain is all real numbers.

7. True   8. True   9. True

10. False, since $\dfrac{3}{8} = .375$

2.2 Exercises

1. Function

3. Not a function since 25 has two different second coordinates.

5. Not a function since 3 has two different second coordinates.

7. Function

9. Not a function since 2 has two different second coordinates.

11. Function   13. Function

15. function   17. function

19. Not a function since $(1, 1)$ and $(1, -1)$ are two points, with different second coordinates, satisfying the equation.

21. Function   23. Function

25. Since $a = 2\pi r$ or equivalently $r = \dfrac{a}{2\pi}$, then a is a function of b and b is a function of a.

27. a is a function of b since a given denomination has a unique length. Since a dollar bill and a five-dollar bill have the same length, then b is not a function of a.

29. Since an item has only one price, b is a function of a. Since two items may have the same price, a is not a function of b.

31. a is not a function of b since it is possible that two different students can obtain the same final exam score but the times spent on studying are different.

    b is not a function of a since it is possible that two different students can spend the same time studying but obtain different final exam scores.

33. Since 1 in. $\approx$ 2.54 cm., a is a function of b and b is a function of a.

35. Domain $\{-3, 5, \pi\}$, range $\{1, 6, \sqrt{2}\}$

37. Domain $(-\infty, \infty)$, range $[5, \infty)$

39. Domain $(-\infty, 0]$, range $(-\infty, \infty)$

41. Domain $[0, \infty)$, Range $[5, \infty)$

43. Domain $[1/2, \infty)$, range $[0, \infty)$

45. Domain and range are both $(-\infty, \infty)$

47. Domain $(0, 3)$, range $(2, 5)$

49. $3 \cdot 4 - 2 = 10$

51. $-4 - 2 = -6$ 53. $|8| = 8$

55. $0.9408 - 0.56 = 0.3808$ 57. $4 + (-6) = -2$

59. $-22 - 2 = -24$ 61. $2/2 = 1$ $3a^2 - a$

63. $3(a^2 + 6a + 9) - a - 3 = 3a^2 + 17a + 24$

65. $3(x^2 + 2xh + h^2) - x - h = 3x^2 + 6xh + 3h^2 - x - h$

67. $(3x^2 + 6xh + 3h^2 - x - h) - 3x^2 + x = 3h^2 + 6xh - h$

69. $3x^2 + 3x - 2$

71. $(3x^2 - x)(4x - 2) = 12x^3 - 10x^2 + 2x$

73. Factor and get $x(3x - 1) + 0$. So $x = 0, 1/3$.

75. Since $|a + 3| = 4$ is equivalent to $a + 3 = 4$ or $a + 3 = -4$, we have $a = 1, -7$.

77. $f(-6) = 3(36) + 6 = 114$

79. $f(1) = 3 - 1 = 2$

81.
$$\frac{f(x+h) - f(x)}{h} = \frac{3(x+h) + 5 - 3x - 5}{h}$$
$$= \frac{3h}{h}$$
$$= 3$$

83.
$$\frac{g(x+h) - g(x)}{h} = \frac{3(x+h)^2 + 1 - 3x^2 - 1}{h}$$
$$= \frac{6xh + 3h^2}{h}$$
$$= 6x + 3h$$

85. Difference quotient is
$$= \frac{-(x+h)^2 + (x+h) - 2 + x^2 - x + 2}{h}$$
$$= \frac{-2xh - h^2 + h}{h}$$
$$= -2x - h + 1$$

87. Difference quotient is
$$= \frac{\sqrt{x+h+2} - \sqrt{x+2}}{h} \cdot \frac{\sqrt{x+h+2} + \sqrt{x+2}}{\sqrt{x+h+2} + \sqrt{x+2}}$$
$$= \frac{(x+h+2) - (x+2)}{h(\sqrt{x+h+2} + \sqrt{x+2})}$$
$$= \frac{h}{h(\sqrt{x+h+2} + \sqrt{x+2})}$$
$$= \frac{1}{\sqrt{x+h+2} + \sqrt{x+2}}$$

89. Difference quotient is
$$= \frac{\frac{1}{x+h} - \frac{1}{x}}{h} \cdot \frac{x(x+h)}{x(x+h)}$$
$$= \frac{x - (x+h)}{xh(x+h)}$$
$$= \frac{-h}{xh(x+h)}$$
$$= \frac{-1}{x(x+h)}$$

91. $A = s^2$ 93. $s = \dfrac{\sqrt{2}d}{2}$

95. $P = 4s$ 97. $A = P^2/16$

99. $C = 353n$ 101. $C = 35n + 50$

103.

(a) In 1999, the combined payments by Medicare and Medicaid is $E(9) + P(9) = (15.6(9) + 99.8) + (12.6(9) + 81.3) = \$434.9$ billions.

(b) The year when Medicaid spending will reach \$250 billion can found by solving $12.6n + 81.3 = 250$. The solution of which is $n = \dfrac{250 - 81.3}{12.6} \approx 13.4$. In the year 2003, Medicaid payments will reach \$250 billion.

105. The average rate of change is $\dfrac{4,000 - 16,000}{5} = -\$2,400$. The car depreciates \$2,400 per year.

107. The average rate of change as income varies from \$100 to \$2,000 per capita is $\dfrac{50 - 10}{2,000 - 100} \approx .02$.

The average rate of change as income varies from \$2,000 to \$31,000 is $\dfrac{10 - 50}{31,000 - 2,000} \approx -.001$.

The average rate of change as income varies from $100 to $31,000 per capita is $\frac{10-10}{31,000-2,000} = 0$.
Among the three average rates of change, the second one (which is -.001 micro grams/m³ per dollar) makes the most sense. This implies that for every $1,000 increase in income per capita, the concentration of sulfur dioxide will decrease by 1 micro gram per cubic meter.

109. The average rate of change is $\frac{982-1056}{1996-1984} \approx -6.2$ millions of hectares per year.

111. The difference quotient when $x = 18$ and $h = 0.1$ is $\frac{R(18.1) - R(18)}{0.1} = 1,950$. The revenue from the concert will increase by approximately $1,950 if the price of a ticket is raised from $18 to $19.

The difference quotient when $x = 22$ and $h = 0.1$ is $\frac{R(18.1) - R(18)}{0.1} = -2,050$. The revenue from the concert will decrease by approximately $2,050 if the price of a ticket is raised from $22 to $23.

113. $MC(x) = 0.03(x+1)^2 + 40(x+1) + 6000 - 0.03x^2 - 40x - 6000 = 0.06x + 40.03$ and $MC(100) = \$46.03$

For Thought

1. True

2. False, since the center is $(3, 2)$.

3. False, it is not a circle becaue of -16.

4. True

5. False, since the range is $\{\pm 1\}$.

6. True

7. True **8.** True

9. False, since the range is the interval $[0, 4]$.

10. True

2.3 Exercises

1. Center$(0, 0)$, radius 4

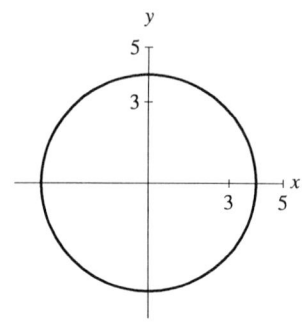

3. Center $(-6, 0)$, radius 6

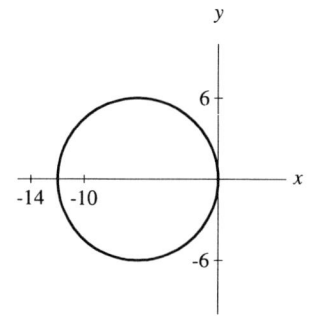

5. Center $(2, -2)$, radius $2\sqrt{2}$

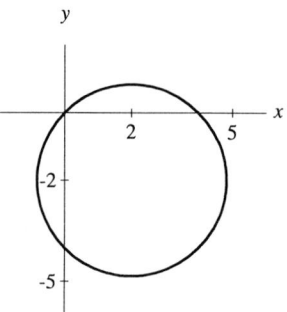

7. $x^2 + y^2 = 7$

9. $(x+2)^2 + (y-5)^2 = 1/4$

11. $(x-3)^2 + (y-5)^2 = 34$

13. $(x-5)^2 + (y+1)^2 = 32$

15. Completing the square, we obtain
$$x^2 + (y^2 + 6y + 9) = 0 + 9$$
$$x^2 + (y+3)^2 = 9.$$
Center $(0,-3)$, radius 3

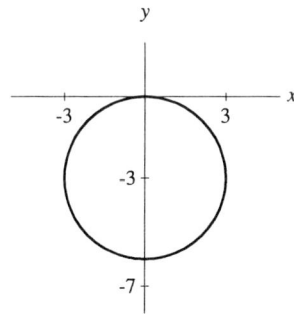

17. Completing the square, we obtain
$$(x^2 - 6x + 9) + (y^2 - 8y + 16) = 9 + 16$$
$$(x-3)^2 + (y-4)^2 = 25.$$
Center $(3,4)$, radius 5

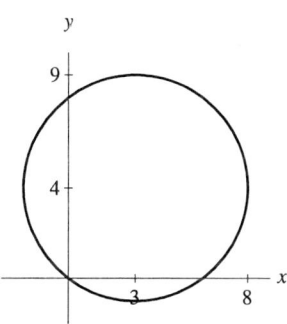

19. Completing the square, we obtain
$$(x^2 - 4x + 4) + (y^2 - 3y + \frac{9}{4}) = 4 + \frac{9}{4}$$
$$(x-2)^2 + (y-\frac{3}{2})^2 = \frac{25}{4}.$$
Center $(2, 3/2)$, radius $5/2$

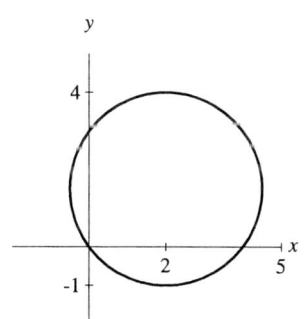

21. Completing the square, we obtain
$$(x^2 - \frac{1}{2}x + \frac{1}{16}) + (y^2 + \frac{1}{3}y + \frac{1}{36}) = \frac{1}{36}$$
$$(x - \frac{1}{4})^2 + (y + \frac{1}{6})^2 = \frac{1}{36}.$$
Center $(1/4, -1/6)$, radius $1/6$

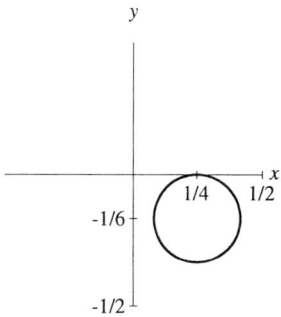

23. Function $x = \sqrt{y}$ goes through $(0,0), (2,4), (3,9)$, domain and range is $[0, \infty)$

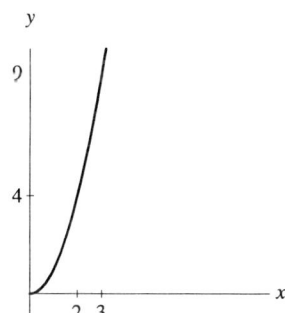

25. Function $y = x^2 - 1$ goes through $(0, -1), (\pm 1, 0)$, domain is $(-\infty, \infty)$, range is $[-1, \infty)$

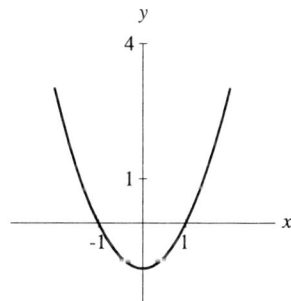

27. Function $y = 5$ includes the points $(0, 5), (\pm 2, 5)$, domain is R, range is $\{5\}$

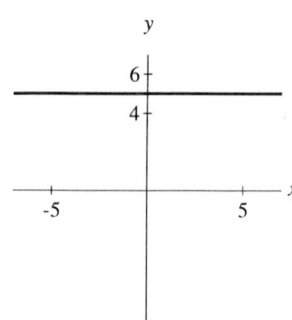

29. Function $x = 2y$ includes the points $(0, 0), (2, 1), (-2, -1)$, domain and range are both R

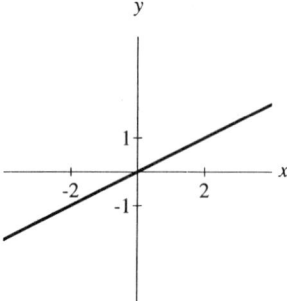

31. Function $x - y = 2$ includes the points $(2, 0), (0, -2), (-2, -4)$, domain and range are both R

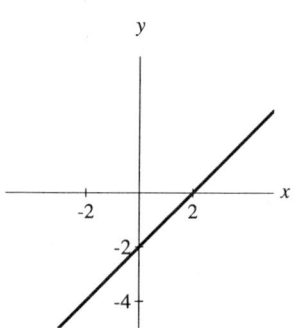

33. Function $y = 2|x|$ includes the points $(0, 0), (\pm 1, 2)$, domain is R, range is $[0, \infty)$

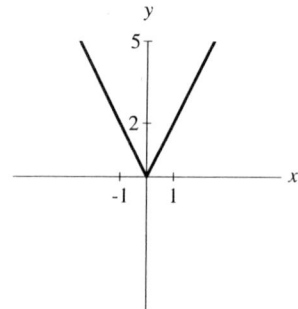

35. $x = y^2 + 1$ is not a function and includes the points $(0, 1), (2, \pm 1)$, domain is $[1, \infty)$, range is R

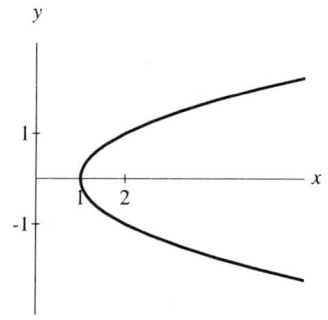

37. Function $y = |x - 1|$ includes the points $(0, 1), (1, 0), (2, 1)$, domain is R, range is $[0, \infty)$

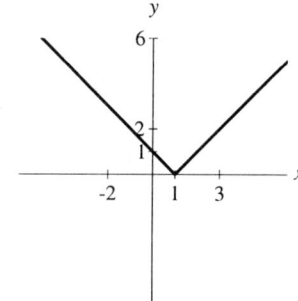

2.3 GRAPHS OF RELATIONS AND FUNCTIONS

39. $x = |y+2|$ is not a function and includes $(1, -3)$, $(0, -2)$, $(1, -1)$, domain is $[0, \infty)$, range is R

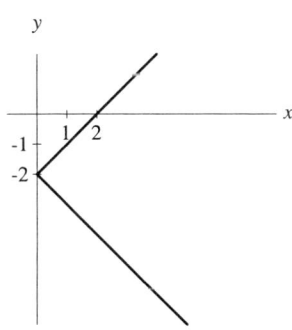

41. Function $y = \sqrt{1-x^2}$ includes the points $(0, 1)$, $(\pm 1, 0)$, domain is $[-1, 1]$, range is $[0, 1]$

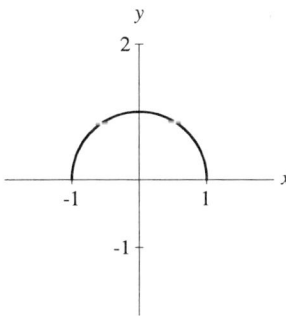

43. $y^2 = 1 - x^2$ is not a function and includes $(\pm 1, 0)$, $(0, \pm 1)$, domain and range are both $[-1, 1]$

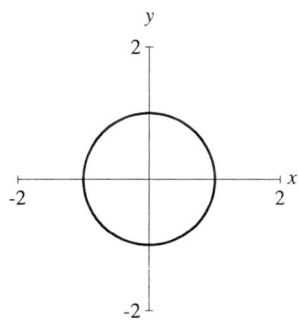

45. $x + y^2 = 0$ is not a function and includes the points $(0, 0), (-4, \pm 2)$, domain is $(-\infty, 0]$, range is R

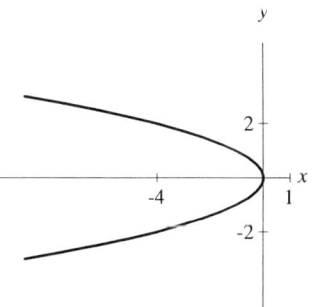

47. Function $y = 1 - x^2$ includes the points $(0, 1)$, $(\pm 1, 0)$, domain is R, range is $(-\infty, 1]$

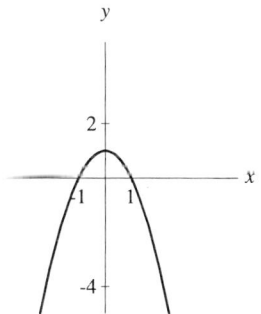

49. Function $y = -|x|$ includes the points $(0, 0)$, $(\pm 1, -1)$, domain is R, range is $(-\infty, 0]$

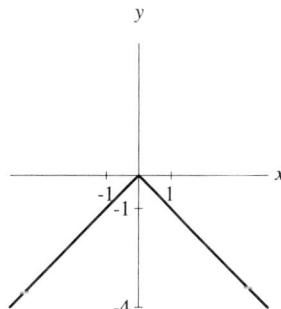

51. No **53.** Yes **55.** Yes

57. Domain is $(-\infty, \infty)$, range is $\{\pm 2\}$

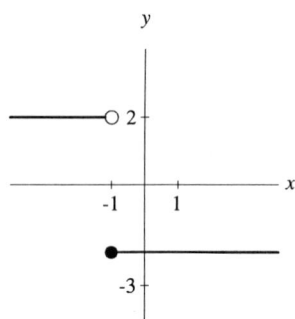

59. Domain is $[-2, \infty)$, range is $(-\infty, 2]$

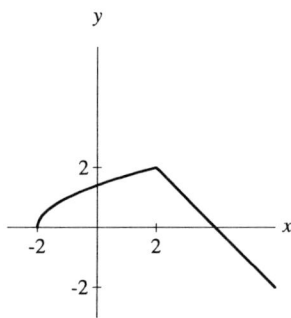

61. Domain is R, range is $[0, \infty)$

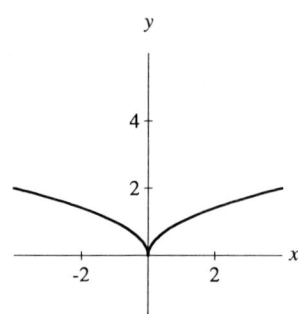

63. Domain is $[-2, \infty)$, range is $[0, \infty)$

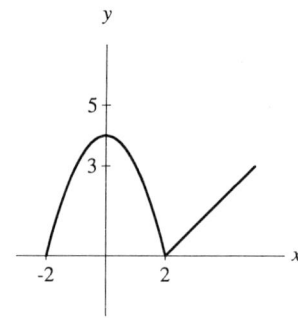

65. Domain is R, range is the set of integers

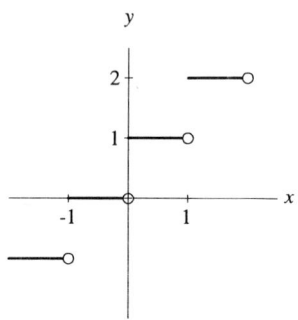

67. Domain $[0, 4)$, range is $\{2, 3, 4, 5\}$

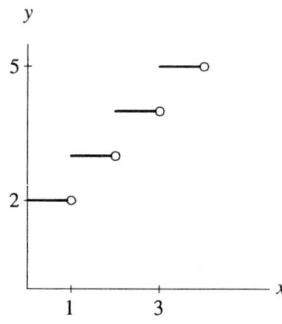

69. Domain is $(-\infty, \infty)$, range is $(-\infty, 4)$
increasing on $(-\infty, 0)$, decreasing on $(0, \infty)$

71. Domain $(-\infty, 2]$, range $(-\infty, 2]$,
increasing on $(-\infty, -2)$, constant on $(-2, 2)$

73. Domain is $[-4, 4]$, range is $[0, 4]$
increasing on $(-4, 0)$, decreasing on $(0, 4)$

75. c, graph was increasing at first, then suddenly dropped and became constant, then increased slightly

77. d, graph was decreasing at first, then fluctuated between increases and decreases, then the market increased

79. The independent variable is time t where t is the number of minutes after 7:45 and the dependent variable is distance D from the holodeck.

D is increasing on the intervals $(0, 3)$ and $(6, 15)$, decreasing on $(3, 6)$ and $(30, 39)$, and constant on $(15, 30)$.

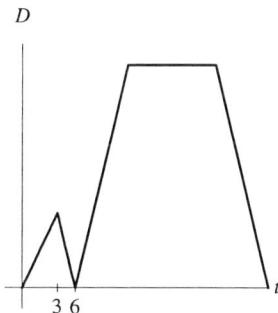

81. Independent variable is time t in years, dependent variable is savings s in dollars

s is increasing on the interval $(0, 2)$; s is constant on $(2, 2.5)$; s is decreasing on $(2.5, 3.5)$.

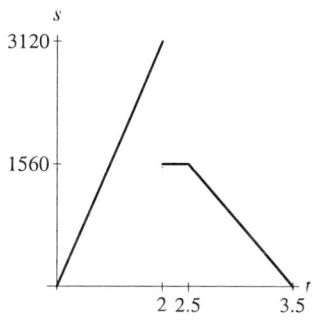

83. Domain and range are both $(-\infty, \infty)$ increasing on $(-\infty, \infty)$

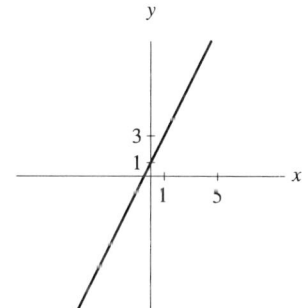

85. Domain is $(-\infty, \infty)$, range is $[0, \infty)$, increasing on $(1, \infty)$, decreasing on $(-\infty, 1)$

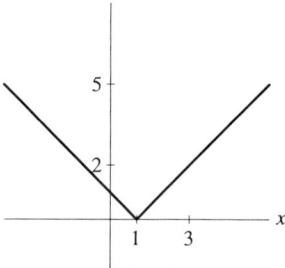

87. Domain is $(-\infty, 0) \cup (0, \infty)$, range is $\{\pm 2\}$, constant on $(-\infty, 0)$ and $(0, \infty)$

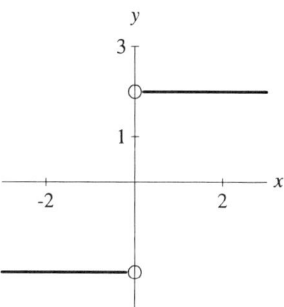

89. Domain is $[-1, 1]$, range is $[-1, 0]$, Increasing on $(0, 1)$, decreasing on $(-1, 0)$

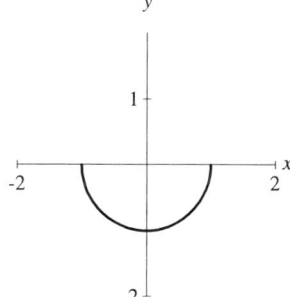

91. Domain is R, range is R,
increasing on $(-\infty, 3)$ and $(3, \infty)$

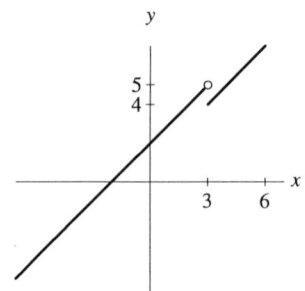

93. Domain is R, range is $(-\infty, 2]$, increasing on $(-\infty, -2)$ and $(-2, 0)$, decreasing on $(0, 2)$ and $(2, \infty)$

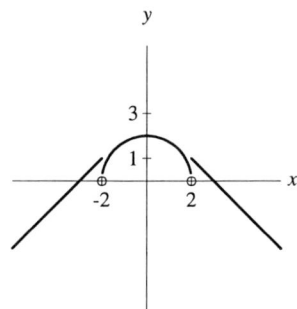

95. increasing on the interval $(0.83, \infty)$, decreasing on $(-\infty, 0.83)$

97. increasing on the interval $(-\infty, -1) \cup (1, \infty)$, decreasing on $(-1, 1)$

99. increasing on the interval $(-1.73, 0) \cup (1.73, \infty)$, decreasing on $(-\infty, -1.73) \cup (0, 1.73)$

101. Since the midpoint (=center) and distance (=diameter) between $(4, 1)$ and $(-6, 5)$ is $(-1, 3)$ and $2\sqrt{29}$, respectively, the standard equation of the said circle is $(x+1)^2 + (y-3)^2 = 29$

103. $f(x) = \begin{cases} -4[-x] & \text{if } 0 < x \le 3 \\ 15 & \text{if } 3 < x \le 8 \end{cases}$

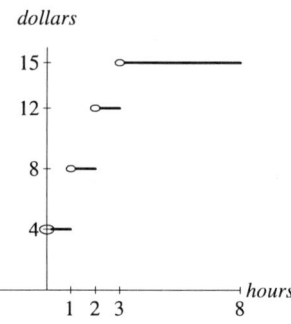

105. Constant on $(0, 10000)$, increasing on $(10000, \infty)$

107. The cost is over \$235 for $t \ge 5$.

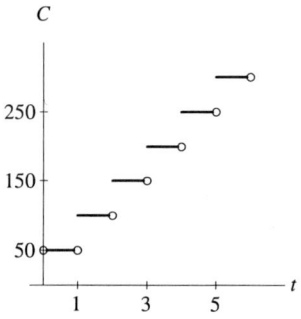

109. In 1988, there were $M(18) = 565$ million cars. In 2002, it is projected that there will be $M(32) = 720$ million cars. The average rate of change from 1984 to 1994 is $\dfrac{M(24) - M(14)}{10} = 14.5$ millions of car per year.

111. A mechanic's fee is \$20 for each half-hour of work with any fraction of a half-hour charged as a half-hour. If y is the fee in dollars and x is the number of hours, then $y = -20[-2x]$.

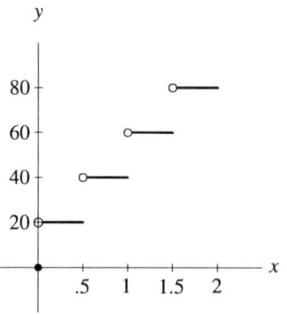

113.

(a) Let x be the number of days. For the first 50 days, the sales y (in dollars) was increasing at an increasing rate. At $x = 50$, sales dropped suddenly due to a product shortage. When $50 < x \leq 75$, y was increasing at a constant rate. Another product shortage occured at $x = 75$. For $75 < x \leq 100$, the sales were the same for each day.

(b) Suppose an antibiotic is being taken by a patient every 7 hours. The graph given can represent the amount, y, of antibiotic in the body as a function of the number of hours, x, that has passed since the antibiotic was first taken.

(c) The price of a stock was decreasing in 1990-95, with the price bottoming out in 1995, and then its price has been increasing and is still projected to increase faster.

For Thought

1. False, it is a reflection in the y-axis.

2. True

3. False, rather it is a left translation.

4. True   5. True

6. False, the down shift should comes after the reflection.

7. True

8. False, since their domains are different.

9. True   10. True

2.4 Exercises

1. $f(x) = \sqrt{x}, g(x) = -\sqrt{x}$

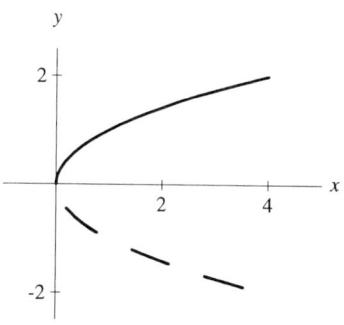

3. $y = x, y = -x$

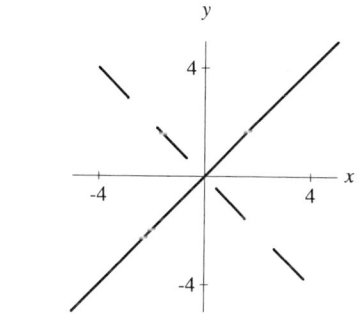

5. $f(x) = |x|, g(x) = |x| - 4$

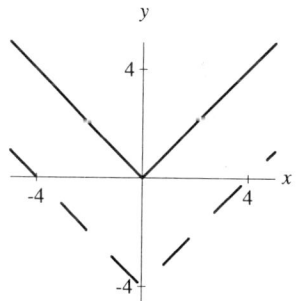

7. $f(x) = x, g(x) = x + 3$

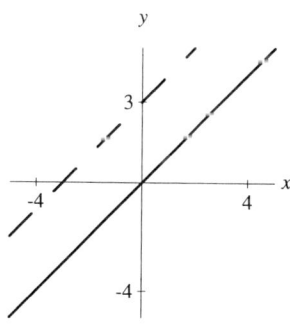

9. $y = x^2, y = (x-3)^2$

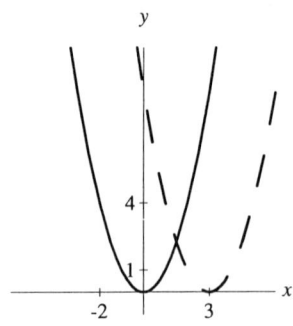

11. $y = \sqrt{x}, y = 3\sqrt{x}$

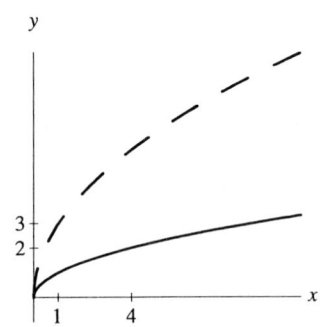

13. $y = x^2, y = \frac{1}{4}x^2$

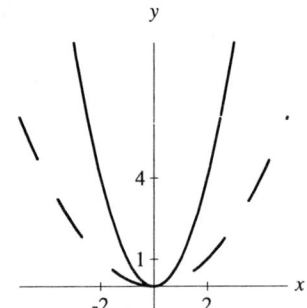

15. g **17.** b

19. c **21.** f

23. $y = (x-10)^2 + 4$

25. $y = -3|x-7| + 9$

27. $y = -(3\sqrt{x} + 5)$

29. $y = (x-1)^2 + 2$; right by 1, up by 2

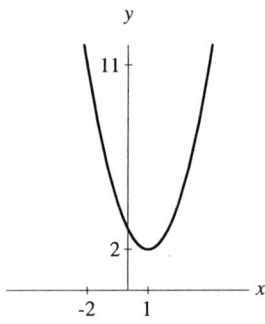

31. $y = |x-1| + 3$; right by 1, up by 3

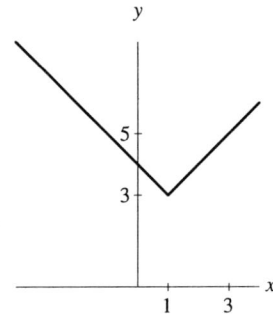

33. $y = 3x - 40$

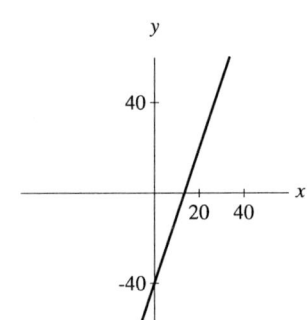

35. $y = \frac{1}{2}x - 20$

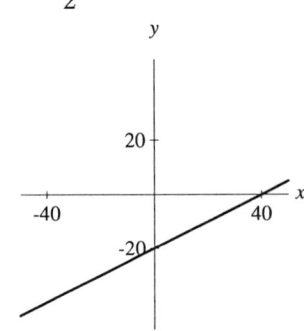

37. $y = -\dfrac{1}{2}|x| + 40$, shrink by 1/2, reflect about x-axis, up by 40

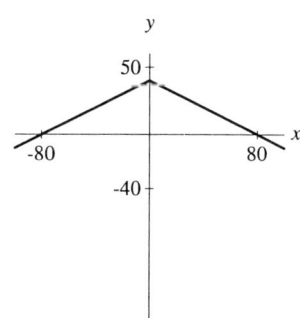

39. $y = -\dfrac{1}{2}|x+4|$, left by 4, reflect about $x-axis$, shrink by 1/2

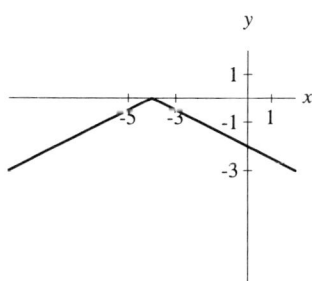

41. $y = -\sqrt{x-3} + 1$, right by 3, reflect about x-axis, up by 1

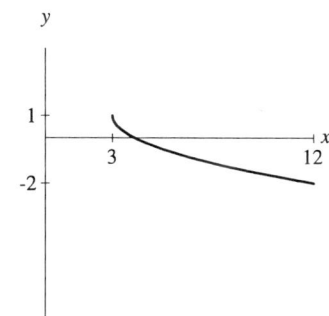

43. $y = -2\sqrt{x+3} + 2$, left by 3, stretch by 2, reflect about x-axis, up by 2

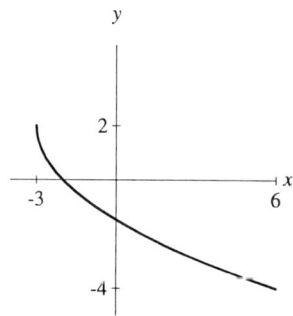

45. Symmetric about y-axis, even function

47. No symmetry, neither even nor odd

49. Symmetric about $x = -3$, neither even nor odd

51. No symmetry, not an even or odd function

53. Symmetric about the origin, odd function

55. No symmetry, not an even or odd function

57. No symmetry, not an even or odd function

59. Symmetry about $x = 2$, not an even or odd function

61. Symmetric about the y-axis, even function

63. Symmetric about the y-axis, even function

65. e   67. g

69. b   71. c

73. $(-\infty, -1] \cup [1, \infty)$

75. $(-\infty, -1) \cup (5, \infty)$

77. Graph of $y=(x-1)^2-9$ shows solution is $(-2,4)$

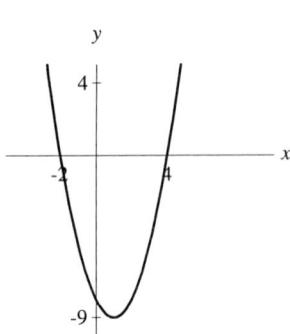

79. Graph of $y=5-\sqrt{x}$ shows solution is $[0,25]$

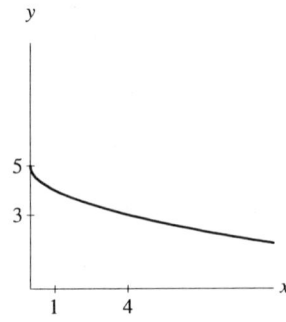

81. Solution is $\left(-\infty, 2-\sqrt{3}\right) \cup \left(2+\sqrt{3}, \infty\right)$

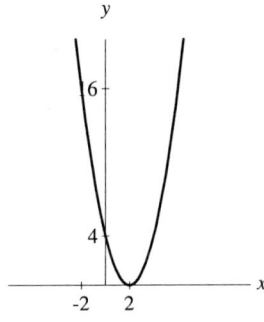

83. Graph of $y=\sqrt{25-x^2}$ shows solution is $(-5,5)$

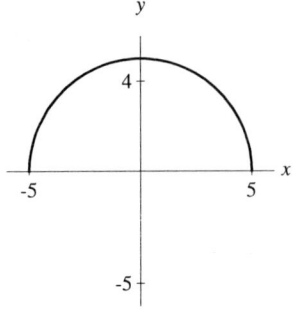

85. From the graph of $y=\sqrt{3}x^2+\pi x-9$,

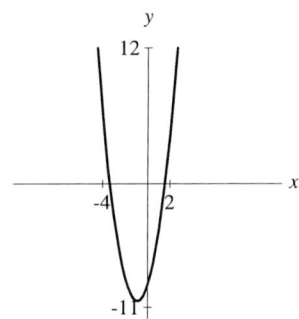

the solution set of $\sqrt{3}x^2+\pi x-9<0$ is $(-3.36, 1.55)$.

87. $N(x)=x+2000$

89. If inflation rate is less than 50% then $1-\sqrt{x}<\dfrac{1}{2}$. This simplifies to $\dfrac{1}{2}<\sqrt{x}$. After squaring we have $\dfrac{1}{4}<x$ and so $x>25\%$.

91. By solving
$$.005x^2+32=46$$
$$x=\sqrt{\dfrac{46-32}{.005}}$$
$$x\approx 52.9$$

and since $y=.005x^2+32$ is a parabola opening up with its vertex at the origin then life expectancy in Lyon surpassed 46 years in the year 1893 ($\approx 1840+52.9$).

93.

(a) Both functions are even functions and the graphs are identical

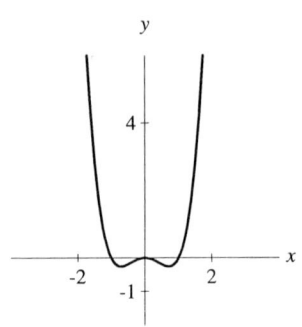

(b) One graph is a reflection of the other about the y-axis. Both functions are odd.

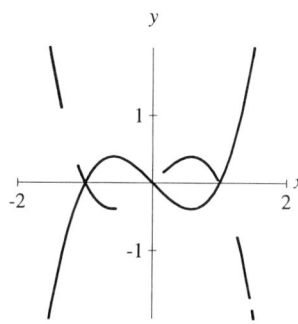

(c) The second graph is obtained by shifting the first one to the left by 1 unit.

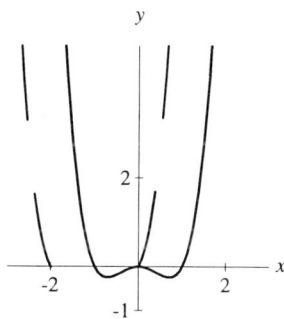

(d) The second graph is obtained by translating the first one to the right by 2 units and 3 units up.

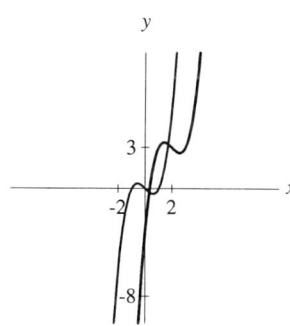

For Thought

1. False, since $f + g$ has an empty domain.

2. True   3. True   4. True

5. True, since $A = P^2/16$.   6. True

7. False, since $(f \circ g)(x) = \sqrt{x} - 2$ 8. True

9. False, since $(h \circ g)(x) = x^2 - 9$.

10. True, since x belongs to the domain if $\sqrt{x-2}$ is a real number, i.e., if $x \geq 2$.

2.5 Exercises

1. $-1 + 2 = 1$ 3. $-5 - 6 = -11$

5. $(-4) \cdot 2 = -8$ 7. $1/12$

9. $(a - 3) + (a^2 - a) = a^2 - 3$

11. $(a-3)(a^2 - a) = a^3 - 4a^2 + 3a$

13. $y = 2(3x + 1) - 3 = 6x - 1$

15. $y = (x^2 + 6x + 9) - 2 = x^2 + 6x + 7$

17. $y = 3 \cdot \dfrac{x+1}{3} - 1 = x + 1 - 1 = x$

19. $y = 2 \cdot \dfrac{x+1}{2} - 1 = x + 1 - 1 = x$

21. $\{(-3,3),(2,6)\}$ 23. $\{(-3,2)\}$

25. $\{(-3,2),(2,0)\}$ 27. $\{(-3,0),(1,0),(4,4)\}$

29. $\{(1,4)\}$ 31. $\{(-3,4),(1,4)\}$

33. $(f+g)(x) = \sqrt{x} + x - 4$, domain is $[0, \infty)$

35. $(g \cdot h)(x) = \dfrac{x-4}{x-2}$, domain is $(-\infty, 2) \cup (2, \infty)$

37. $\left(\dfrac{f}{g}\right)(x) = \dfrac{\sqrt{x}}{x-4}$, domain is $[0,4) \cup (4, \infty)$

39. $\left(\dfrac{g}{h}\right)(x) = \dfrac{x-4}{\frac{1}{x-2}} = (x-4)(x-2) = x^2 - 6x + 8$, domain is $(-\infty, 2) \cup (2, \infty)$

41. $f(2) = 5$ 43. $f(2) = 5$

45. $f(20.2721) = 59.8163$

47. $3(x^2+1) - 1 = 3x^2 + 2$

49. $(3x-1)^2 + 1 = 9x^2 - 6x + 2$

51. $\dfrac{x^2 + 2}{3}$

53. $3(3x-1) - 1 = 9x - 4$

55. $h\left((3x-1)^2 + 1\right) = \dfrac{(9x^2 - 6x + 2) + 1}{3} = 3x^2 - 2x + 1$

57. $(f \circ g)(x) = \sqrt{2x-1}$, domain is $[1/2, \infty)$

59. $(h \circ f)(x) = \dfrac{1}{\sqrt{x}-3}$, domain is $[0,9) \cup (9, \infty)$

61. $(h \circ h)(x) = \dfrac{1}{\dfrac{1}{x-3} - 3} = \dfrac{x-3}{10-3x}$, domain is $(-\infty, 3) \cup (3, 10/3) \cup (10/3, \infty)$

63. $(f \circ f)(x) = \sqrt{\sqrt{x}} = \sqrt[4]{x}$, domain is $[0, \infty)$

65. $(f \circ g)(x) = 2\left(\dfrac{x+1}{2}\right) - 1 = (x+1) - 1 = x$,
$(g \circ f)(x) = \dfrac{(2x-1) + 1}{2} = \dfrac{2x}{2} = x$

67. Multiply by the LCD in both cases.
$(f \circ g)(x) = \dfrac{\dfrac{x+1}{1-x} - 1}{\dfrac{x+1}{1-x} + 1}$
$= \dfrac{x + 1 - (1-x)}{(x+1) + (1-x)}$
$= \dfrac{2x}{2} = x$ and
$(g \circ f)(x) = \dfrac{\dfrac{x-1}{x+1} + 1}{1 - \dfrac{x-1}{x+1}}$
$= \dfrac{(x-1) + (x+1)}{(x+1) - (x-1)}$
$= \dfrac{2x}{2} = x$

69. $(f \circ g)(x) = \sqrt[3]{2 \cdot \dfrac{x^3+1}{2} - 1} = \sqrt[3]{(x^3+1) - 1} = \sqrt[3]{x^3} = x$,
$(g \circ f)(x) = \sqrt[5]{\dfrac{(2x^5 - 1) + 1}{2}} = \sqrt[5]{\dfrac{2x^5}{2}} = x$

71. After multiplying y by $\dfrac{x+1}{x+1}$ we have
$y = \dfrac{\dfrac{x-1}{x+1} + 1}{\dfrac{x-1}{x+1} - 1} = \dfrac{(x-1) + (x+1)}{(x-1) - (x+1)} = -x$

The domain of the original function is $(-\infty, -1) \cup (-1, \infty)$ while the domain of the simplified function is $(-\infty, \infty)$. The two functions are not the same.

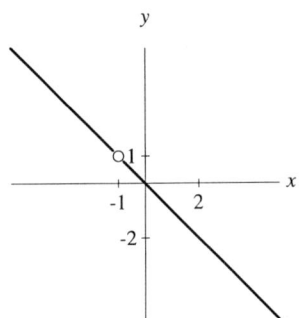

73. Domain $[-1, \infty)$, range $[-7, \infty)$

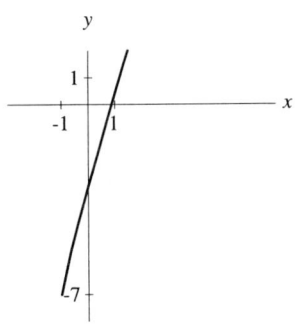

75. Domain $[1, \infty)$, range $[0, \infty)$,

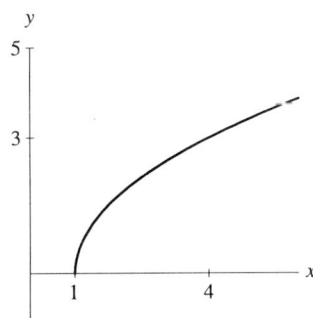

77. Domain $[0, \infty)$, range $[4, \infty)$,

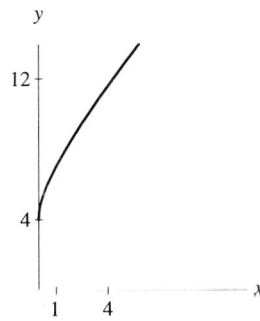

79. Since $m = n - 4$ and $y = m^2$, $y = (n-4)^2$.

81. Since $w = x + 16$, $z = \sqrt{w}$, and $y = \dfrac{z}{8}$, then $y = \dfrac{\sqrt{x+16}}{8}$.

83. $F = g \circ h$ **85.** $H = f \circ g \circ h$

87. $N = h \circ g \circ f$ **89.** $P = g \circ f \circ g$

91. $A = d^2/2$

93. $(f + g)(x) = 0.55x$ is the area of rain forest destructed by waste and harvest combined.

95. $F(x) = \left(\dfrac{x}{3} + 150\right) + \left(\dfrac{5x}{3} + 40\right) + \left(\dfrac{x}{15} + 48\right) = \dfrac{31x}{15} + 238.$

$F(x)$ represents the combined energy consumption of OECD, transition, and developing countries. $F(20) = \dfrac{31x}{15} + 238 \approx 279.3$, energy consumption in 1993 (20 years after 1973).

97. Note, $D = \dfrac{d/2240}{x} = \dfrac{d/2240}{L^3/100^3} = \dfrac{100^3 d}{2240 L^3} = \dfrac{100^3(26000)}{2240 L^3} = \dfrac{100^4(26)}{224 L^3} = \dfrac{100^4(13)}{112 L^3}.$

Expressing D as a function of L we write $D = \dfrac{(13)100^4}{112 L^3}$ or $D = \dfrac{1.16 \times 10^7}{L^3}$.

99. The area of a semicircle with radius $s/2$ is $(1/2)\pi(s/2)^2 = \pi s^2/8$. The area of the square is s^2. The area of the window is
$$W = \dfrac{\pi s^2}{8} + s^2 = \dfrac{s^2(\pi + 8)}{8}.$$

101. Form a right triangle with two sides of length $s/2$ and a hypotenuse of length $d/2$. By the Pythagorean Theorem we have $d^2/4 = s^2/4 + s^2/4$. Solving for s we have $s = \dfrac{\sqrt{2}d}{2}$.

103.

(a) $s = 0.7x$

(b) $c = 0.75s$

(c) $c = 0.75(0.7x)$

(d) Since $0.75(0.7) = 0.525$, at the final clearance sale one is getting 47.5% off and not 55% off.

For Thought

1. False, since the inverse function is $\{(3,2), (5,5)\}$.

2. False, since it is not one-to-one.

3. False, $g^{-1}(x)$ does not exist since g is not one-to-one.   **4.** True   **5.** False, a function that fails the horizontal line test has no inverse.

6. False, since it fails the horizontal line test.

7. False, since $f^{-1}(x) = \left(\dfrac{x}{3}\right)^2 + 2$ where $x \geq 0$.

8. False, $f^{-1}(x)$ does not exist since f is not one-to-one.

9. False, since $y = |x|$ is V shaped.   **10.** True

2.6 Exercises

1. $f^{-1} = \{(1,2),(5,3)\}$, $f^{-1}(5) = 3$,
   $(f \circ f^{-1})(2) = 2$

3. $f^{-1} = \{(-3,-3),(5,0),(-7,2)\}$, $f^{-1}(5) = 0$,
   $(f \circ f^{-1})(2) = 2$

5. Not invertible

7. Invertible, inverse is $\{(0,3),(5,2),(6,4),(9,7)\}$

9. Invertible, inverse is $\{(1,1),(2,2),(4.5,4.5)\}$

11. Invertible and the inverse is
    $\{(x,y) \mid x = y+2\} = \{(x,y) \mid y = x-2\}$

13. Invertible and the inverse is
    $\{(x,y) \mid x = 2y+7\} = \left\{(x,y) \mid y = \dfrac{x-7}{2}\right\}$

15. Not invertible

17. Not invertible

19. Invertible

21. Not invertible

23. Not one-to-one

25. Not one-to-one since it fails the Horizontal Line Test

27. One-to-one

29. Not one-to-one since it fails the Horizontal Line Test

31. Not one-to-one since it fails the Horizontal Line Test

33. Not invertible, there can be two different items with the same price.

35. Invertible, since the playing time is a function of the length of the VCR tape.

37. Invertible, assuming that cost is simply a multiple of the number of days. If cost includes extra charges, then the function may not be invertible.

39. Not invertible since it fails the Horizontal Line Test.

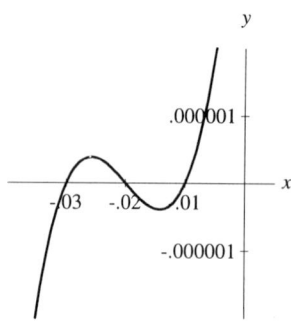

41. Not invertible since it fails the Horizontal Line Test.

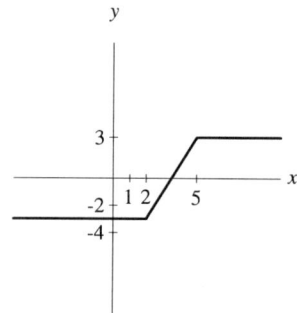

43. Interchange x and y then solve for y

    $$x = 3y - 7$$
    $$\dfrac{x+7}{3} = y$$
    $$\dfrac{x+7}{3} = f^{-1}(x)$$

45. Interchange x and y then solve for y

    $$x = 2 + \sqrt{y-3} \quad x \geq 2$$
    $$(x-2)^2 = y - 3 \quad x \geq 2$$
    $$f^{-1}(x) = (x-2)^2 + 3 \quad x \geq 2$$

47. Interchange x and y then solve for y

    $$x = -y - 9$$
    $$y = -x - 9$$
    $$f^{-1}(x) = -x - 9$$

49. Interchange x and y then solve for y
$$x = \frac{y+3}{y-5}$$
$$xy - 5x = y + 3$$
$$xy - y = 5x + 3$$
$$y(x-1) = 5x + 3$$
$$f^{-1}(x) = \frac{5x+3}{x-1}$$

51. Interchange x and y then solve for y
$$x = -\frac{1}{y}$$
$$xy = -1$$
$$f^{-1}(x) = -\frac{1}{x}$$

53. Interchange x and y then solve for y
$$x = \sqrt[3]{y-9} + 5$$
$$x - 5 = \sqrt[3]{y-9}$$
$$(x-5)^3 = y - 9$$
$$f^{-1}(x) = (x-5)^3 + 9$$

55. Interchange x and y then solve for y
$$x = (y-2)^2 \quad x \geq 0$$
$$\sqrt{x} = y - 2$$
$$f^{-1}(x) = \sqrt{x} + 2$$

57. g is the inverse of f since they are both one-to-one and $(g \circ f)(x) = 0.25(4x+4) - 1 = x$

59. g is not the inverse of f since $(g \circ f)(x) = \sqrt{x^2} = |x|$

61. g is the inverse of f since they both satisfy the Horizontal Line Test and
$$(g \circ f)(x) = \frac{1}{\left(\frac{1}{x}+3\right)-3}$$
$$= \frac{1}{1/x}$$
$$(g \circ f)(x) = x$$

63. g is the inverse of f since they both satisfy the Horizontal Line Test and
$$(g \circ f)(x) = 5\left(\sqrt[3]{\frac{x-2}{5}}\right)^3 + 2$$
$$= 5\left(\frac{x-2}{5}\right) + 2$$
$$= (x-2) + 2$$
$$(g \circ f)(x) = x$$

65. The first two functions are inverses of each other and the third one is the composition of the two.

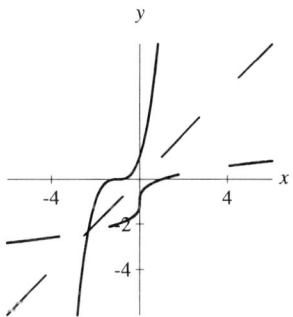

67. $f^{-1}(x) = x/5$ **69.** $f^{-1}(x) = x + 88$

71. $f^{-1}(x) = (x+7)/3$ **73.** $f^{-1}(x) = \dfrac{x-4}{-3}$

75. $f^{-1}(x) = 2(x+9) = 2x + 18$

77. $f^{-1}(x) = -x$ **79.** $f^{-1}(x) = (x+9)^3$

81. $f^{-1}(x) = \sqrt[3]{\dfrac{x+7}{2}}$

83. No, since they fail the Horizontal Line Test.

85. Yes since $(f \circ g)(x) = x$ and $(g \circ f)(x) = x$.

87. $f^{-1}(x) = \dfrac{x-2}{3}$

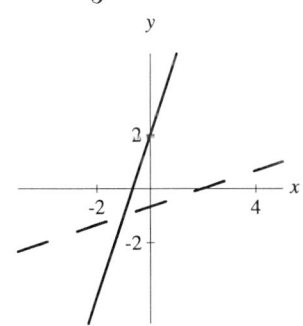

89. $f^{-1}(x) = \sqrt{x+4}$

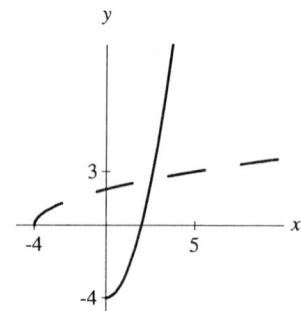

91. $f^{-1}(x) = \sqrt[3]{x}$

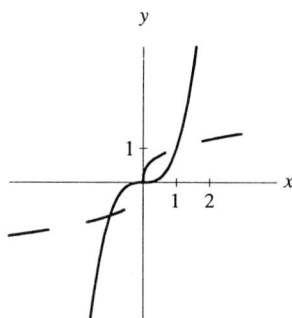

93. $f^{-1}(x) = (x+3)^2, x \geq -3$

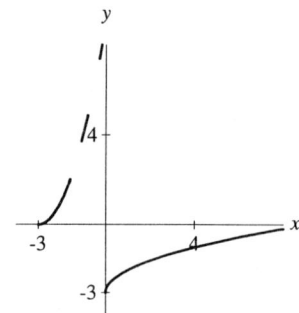

95. $C = 1.08P$ expresses the total cost as a function of the purchase price; and $P = C/1.08$ is the purchase price as a function of the total cost.

97. The graph of t as a function of r satisfies the Horizontal Line Test and is invertible. Solving for r we find,
$$t - 7.89 = -.39r$$
$$r = \frac{t - 7.89}{-.39}$$
and the inverse function is $r = \dfrac{t - 7.89}{-.39}$. If $t = 5.55$ min., there are $r = \dfrac{5.55 - 7.89}{-.39} = 6$ rowers.

99. Solving for w, we obtain
$$1.496w = V^2$$
$$w = \frac{V^2}{1.496}$$
and the inverse function is $w = \dfrac{V^2}{1.496}$. If $V = 115$ ft./sec., then $w = \dfrac{115^2}{1.496} \approx 8,840.2$ lbs.

101. If a car is worth $28,000 after 5 years, its depreciation rate is $r = 1 - \left(\dfrac{28,000}{50,000}\right)^{1/5} \approx .109$ or $r \approx 10.9\%$.

Writing V as a function of r we find
$$1 - r = \left(\frac{V}{50,000}\right)^{1/5}$$
$$(1-r)^5 = \frac{V}{50,000}$$
and $V = 50000(1-r)^5$.

103. One can easily see that the slope of the line joining (a,b) to (b,a) is -1 and that their midpoint is $\left(\dfrac{a+b}{2}, \dfrac{a+b}{2}\right)$. This midpoint lies on the line $y = x$ whose slope is 1. Then $y = x$ is the perpendicular bisector of the line segment joining the points (a,b) and (b,a)

105. Dividing we get $\dfrac{x-3}{x+2} = 1 - \dfrac{5}{x+2}$

107. Since $g^{-1}(x) = \dfrac{x+5}{3}$ and $f^{-1}(x) = \dfrac{x-1}{2}$ then
$g^{-1} \circ f^{-1}(x) = \dfrac{\frac{x-1}{2} + 5}{3} = \dfrac{x+9}{6}$. Since $f \circ g(x) = 6x - 9$ then $(f \circ g)^{-1}(x) = \dfrac{x+9}{6}$.
So $(f \circ g)^{-1} = g^{-1} \circ f^{-1}$.

2.7 VARIATION

For Thought

1. False

2. False, since cost varies directly with the number of pounds purchased.   3. True

4. True   5. True, since the area of a circle varies directly with the square of its radius.

6. False, since $y = k/x$ is undefined when $x = 0$.

7. True   8. True   9. True

10. False, the surface area is not equal to $k \cdot length \cdot width \cdot height$ for any constant k.

2.7 Exercises

1. $G = kn$ 3. $m_1 = k/m_2$ 5. $C = khr$

7. $Y = \dfrac{kx}{\sqrt{z}}$

9. A varies directly with the square of r

11. a varies jointly with z and w

13. Not a variation expression

15. y varies inversely with x

17. H varies directly with the square root of l and inversely with s

19. D varies jointly with L and J and inversely with W

21. Since $y = kx$ and $5 = k \cdot 9$, $k = 5/9$.
So $y = 5x/9$.

23. Since $T = k/y$ and $-30 = k/5$, $k = -150$.
So $T = -150/y$.

25. Since $m = kt^2$ and $54 = k \cdot 18$, $k = 3$.
So $m = 3t^2$.

27. Since $y = kx/\sqrt{z}$ and $2.192 = k(2.4)/\sqrt{2.25}$,
$k = 1.37$. So $y = 1.37x/\sqrt{z}$.

29. Since $y = kx$ and $9 = k(2)$,
$$y = \dfrac{9}{2} \cdot (-3) = -27/2.$$

31. Since $P = k/w$ and $2/3 = \dfrac{k}{1/4}$, $P = \dfrac{1/6}{1/6} = 1$.

33. Since $A = kLW$ and $30 = k(3)(5\sqrt{2})$,
$A = \sqrt{2}(2\sqrt{3})(1/2) = \sqrt{6}$.

35. Since $y = ku/v^2$ and $7 = k \cdot 9/64$,
$y = 28 \cdot 4/64 = 7/4$.

37. Let L_i and L_f be the length in inches and feet, respectively. Then $L_i = 12L_f$ is a direct variation.

39. Let P and n be the cost per person and the number of persons, respectively. Then $P = 20/n$ is an inverse variation.

41. Let S_m and S_k be the speeds of the car in mph and kph, respectively. Then $S_m \approx S_k/1.6 \approx 0.6 \cdot S_k$ is a direct variation.

43. Not a variation

45. Let A and W be the area and width, respectively. Then $A = 30W$ is a direct variation.

47. Let n and p be the number of gallons and price per gallon, respectively. Then $n = 5/p$ is an inverse variation.

49. If p is the pressure at depth d, then $p = kd$. Since $4.34 = k(10)$, $k = .434$. At $d = 6,000$ ft., the pressure is $p = .434(6000) = 2,604$ lbs. per square inch.

51. If h is the number of hours, p is the number of pounds, and w is the number of workers then $h = kp/w$. Since $8 = k(3000)/6$, $k = 0.016$. Five workers can process 4000 pounds in $h = (0.016)(4000)/5 = 12.8$ hrs.

53. Since $I = kPt$ and $20.80 = k(4000)(16)$, $k = 0.000325$. The interest from a deposit of \$6500 for 24 days is $I = (0.000325)(6500)(24) \approx \$50.70$

55. Since $C = kDL$ and $18.60 = k(6)(20)$, $k = 0.155$. The cost of a 16 ft pipe with a diameter of 8 inches is
$C = 0.155(8)(16) = \$19.84$

57. Since $w = khd^2$ and $14.5 = k(4)(6^2)$,
$k = \dfrac{14.5}{144}$. Then a 5-inch high can with a diameter of 6 inches is $w = \dfrac{14.5}{144}(5)(6^2) = 18.125$ ounces.

59. Since $V = kh/l$ and $10 = k(50)/(200)$, $k = 40$. The velocity if the head is 60 ft and the length is 300 ft is $V = (40)(60)/(300) = 8$ ft/year.

61. No, it is not directly proportional otherwise the following ratios $\dfrac{42,506}{1.34} \approx 31,720$, $\dfrac{59,085}{0.295} \approx 200,288$, and $\dfrac{738,781}{0.958} \approx 771,170$ would be constant but they are not.

63. Since $g = ks/p$ and $76 = k(12)/(10)$, $k = \dfrac{190}{3}$.

If Calvin studies for 9 hours and plays for 15 hours then his score is $g = \dfrac{190}{3} \cdot (9)/(15) = 38$.

65. Since $h = kv^2$ and $16 = k(32)^2$ then $k = \dfrac{1}{64}$. To reach a height of 20'2.5" then the velocity v must satisfy $20 + \dfrac{2.5}{12} = \dfrac{1}{64}v^2$. The necessary speed is $v \approx 35.96$ ft/sec.

Review Exercises

1. Function, domain and range are both $\{0, 1, -2\}$

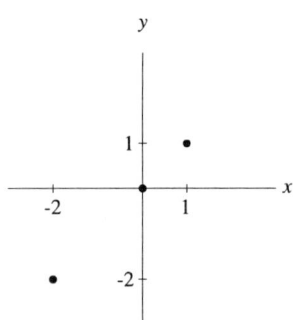

3. $y = 3 - x$ is a function, domain and range are both $(-\infty, \infty)$

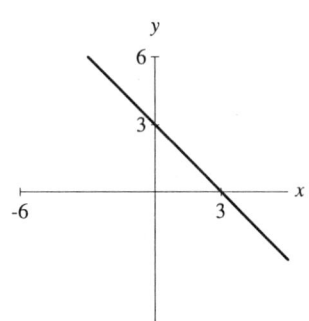

5. Not a function, domain is $\{2\}$, range is $(-\infty, \infty)$

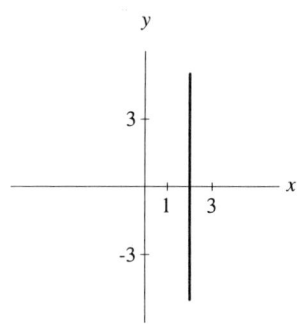

7. $x^2 + y^2 = 0.01$ is not a function, domain and range are both $[-0.1, 0.1]$

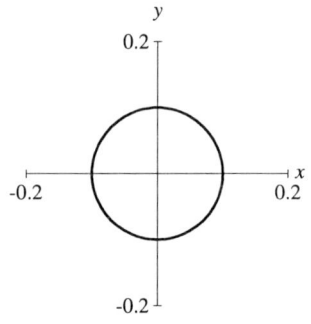

9. $x = y^2 + 1$ is not a function, domain is $[1, \infty)$, range is $(-\infty, \infty)$

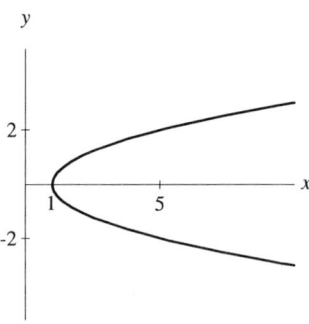

11. $y = \sqrt{x} - 3$ is a function, domain is $[0, \infty)$, range is $[-3, \infty)$

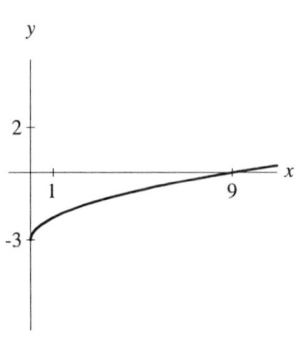

13. $9 + 3 = 12$ 15. $24 - 7 = 17$

17. If $x^2 + 3 = 19$, then $x^2 = 16$ or $x = \pm 4$.

19. $g(12) = 17$ 21. $7 + (-3) = 4$

23. $(4)(-9) = -36$ 25. $f(-3) = 12$

27.
$$\begin{aligned} f(g(x)) &= f(2x - 7) \\ &= (2x - 7)^2 + 3 \\ &= 4x^2 - 28x + 52 \end{aligned}$$

29. $(x^2 + 3)^2 + 3 = x^4 + 6x^2 + 12$

31. $(a + 1)^2 + 3 = a^2 + 2a + 4$

33.
$$\begin{aligned} \frac{f(3 + h) - f(3)}{h} &= \frac{(9 + 6h + h^2) + 3 - 12}{h} \\ &= \frac{6h + h^2}{h} \\ &= 6 + h \end{aligned}$$

35.
$$\begin{aligned} \frac{f(x + h) - f(x)}{h} &= \frac{(x^2 + 2xh + h^2) + 3 - x^2 - 3}{h} \\ &= \frac{2xh + h^2}{h} \\ &= 2x + h \end{aligned}$$

37. $g\left(\dfrac{x + 7}{2}\right) = (x + 7) - 7 = x$

39. $g^{-1}(x) = \dfrac{x + 7}{2}$

41. $f(x) = \sqrt{x}, g(x) = 2\sqrt{x + 3}$; left by 3, stretch by 2

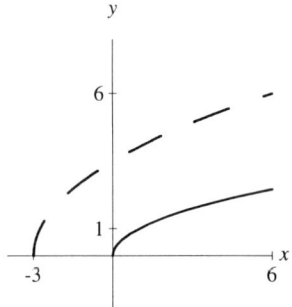

43. $f(x) = |x|, g(x) = -2|x + 2| + 4$; left by 2, stretch by 2, reflect about x-axis, up by 4

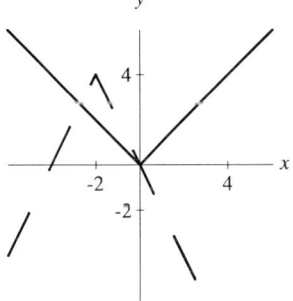

45. $f(x) = x^2, g(x) = \dfrac{1}{2}(x - 2)^2 + 1$; right by 2, stretch by $\dfrac{1}{2}$, up by 1

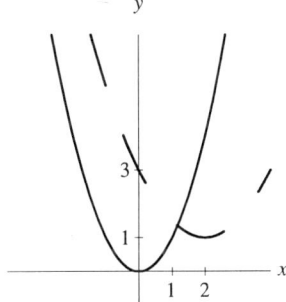

47. $F = f \circ g$ 49. $H = f \circ h \circ q \circ j$

51. $N = h \circ f \circ j$ 53. $R = g \circ h \circ j$

55. distance is $\sqrt{25 + 121} = \sqrt{146}$, midpoint is $(-1/2, -1/2)$

57. distance is $\sqrt{1/16 + 4/9} = \sqrt{73}/12$, midpoint is $(3/8, 2/3)$

59.
$$\frac{f(x+h)-f(x)}{h} = \frac{-5(x+h)+9+5x-9}{h}$$
$$= \frac{-5h}{h}$$
$$= -5$$

61.
$$\frac{f(x+h)-f(x)}{h}$$
$$= \frac{\frac{1}{2x+2h} - \frac{1}{2x}}{h} \cdot \frac{(2x+2h)(2x)}{(2x+2h)(2x)}$$
$$= \frac{(2x)-(2x+2h)}{h(2x+2h)(2x)}$$
$$= \frac{-2}{(2x+2h)(2x)}$$
$$= \frac{-1}{(x+h)(2x)}$$

63. Domain is $[-10, 10]$, range is $[0, 10]$, increasing on $(-10, 0)$, decreasing on $(0, 10)$

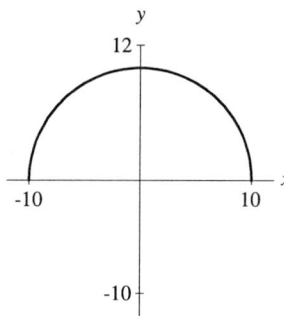

65. Domain and range are both R, increasing on $(-\infty, \infty)$

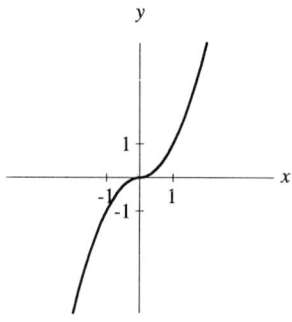

67. Domain is R, range is $[-2, \infty)$, increasing on $(-2, 0)$ and $(2, \infty)$, decreasing on $(-\infty, -2)$ and $(0, 2)$

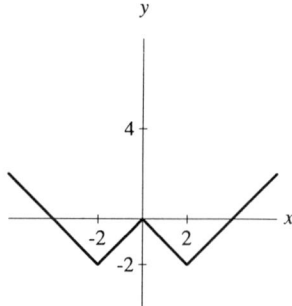

69. $y = |x| - 3$, domain is $(-\infty, \infty)$, range is $[-3, \infty)$

71. $y = -2|x| + 4$, domain is $(-\infty, \infty)$, range is $(-\infty, 4]$

73. $y = |x+2|+1$, domain is $(-\infty, \infty)$, range is $[1, \infty)$

75. Symmetry: y-axis   **77.** Symmetric about origin

79. Neither symmetry   **81.** Symmetric about y-axis

83. From the graph of $y = |x-3|-1$ the solution set is $(-\infty, 2] \cup [4, \infty)$

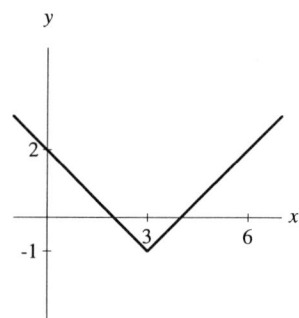

85. From the graph of $y = -2x^2 + 4$ the solution set is $(-\sqrt{2}, \sqrt{2})$

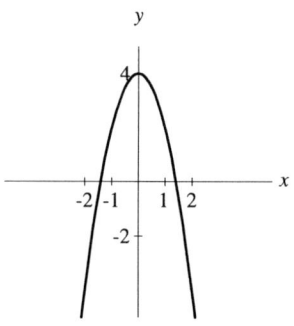

87. No solution since $-\sqrt{x+1} - 2 \leq -2$

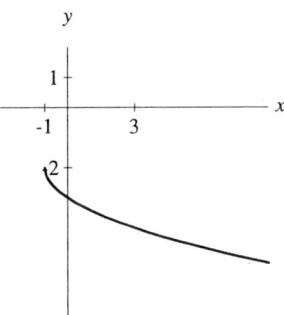

89. $f(x) = \sqrt{x+3}, g(x) = x^2 - 3$ for $x \geq 0$

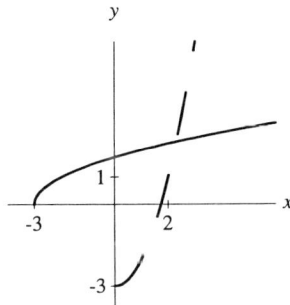

91. $f(x) = 2x - 4, g(x) = \frac{1}{2}x + 2$

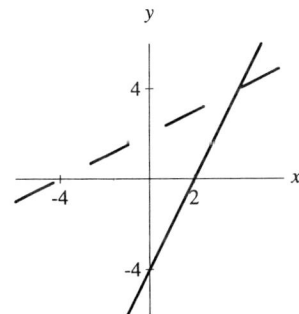

93. Not invertible

95. Inverse is $f^{-1}(x) = \dfrac{x + 21}{3}$ with domain and range both $(-\infty, \infty)$

97. Not invertible

99. Inverse is $f^{-1}(x) = x^2 + 9$ for $x \geq 0$ with domain $[0, \infty)$ and range $[9, \infty)$

101. Inverse is $f^{-1}(x) = \dfrac{5x + 7}{1 - x}$ with domain $\{x | x \neq 1\}$ and range $\{y | y \neq -5\}$

103. Inverse is $f^{-1}(x) = \sqrt{x - 1}$ with domain $[1, \infty)$ and range $(-\infty, 0]$

105. x-intercept is $(3, 0)$, y-intercept is $(0, -9/4)$

107. Since $t = ku/v$ and $6 = k \cdot 8/2$, $k = 1.5$. So $t = (1.5)(19)/3 = 19/2$.

109. Since $\left(x^2 - x + \dfrac{1}{4}\right) + (y+1)^2 = 1 + \dfrac{1}{4} + 1 = \dfrac{9}{4}$, $\left(x - \dfrac{1}{2}\right)^2 + (y+1)^2 = 9/4$.

The center is $\left(\dfrac{1}{2}, -1\right)$ and the radius is $3/2$.

111. If A is the area of the square then the length of one side is $\sqrt{A}$. Since one side of the square is twice the radius r then $\sqrt{A} = 2r$. So $A = 4r^2$.

113. The average change is $\dfrac{130 - 40}{3} = 30$ mph/sec.

115. Since $R = k/p$ and $21 = k/240$, $k = 5040$. If $p = 224$ then he needs $R = 5040/224 = 22.5$ rows.

117. $F = k\dfrac{m_1 m_2}{d^2}$ where m_1, m_2 are the masses and d is the distance between the centers of the objects.

Chapter 2 Test

1. No 2. Yes 3. No 4. Yes

5. Domain is $\{2,5\}$, range is $\{-3,-4,7\}$

6. Domain is $[9,\infty)$, range is $[0,\infty)$

7. Domain is $[0,\infty)$, range is $(-\infty,\infty)$

8. Graph of $3x - 4y = 12$ includes the points $(4,0), (0,-3)$

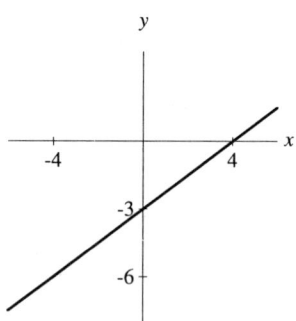

9. Graph of $y = 2x - 3$ includes the points $(3/2, 0), (0, -3)$

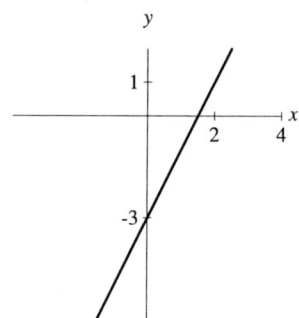

10. $y = \sqrt{25 - x^2}$ is a semicircle with radius 5

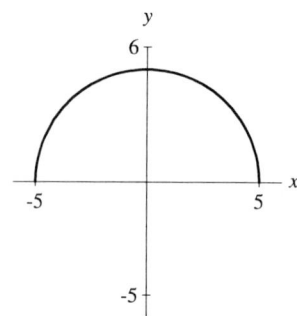

11. $y = -(x-2)^2 + 5$ is a parabola with vertex $(2,5)$

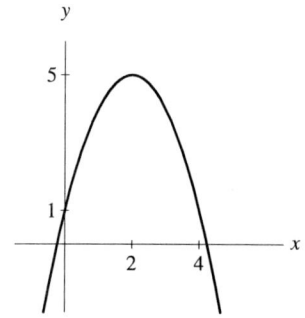

12. $y = 2|x| - 4$ includes the points $(0,-4), (\pm 3, 2)$

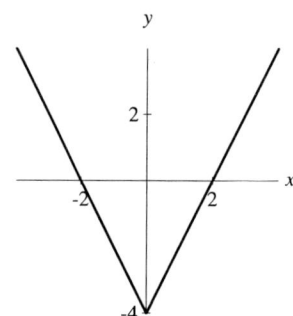

13. $y = \sqrt{x+3} - 5$ includes the points $(1,-3), (6,-2)$

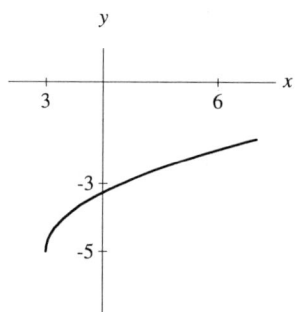

14. Graph includes the points $(-2,-2),(0,2),(3,2)$

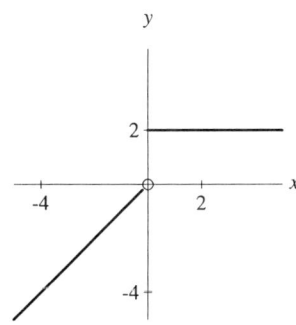

15. $\sqrt{9}=3$ **16.** $f(5)=\sqrt{7}$

17. $f(3x-1)=\sqrt{(3x-1)+2}=\sqrt{3x+1}$.

18. $g^{-1}(x)=\dfrac{x+1}{3}$ **19.** $\sqrt{16}+41=45$

20.
$$\dfrac{g(x+h)-g(x)}{h}=\dfrac{3(x+h)-1-3x+1}{h}$$
$$=\dfrac{3h}{h}$$
$$=3$$

21. $(0,11/2)$ **22.** $\sqrt{25+25}=5\sqrt{2}$

23. $(x+4)^2+(y-1)^2=7$

24. Increasing on $(3,\infty)$, decreasing on $(-\infty,3)$

25. Symmetric about y-axis

26. Add 1 to $-3<x-1<3$ and we have the solution set $(-2,4)$.

27. $g^{-1}(x)=(x-3)^3+2$

28. $\dfrac{60-35}{200}=\$0.125$ per envelope

29. Since $I=k/d^2$ and $300=k/4$, $k=1200$. If $d=10$ then $I=1200/100=12$ candlepower.

30. Let s be the length of one side of the cube. By the Pythagorean Theorem we have $s^2+s^2=d^2$. So $s=\dfrac{d}{\sqrt{2}}$ and the volume is $V=\left(\dfrac{d}{\sqrt{2}}\right)^3=\dfrac{\sqrt{2}d^3}{4}$.

Tying It All Together

1. $5x$ **2.** $9-6x+x^2$

3. $|x|$ **4.** x^2-4

5. $x^4+x^3-6x^2$ **6.** $9+4=13$

7. $\dfrac{-4\pm\sqrt{8}}{2}=\dfrac{-4\pm 2\sqrt{2}}{2}=-2\pm\sqrt{2}$

8. $\dfrac{-6\pm\sqrt{-24}}{6}=\dfrac{-6\pm 2i\sqrt{6}}{6}=\dfrac{-3\pm i\sqrt{6}}{3}$

9. An identity with solution set $(-\infty,\infty)$

10. Squaring both sides, we obtain
$$9-x^2=4$$
$$5=x^2$$
$$\pm\sqrt{5}=x$$
The solution set is $\{\pm\sqrt{5}\}$.

11. $\{\pm 1\}$ **12.** $\{\pm 2\}$

13. Complete the square: $x^2-4x+4=-5+1$ Then $(x-2)^2=-1$ and the solution set is $\{2\pm i\}$.

14. Add 2 to both sides and get $4x^2-4x+1=2$. Factor as $(2x-1)^2=2$. By the square root property, solution set is $\left\{\dfrac{1\pm\sqrt{2}}{2}\right\}$.

15. Since $x(x-9)=0$, the solution set is $\{0,9\}$.

16. Multiplying the equation by $x(x-1)$, we find
$$(x-1)+x=3$$
$$2x=4$$
The solution set is $\{2\}$.

17. Line $y = 5x$ includes the points $(0,0), (1,5)$

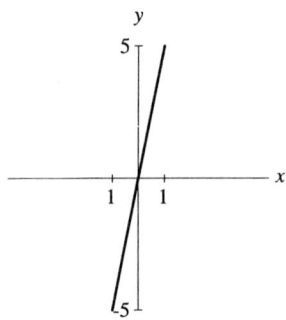

18. $y = \sqrt{9 - x^2}$ is a semicircle with radius 3

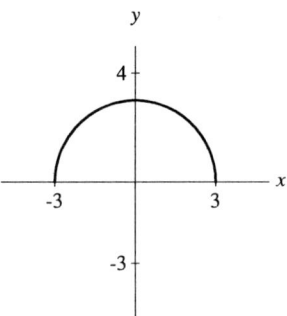

19. Graph of $y = |x| - 1$ includes the points $(0, -1), (\pm 2, 1)$

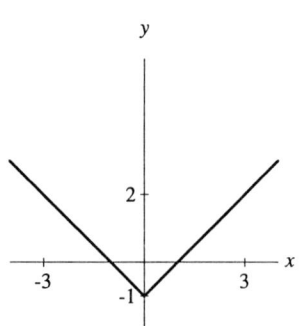

20. $y = x^2 - 4$ is a parabola with vertex $(0, -4)$

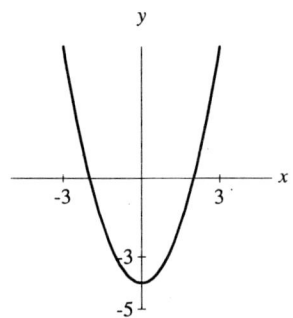

21. Line $y = 3 - x$ includes the points $(3, 0), (0, 3)$

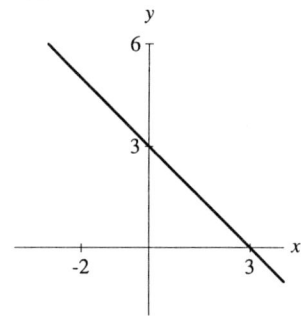

22. Graph of $y = 3 - \sqrt{x}$ includes the points $(0, 3), (4, 1)$

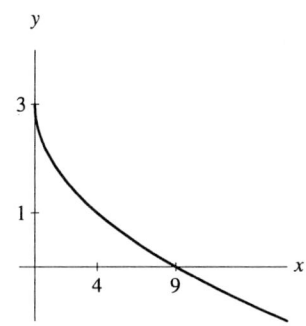

23. Graph of $y = 2(x + 2)^2 - 5$ includes the points $(-2, -5), (-1, -3)$

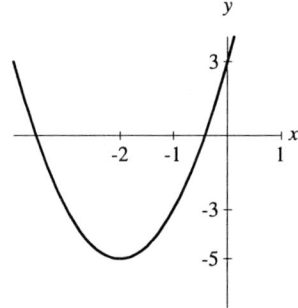

24. Graph of $y = x^3 - 1$ includes the points $(-1, -2), (0, -1), (2, 7)$

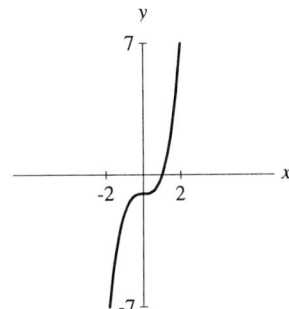

25. $\dfrac{x^2 + 1}{x}$

26. $\dfrac{x^2 - 1}{x - 1} + \dfrac{1}{x - 1} = \dfrac{x^2}{x - 1}$

27. $\dfrac{3(x - 4) + 2}{x - 4} = \dfrac{3x - 10}{x - 4}$

28. $\dfrac{-2(x + 3) - 1}{x + 3} = \dfrac{-2x - 7}{x + 3}$

For Thought

1. False, the slope is 1. 2. True

3. False, slopes of vertical lines are undefined.

4. False, it is a vertical line. 5. True

6. False, $x = 1$ can not be written in the slope-intercept form. 7. False, since the leading coeficient is -1. 8. True

9. False 10. True

3.1 Exercises

1. $\dfrac{5-3}{4+2} = \dfrac{1}{3}$ 3. $\dfrac{3+5}{1-3} = -4$

5. $\dfrac{2-2}{5+3} = 0$

7. $\dfrac{1+3}{\sqrt{2}-3\sqrt{2}} = \dfrac{4\cdot\sqrt{2}}{-2\sqrt{2}\cdot\sqrt{2}} = \dfrac{4\sqrt{2}}{-4} = -\sqrt{2}$

9. $\dfrac{\pi/2 - \pi/4}{3-1} = \dfrac{\pi/4}{2} = \dfrac{\pi}{8}$

11. $y = \dfrac{3}{5}x - 2$, slope is $\dfrac{3}{5}$, y-intercept is $(0,-2)$

13. Since $y - 3 = 2x - 8$, $y = 2x - 5$. The slope is 2 and y-intercept is $(0,-5)$.

15. Since $y + 1 = \dfrac{1}{2}x + \dfrac{3}{2}$, $y = \dfrac{1}{2}x + \dfrac{1}{2}$. The slope is $\dfrac{1}{2}$ and y-intercept is $\left(0, \dfrac{1}{2}\right)$.

17. Since $y = 4$, the slope is $m = 0$ and the y-intercept is $(0,4)$.

19. Since $y - 0.4 = 0.03x - 3$, $y = 0.03x - 2.6$. The slope is 0.03 and y-intercept is $(0,-2.6)$

21. $y = \dfrac{1}{2}x - 2$ goes through the points $(0,-2), (2,-1)$, and $(4,0)$.

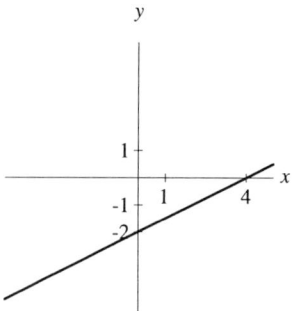

23. $f(x) = -3x + 1$ goes through $(0,1), (1,-2)$

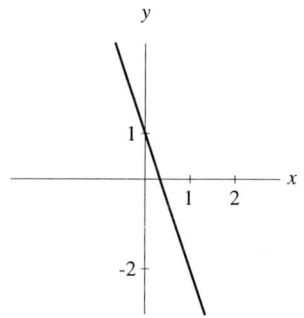

25. $y = -\dfrac{3}{4}x - 1$ goes through $(0,-1), (-4/3, 0)$

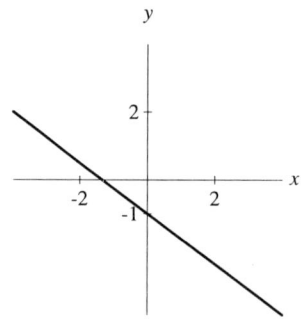

27. $x - y = 3$ goes through $(0,-3), (3,0)$

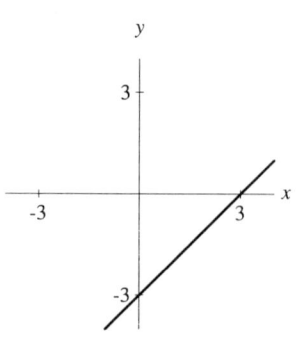

29. $y = 5$ is a horizontal line

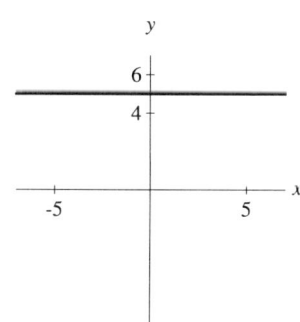

31. $y = 3x - 4$ goes through $(0, -4), (4/3, 0)$

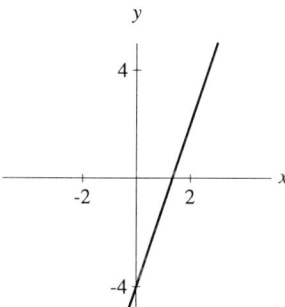

33. $y = \sqrt{2}x - \sqrt{2}$ goes through $(0, -\sqrt{2}), (1, 0)$

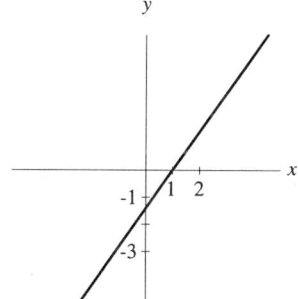

35. $y = \dfrac{2}{3}x - 1$

37. $y = \dfrac{5}{2}x + \dfrac{3}{2}$

39. $y = -2x + 4$

41. Since $m = 4/3$ and $y - 0 = \dfrac{4}{3}(x - 3)$,
$4x - 3y = 12$.

43. Since $m = 4/5$ and $y - 3 = \dfrac{4}{5}(x - 2)$,
$5y - 15 = 4x - 8$ and $4x - 5y = -7$.

45. $x = -4$ is a vertical line.

47. $y = -6$ is a horizontal line.

49. 0.5 **51.** -1

53. 0

55. Since $y + 2 = 2(x - 1)$, $2x - y = 4$

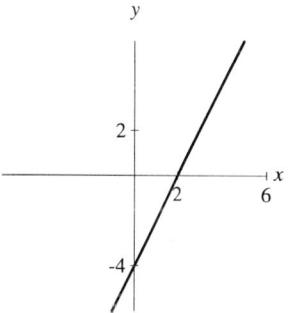

57. Since slope of $y = -3x$ is -3 and $y - 4 = -3(x - 1)$,
$3x + y = 7$

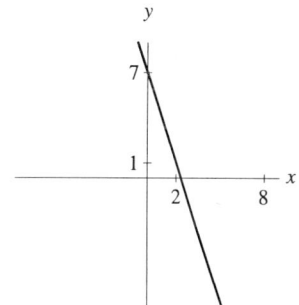

59. Since slope of $y = \dfrac{1}{2}x - \dfrac{3}{2}$ is $1/2$ and
$y - 1 = -2(x + 3)$, $2x + y = -5$

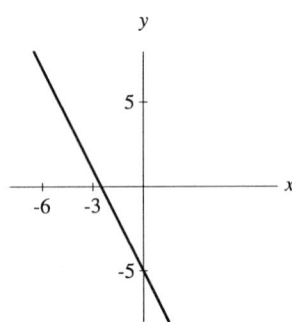

61. Since $x = 4$ is a vertical line, a horizontal line through $(2, 5)$ is $y = 5$

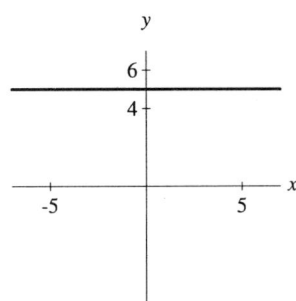

63. Since $m = \dfrac{1+2}{4+1} = \dfrac{3}{5}$ and $y - 3 = \dfrac{3}{5}(x - 0)$,
$3x - 5y = -15$

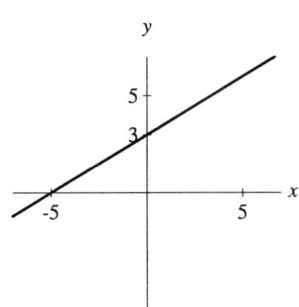

65. Since $\dfrac{a-4}{7-3} = \dfrac{a-4}{4} = \dfrac{2}{3}$, $a = \dfrac{20}{3}$.

67. Since $\dfrac{5-3}{8+2} = \dfrac{1}{5} = -\dfrac{1}{a}$, $a = -5$.

69. Plot the points $A(-1, 2)$, $B(2, -1)$, $C(3, 3)$, and $D(-2, -2)$, respectively. The slopes of the opposite sides are $m_{AC} = m_{BD} = 1/4$ and $m_{AD} = m_{BC} = 4$. Since the opposite sides are parallel, it is a parallelogram.

71. Plot the points $A(-5, -1)$, $B(-3, -4)$, $C(3, 0)$, and $D(1, 3)$, respectively. The slopes of the opposite sides are $m_{AB} = m_{CD} = -3/2$ and $m_{AD} = m_{BC} = 2/3$. Since the adjacent sides are perpendicular, it is a rectangle.

73. Plot the points $A(-5, 1)$, $B(-2, -3)$, and $C(4, 2)$, respectively. The slopes of the sides $m_{AB} = -4/3$, $m_{BC} = 5/6$ and $m_{AC} = 1/9$. It is not a right triangle since no two sides are perpendicular.

75. Plot the points $A(-3, 2)$, $B(-1, -2)$, $C(6, -1)$, and $D(1, 4)$, respectively. Since the slopes of the diagonals are $m_{BD} = 3$ and $m_{AC} = -1/3$, the diagonals are perpendicular.

77. No, they are not parallel since their slopes are not equal, i.e., $\dfrac{1}{3} \neq .33$. Although, they may appear parallel.

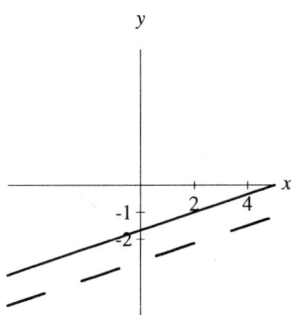

79. Since $x^3 - 8 = (x-2)(x^2 + 2x + 4)$, a linear function for the graph is $y = x - 2$.

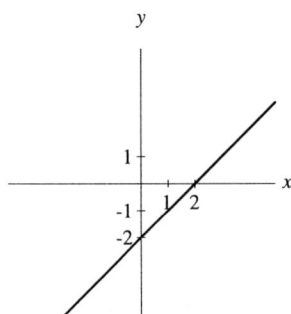

81. The slope is $\dfrac{212 - 32}{100 - 0} = \dfrac{9}{5}$.
Since $F - 32 = \dfrac{9}{5}(C - 0)$ then $F = \dfrac{9}{5}C + 32$.
When $C = 150$, $F = \dfrac{9}{5}(150) + 32 = 302°$F.

83. Linear function through $(1, 49)$ and $(2, 48)$ is $c = 50 - n$. With 40 people in a tour she would charge $10 each and make $400.

85. Since the slope is $m = 0.1$ for $35 \leq a \leq 50$ and from $y = mx + b$ we derive $0.5 = 0.1(35) + b$ or $b = -3$. Then $y = 0.1a - 3$ expresses y as a function of a for $35 \leq a \leq 50$.

    For $50 < a \leq 65$, $m = .2$ and $y = .2x + b$. Then $4 = 0.2(60) + b$ or $-8 = b$; and $y = 0.2a - 8$ for $50 < a \leq 65$.

    If $a = 47$, then $y = 0.1(47) - 3 = 1.7$ years.
    If $a = 63$, then $y = 0.2(63) - 8 = 4.6$ years.

87. The slope is $\dfrac{75 - 95}{4000} = -0.005$.
    Since $S - 95 = -0.005(D - 0)$, $S = -0.005D + 95$.

89. Let c and p be the number of computers and printers, respectively. Since $60000 = 2000c + 1500p$, then
    $$2000c = -1500p + 60000$$
    $$c = -\dfrac{3}{4}p + 30.$$
    The slope is $-\dfrac{3}{4}$, i.e., if 4 more printers are purchased then 3 less computers must be bought.

91. (a) $y = -1,645x + 16,203$ where x is the age in years and y is the retail price.

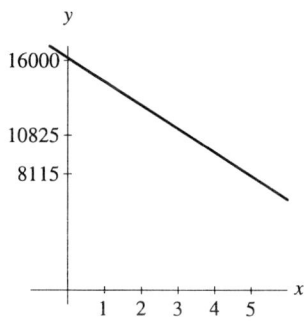

   (b) The slope is $-1,645$, i.e., the car's retail price decreases by $1,645 every year.
   (c) $y = -1,645(7) + 16,203 = \$4,688$.
   (d) By solving $-1,645x + 16,203 = 0$, one finds $x \approx 10$ or after 10 years a Z28 becomes worthless.

93. Since $C(x) = 50x + 8000$, then
    $$MC(x) = C(x + 1) - C(x)$$
    $$= 50(x + 1) + 8000 - 50x - 8000$$
    $$= 50.$$
    and MC is constant $(0, \infty)$.

95. Consider the isosceles right triangle with vertices at the points $S(0, 0)$, $T(1, 2)$, and $U(3, 1)$. Then the angle at U is $45°$, i.e., $\angle TUS = 45°$ Let V be the point $(0, 1)$. Note, $\angle TUV = B$ and $\angle SUV = A$. Then $A + B = 45°$. Since C is an angle of the isosceles right triangle with vertices at $(2, 0)$, $(3, 0)$, and $(3, 1)$, then $C = 45°$. Thus, $C = A + B$.

97. Let $b_1 \neq b_2$. If $y = mx + b_1$ and $y = mx + b_2$ have a point (s, t) in common, then $ms + b_1 = ms + b_2$. After subtracting ms from both sides, we get $b_1 = b_2$; a contradiction. Thus, $y = mx + b_1$ and $y = mx + b_2$ have no points in common if $b_1 \neq b_2$.

For Thought

1. False, the range of $y = x^2$ is $[0, \infty)$.
2. False, vertex is the point $(3, -1)$.
3. True   4. True   5. True, since $\dfrac{-b}{2a} = \dfrac{6}{2 \cdot 3} = 1$.
6. True, x-intercept of $y = (3x + 2)^2$ is the vertex $(-2/3, 0)$ and the y-intercept is $(0, 4)$.
7. True   8. True, since $(x - \sqrt{3})^2$ is always nonnegative.
9. True, since if x and $\dfrac{p - 2x}{2}$ are the length and the width, respectively, of a rectangle with perimeter p, the area is $y = x \cdot \dfrac{p - 2x}{2}$. This is a parabola opening down with vertex $\left(\dfrac{p}{4}, \dfrac{p^2}{16}\right)$.
   So the maximum area is $\dfrac{p^2}{16}$.
10. False

3.2 Exercises

1. $y = \left(x^2 - 3x + \frac{9}{4}\right) - \frac{9}{4} = \left(x - \frac{3}{2}\right)^2 - \frac{9}{4}$

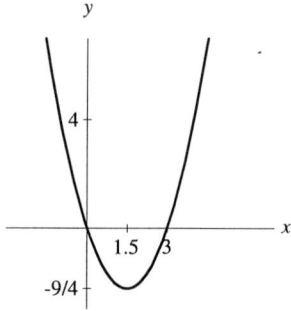

3. $y = 2\left(x^2 - 6x + 9\right) - 18 + 22 = 2(x-3)^2 + 4$

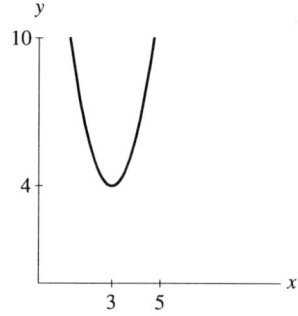

5. $y = -\frac{1}{2}\left(x^2 - 2x + 1\right) + \frac{1}{2} + \frac{5}{2} = -\frac{1}{2}(x-1)^2 + 3$

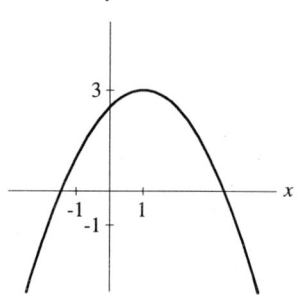

7. $y = \left(x^2 + 3x + \frac{9}{4}\right) - \frac{9}{4} + \frac{5}{2} = \left(x + \frac{3}{2}\right)^2 + \frac{1}{4}$

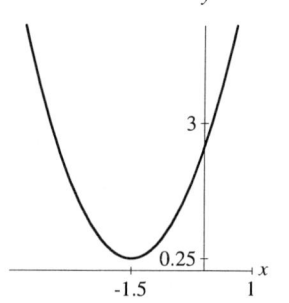

9. $y = -2\left(x^2 - \frac{3}{2}x + \frac{9}{16}\right) + \frac{9}{8} - 1 =$
$-2\left(x - \frac{3}{4}\right)^2 + \frac{1}{8}$

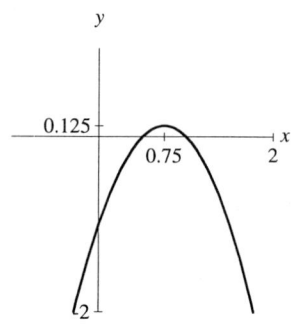

11. $y = -3\left(x^2 - \frac{2}{3}x + \frac{1}{9}\right) + \frac{1}{3} = -3\left(x - \frac{1}{3}\right)^2 + \frac{1}{3}$

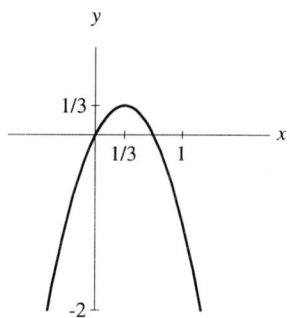

13. Since $\dfrac{-b}{2a} = \dfrac{12}{6} = 2$ and $f(2) = 12 - 24 + 1 = -11$, the vertex is $(2, -11)$.

15. Vertex: $(4, 1)$

17. Since $\dfrac{-b}{2a} = \dfrac{1/3}{-1} = -\dfrac{1}{3}$ and $f(-1/3) = -\dfrac{1}{18} + \dfrac{1}{9} = \dfrac{1}{18}$, the vertex is $\left(-\dfrac{1}{3}, \dfrac{1}{18}\right)$.

19. Up, vertex $(1, -4)$, axis of symmetry $x = 1$, range $[-4, \infty)$, minimum value -4, decreasing on $(-\infty, 1)$, inreasing on $(1, \infty)$.

21. Since it opens up and vertex $(1, -1)$, range is $[-1, \infty)$, minimum value is -1, decreasing on $(-\infty, 1)$, and increasing on $(1, \infty)$.

23. Since it opens down and vertex $(0, \sqrt{3})$, range is $(-\infty, \sqrt{3}]$, maximum value is $\sqrt{3}$, decreasing on $(0, \infty)$, and increasing on $(-\infty, 0)$.

25. Since it opens up and vertex is $(3, 4)$, range is $[4, \infty)$, minimum value is 4, decreasing on $(-\infty, 3)$, and increasing on $(3, \infty)$.

27. Since it opens down and vertex is $(1/2, 9)$, range is $(-\infty, 9]$, maximum value is 9, decreasing on $(1/2, \infty)$, and increasing on $(-\infty, 1/2)$.

29. Since it opens down and vertex $(3/2, 27/2)$, range is $(-\infty, 27/2]$, maximum value is $27/2$, decreasing on $(3/2, \infty)$, and increasing on $(-\infty, 3/2)$.

31. Since it opens up and vertex is $(-1, -2 - \sqrt{2})$, range is $[-2 - \sqrt{2}, \infty)$, minimum value is $-2 - \sqrt{2}$, decreasing on $(-\infty, -1)$, and increasing on $(-1, \infty)$.

33. Vertex $(0, -3)$, axis $x = 0$, y-intercept $(0, -3)$, x-intercepts $(\pm\sqrt{3}, 0)$, opening up

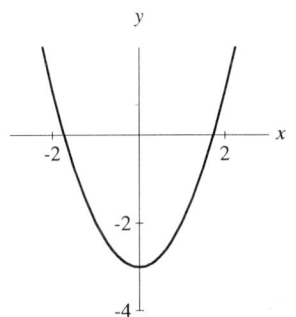

35. Vertex $(\pi/2, -\pi^2/4)$, axis $x = \pi/2$, y-intercept $(0, 0)$, x-intercepts $(0, 0), (\pi, 0)$, opening up

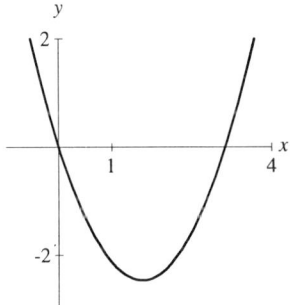

37. Vertex $(-3, 0)$, axis $x = -3$, y-intercept $(0, 9)$, x-intercepts $(-3, 0)$, opening up

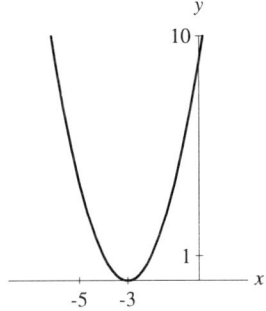

39. Vertex $(3,-4)$, axis $x=3$, y-intercept $(0,5)$, x-intercepts $(1,0),(5,0)$, opening up

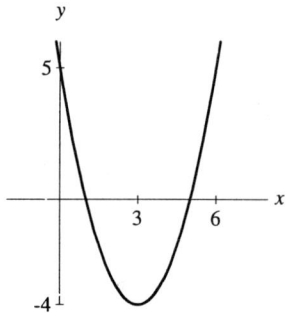

41. Vertex $(2,12)$, axis $x=2$, y-intercept $(0,0)$, x-intercepts $(0,0),(4,0)$, opening down

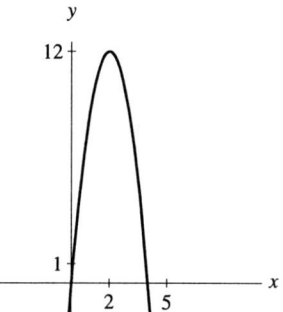

43. Vertex $(1,3)$, axis $x=1$, y-intercept $(0,1)$, x-intercepts $\left(1\pm\dfrac{\sqrt{6}}{2},0\right)$, opening down

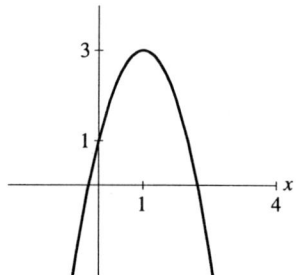

45. $(-\infty,-1]\cup[3,\infty)$

47. $(-3,1)$

49. $[-3,1]$

51. From the x-intercepts of $y=x^2-5x+6$, the solution set is $[2,3]$.

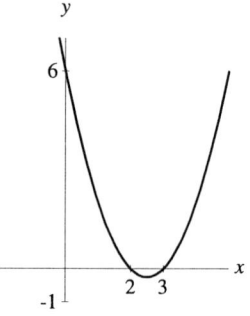

53. From the x-intercepts of $y=x^2-6x+7$, the solution set is $(-\infty,3-\sqrt{2})\cup(3+\sqrt{2},\infty)$.

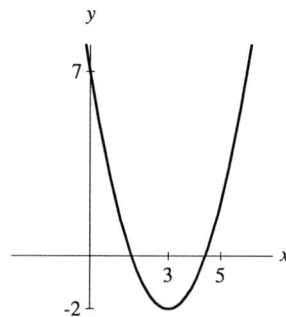

55. Since $y=x^2-6x+10$ has no x-intercepts, the solution set is $(-\infty,\infty)$.

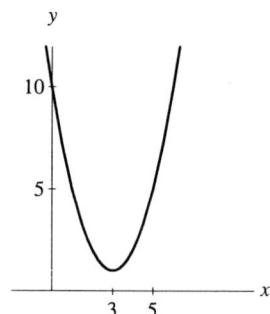

57. From the x-intercepts of $y = x^2 + 0.1x - 0.021$, solution set is approximately $(-0.203, 0.103)$

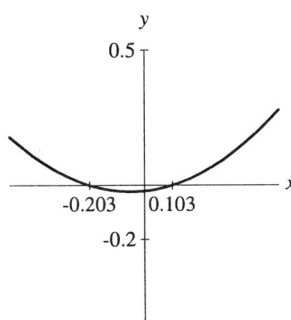

59. From the x-intercepts of $y = x^2 - 24x - 3456$, a factorization is $y = (x - 72)(x + 48)$

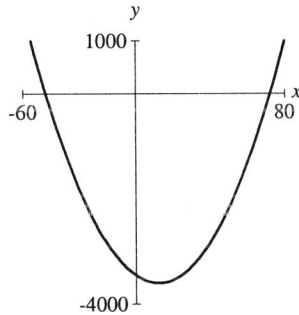

61. From the x-intercepts of $y = 2x^2 - 39x - 2970$, a factorization is $2\left(x - \dfrac{99}{2}\right)(x + 30)$ or $(2x - 99)(x + 30)$

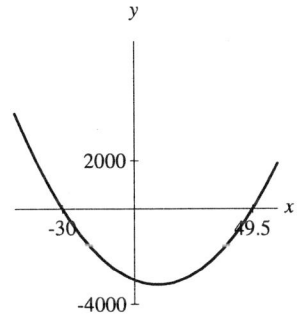

63. Axis $x = 0$, point $(-2, 4)$

65. Axis $x = 1$, point $(0, 8)$

67. Axis $x = 1$, point $(3, 16)$

69. Axis $x = -1/2$, point $(1, -13)$

71. Since $\dfrac{-b}{2a} = \dfrac{-b}{-6} = -2$, $b = -12$.

73. Since $2 = a(0 - 1)^2 - 3$, $a = 5$ and $y = 5(x - 1)^2 - 3$.

75. Since $0 = a(-1 + 3)^2 + 2$, $a = -1/2$ and
$y = -\dfrac{1}{2}(x + 3)^2 + 2$.

77. Let x and $7 - x$ be the numbers. Their product is $y = x(7 - x)$ which is a parabola opening up with
$x = \dfrac{-b}{2a} = \dfrac{7}{2}$. Since $7 - x = 7 - \dfrac{7}{2} = \dfrac{7}{2}$, the numbers are $\dfrac{7}{2}$ and $\dfrac{7}{2}$.

79. Let x be the length of a folded side. The area of the cross-section is $A = x(10 - 2x)$. This is a parabola opening down with $-b/2a = 2.5$. The dimensions of the cross-section are 2.5 in. high and 5 in. wide.

81. The value of A that would maximize M is
$A = \dfrac{-b}{2a} = \dfrac{-0.127}{-0.001306} \approx 97.24$ mph.
In 1 hour, Lindbergh flying at 97 mph would use
$\dfrac{97}{1.2} \approx 80.83$ lbs. of fuel; which weighed
$\dfrac{80.83}{6.12} \approx 13.2$ gallons of fuel.

83. Since $\dfrac{-b}{2a} = \dfrac{-128}{-2(-16)} = 4$, the maximum height is $h(4) = 261$ ft.

85. Let n and p be the number of persons and the price of a tour per person, respectively. The function expressing p as a function of n is $p = 50 - n$. The revenue, R, is $R = (50 - n)n$. This is a parabola opening down with vertex $(25, 625)$. Therefore, 25 persons will give her the maximum revenue of $625.

87. Since $v = 50p - 50p^2$ is a parabola opening down, to maximize v, choose $p = \dfrac{-b}{2a} = \dfrac{50}{100} = 1/2$.

89. (a) A graph of atmospheric pressure versus altitude

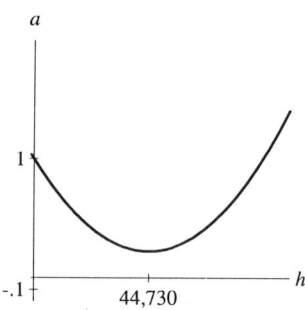

(b) From the graph, one finds the function is decreasing when $h = 29,029$ feet.

(c) Since $\dfrac{-b}{2a} = \dfrac{3.48 \times 10^{-5}}{2(3.89 \times 10^{-10})} \approx 44,730$ feet, the function is decreasing on the interval $(0, 44730)$ and increasing on $(44,730, \infty)$.

(d) It does not make sense to speak of atmospheric pressure at heights that lie in $(44,730, \infty)$ since $44,730$ feet is over 8 miles.

(e) It is valid for altitudes less than $44,730$ feet.

91. Note, $(3 - 2m)^2 = 4m^2 - 12m + 9$, $(6m - 5)^2 = 36m^2 - 60m + 25$, and $d = 40m^2 - 72m + 34$.

To minimize d, let $m = \dfrac{-b}{2a} = \dfrac{72}{2(40)} = 0.9$

93. (a) If I is the index and n is number of years after 1985, a linear regression is $I = 14.6n + 110.9$.

(b) The consumer price index for tuition in 1998 is $I = 14.6(13) + 110.9 \approx 300.7$

(c) Solving $\dfrac{1489}{249.8} = \dfrac{x}{300.7}$, one finds $x = \$1,792$ - the expected amount to be paid in 1998.

For Thought

1. False 2. True, by the Remainder Theorem.

3. True, by the Factor Theorem.

4. True, since $2^5 - 1 = 31$.

5. False, since $P(x) = 1$ has no zero.

6. False, rather $c^3 - c^2 + 4c - 5 = b$.

7. False, since $P(4) = -15$. 8. True

9. True, since 1 is a root.

10. False, since -3 is not root.

3.3 Exercises

1. Quotient $x - 3$, remainder 1

$$\begin{array}{r} x - 3 \\ x - 2 \overline{\smash{)}x^2 - 5x + 7} \\ \underline{x^2 - 2x} \\ -3x + 7 \\ \underline{-3x + 6} \\ 1 \end{array}$$

3. Quotient $-2x^2 + 6x - 14$, remainder 33

$$\begin{array}{r} -2x^2 + 6x - 14 \\ x + 3 \overline{\smash{)}-2x^3 + 0x^2 + 4x - 9} \\ \underline{-2x^3 - 6x^2} \\ 6x^2 + 4x \\ \underline{6x^2 + 18x} \\ -14x - 9 \\ \underline{-14x - 42} \\ 33 \end{array}$$

5. Quotient $s^2 + 2$, remainder 16

$$\begin{array}{r} s^2 + 2 \\ s^2 - 5 \overline{\smash{)}s^4 - 3s^2 + 6} \\ \underline{s^4 - 5s^2} \\ 2s^2 + 6 \\ \underline{2s^2 - 10} \\ 16 \end{array}$$

7. Quotient $x + 6$, remainder 13

$$\begin{array}{r|rrr} 2 & 1 & 4 & 1 \\ & & 2 & 12 \\ \hline & 1 & 6 & 13 \end{array}$$

3.3 ZEROS OF POLYNOMIAL FUNCTIONS

9. Quotient $-x + 7$, remainder -12

```
-3 | -1   4    9
   |      3  -21
   ------------------
     -1   7  -12
```

11. Quotient $4x^2 + 2x - 4$, remainder 0

```
1/2 | 4   0   -5    2
    |     2    1   -2
    --------------------
      4   2   -4    0
```

13. Quotient $2a^2 - 4a + 6$, remainder 0

```
-1/2 | 2  -3    4    3
     |    -1    2   -3
     ---------------------
       2  -4    6    0
```

15. Quotient $x^3 + x^2 + x + 1$, remainder -2

```
1 | 1   0   0   0   -3
  |     1   1   1    1
  -----------------------
    1   1   1   1   -2
```

17. Quotient $x - 5/2$, remainder $-1/4$

```
1/2 | 1   -3      1
    |     1/2  -5/4
    -------------------
      1  -5/2  -1/4
```

19. $f(1) = 0$

```
1 | 1   0   0   0   0   -1
  |     1   1   1   1    1
  ---------------------------
    1   1   1   1   1    0
```

21. $f(-2) = -33$

```
-2 | 1    0    0    0    0     1
   |     -2    4   -8   16   -32
   -------------------------------
     1   -2    4   -8   16   -33
```

23. $g(1) = 5$

```
1 | 1   -4    0    8
  |      1   -3   -3
  ---------------------
    1   -3   -3    5
```

25. $g(-1/2) = 55/8$

```
-1/2 | 1    -4      0      8
     |     -1/2   9/4   -9/8
     ---------------------------
       1   -9/2   9/4   55/8
```

27. $h(-1) = 0$

```
-1 | 2    1   -1    3    3
   |     -2    1    0   -3
   ---------------------------
     2   -1    0    3    0
```

29. $h(1) = 8$

```
1 | 2   1   -1   3   3
  |     2    3   2   5
  ------------------------
    2   3    2   5   8
```

31. Yes, $(x+3)(x^2 + x - 2) = (x+3)(x+2)(x-1)$

```
-3 | 1    4    1   -6
   |     -3   -3    6
   ---------------------
     1    1   -2    0
```

33. Yes, $(x-4)(x^2 + 8x + 15) = (x-4)(x+5)(x+3)$

```
4 | 1    4   -17   -60
  |      4    32    60
  -----------------------
    1    8    15     0
```

35. Yes, since the remainder below is zero

```
3 | 2   -5   -4    3
  |      6    3   -3
  ---------------------
    2    1   -1    0
```

37. No, since the remainder below is not zero

```
2 | 1    2    3    1
  |     -2    0   -6
  ---------------------
    1    0    3   -5
```

39. Yes, since the remainder below is zero

```
-1 | 1    2    4    6    3
   |     -1   -1   -3   -3
   ---------------------------
     1    1    3    3    0
```

41. No, since the remainder below is not zero

$$\begin{array}{c|cccc} 1/2 & 1 & 3 & -5 & 7 \\ & & 1/2 & 7/4 & -13/8 \\ \hline & 1 & 7/2 & -13/4 & 43/8 \end{array}$$

43. $\pm\{1, 2, 3, 4, 6, 8, 12, 24\}$

45. $\pm\{1, 3, 5, 15\}$

47. $\pm\left\{1, 3, 5, 15, \dfrac{1}{2}, \dfrac{1}{4}, \dfrac{1}{8}, \dfrac{3}{2}, \dfrac{3}{4}, \dfrac{3}{8}, \dfrac{5}{2}, \dfrac{5}{4}, \dfrac{5}{8}, \dfrac{15}{2}, \dfrac{15}{4}, \dfrac{15}{8}\right\}$

49. $\pm\left\{1, 2, \dfrac{1}{2}, \dfrac{1}{3}, \dfrac{1}{6}, \dfrac{1}{9}, \dfrac{1}{18}, \dfrac{2}{3}, \dfrac{2}{9}\right\}$

51. Zeros are $2, 3, 4$ since

$$\begin{array}{c|cccc} 2 & 1 & -9 & 26 & -24 \\ & & 2 & -14 & 24 \\ \hline & 1 & -7 & 12 & 0 \end{array}$$

and $x^2 - 7x + 12 = (x-4)(x-3)$

53. Zeros are $-3, 2 \pm i$ since

$$\begin{array}{c|cccc} -3 & 1 & -1 & -7 & 15 \\ & & -3 & 12 & -15 \\ \hline & 1 & -4 & 5 & 0 \end{array}$$

and $x^2 - 4x + 5 = (x-2)^2 + 1 = 0$ or $x - 2 = \pm i$

55. Zeros are $1/2, 3/2, 5/2$ since

$$\begin{array}{c|cccc} 1/2 & 8 & -36 & 46 & -15 \\ & & 4 & -16 & 15 \\ \hline & 8 & -32 & 30 & 0 \end{array}$$

and $8x^2 - 32x + 30 = 2(4x^2 - 16x + 15) = 2(2x-3)(2x-5)$

57. Zeros are $1/2, \dfrac{1 \pm i}{3}$ since

$$\begin{array}{c|cccc} 1/2 & 18 & -21 & 10 & -2 \\ & & 9 & -6 & 2 \\ \hline & 18 & -12 & 4 & 0 \end{array}$$

and the zeros of $18x^2 - 12x + 4$ are
(by the quadratic formula) $\dfrac{1 \pm i}{3}$

59. Zeros are $1, -2, \pm i$ since

$$\begin{array}{c|cccccc} 1 & 1 & 1 & -1 & 1 & -2 \\ & & 1 & 2 & 1 & 2 \\ \hline & 1 & 2 & 1 & 2 & 0 \end{array}$$

$$\begin{array}{c|cccc} -2 & 1 & 2 & 1 & 2 \\ & & -2 & 0 & -2 \\ \hline & 1 & 0 & 1 & 0 \end{array}$$

and the zeros of $x^2 + 1$ are $\pm i$

61. Zeros are $-1, \pm\sqrt{2}$ since

$$\begin{array}{c|ccccc} -1 & 1 & 2 & -1 & -4 & -2 \\ & & -1 & -1 & 2 & 2 \\ \hline & 1 & 1 & -2 & -2 & 0 \end{array}$$

$$\begin{array}{c|cccc} -1 & 1 & 1 & -2 & -2 \\ & & -1 & 0 & 2 \\ \hline & 1 & 0 & -2 & 0 \end{array}$$

and the zeros of $x^2 - 2$ are $\pm\sqrt{2}$

63. Zeros are $1/2, 1/3, 1/4$ since

$$\begin{array}{c|cccc} 1/2 & 24 & -26 & 9 & -1 \\ & & 12 & -7 & 1 \\ \hline & 24 & -14 & 2 & 0 \end{array}$$

and $2(12x^2 - 7x + 1) = 2(4x-1)(3x-1)$

65. Rational zero is $1/16$ since

$$\begin{array}{c|cccc} 1/16 & 16 & -33 & 82 & -5 \\ & & 1 & -2 & 5 \\ \hline & 16 & -32 & 80 & 0 \end{array}$$

and by the quadratic formula $16x^2 - 32x + 80$ has imaginary zeros $1 \pm 2i$.

67. Rational zeros are $7/3, -6/7$ since

$$\begin{array}{r|rrrrr} 7/3 & 21 & -31 & -21 & -31 & -42 \\ & & 49 & 42 & 49 & 42 \\ \hline & 21 & 18 & 21 & 18 & 0 \end{array}$$

$$\begin{array}{r|rrrr} -6/7 & 21 & 18 & 21 & 18 \\ & & -18 & 0 & -18 \\ \hline & 21 & 0 & 21 & 0 \end{array}$$

and $21x^2 + 21 = 0$ has imaginary zeros $\pm i$.

69.
$$\frac{2x+1}{x-2} = 2 + \frac{5}{x-2} \text{ since}$$

$$\begin{array}{r|rr} 2 & 2 & 1 \\ & & 4 \\ \hline & 2 & 5 \end{array}$$

71.
$$\frac{a^2 - 3a + 5}{a - 3} = a + \frac{5}{a - 3} \text{ since}$$

$$\begin{array}{r|rrr} 3 & 1 & -3 & 5 \\ & & 3 & 0 \\ \hline & 1 & 0 & 5 \end{array}$$

73.
$$\frac{c^2 - 3c - 4}{c^2 - 4} = 1 + \frac{-3c}{c^2 - 4} \text{ since}$$

$$\begin{array}{r} 1 \\ c^2 - 4 \overline{\smash{)}c^2 - 3c - 4} \\ \underline{c^2 + 0c - 4} \\ -3c \end{array}$$

75.
$$\frac{4t - 5}{2t + 1} = 2 + \frac{-7}{2t + 1} \text{ since}$$

$$\begin{array}{r} 2 \\ 2t + 1 \overline{\smash{)}4t - 5} \\ \underline{4t + 2} \\ -7 \end{array}$$

77. Note, $\dfrac{P(t)}{t} = -t^3 + 12t^2 - 58t + 132$.

$$\begin{array}{r|rrrr} 6 & -1 & 12 & -58 & 132 \\ & & -6 & 36 & -132 \\ \hline & -1 & 6 & -22 & 0 \end{array}$$

The drug will be eliminated in $t = 6$ hrs.

79. If w is the width, then $w(w+4)(w+9) = 630$. This can be re-written as $w^3 + 13w^2 + 36w - 630 = 0$. Using synthetic division, we find

$$\begin{array}{r|rrrr} 5 & 1 & 13 & 36 & -630 \\ & & 5 & 90 & 630 \\ \hline & 1 & 18 & 126 & 0 \end{array}$$

Since $w^2 + 18w + 126 = 0$ has non-real roots, the width of the HP box is $w = 5$ inches. The dimensions are 5 in. by 9 in by 14 in.

81. Remainder is $c = 10$ since

$$\begin{array}{r|rrr} 3 & 1 & -2 & 7 \\ & & 3 & 3 \\ \hline & 1 & 1 & 10 \end{array}$$

83. If $x - c$ is a factor of $P(x)$ then $P(x) = Q(x) \cdot (x - c)$ for some polynomial $Q(x)$.

So $P(c) = Q(c) \cdot 0 = 0$ and c is a zero of $P(x)$.

For Thought

1. False, since 1 has multiplicity 1.   **2.**   True

3. True   **4.**   False, it factors as $(x-5)^4(x+2)$.

5. False, rather $4+5i$ is also a solution.   **6.**   True

7. False, since they are solutions to a polynomial with real coefficients of degree at least 4.

8. False, 2 is not a solution.

9. True, since $-x^3 - 5x^2 - 6x - 1 = 0$ has no sign changes.   **10.**   True

3.4 Exercises

1. Degree 2, the root is 5 and its multiplicity is 2, since $(x-5)^2 = 0$

3. Degree 5, roots are 0 with multiplicity 3 and ± 3 since $x^3(x-3)(x+3) = 0$

5. Degree 4, roots are $x = 0, 1$ each of multiplicity 2 since $x^2(x^2 - 2x + 1) = x^2(x-1)^2$

7. Degree 4, roots $3/2, -4/3$ each of multiplicity 2

9. Degree 3, roots are $0, 2 \pm \sqrt{10}$ since $x(x^2 - 4x - 6) = x((x-2)^2 - 10) = 0$

11. $x^2 + 9$

13. $\left[(x-1) - \sqrt{2}\right]\left[(x-1) + \sqrt{2}\right] = (x-1)^2 - 2 = x^2 - 2x - 1$

15. $[(x-3) - 2i][(x-3) + 2i] = (x-3)^2 + 4 = x^2 - 6x + 13$

17. $(x-2)[(x-3) - 4i][(x-3) + 4i] = (x-2)[(x-3)^2 + 16] = x^3 - 8x^2 + 37x - 50$

19. $(x+3)(x-5) = 0$ or $x^2 - 2x - 15 = 0$

21. $(x+4i)(x-4i) = 0$ or $x^2 + 16 = 0$

23. $(x - (3-i))(x - (3+i)) = 0$ or $x^2 - 6x + 10 = 0$

25. $(x+2)(x-i)(x+i) = 0$ or $x^3 + 2x^2 + x + 2 = 0$

27. $x(x - i\sqrt{3})(x + i\sqrt{3}) = 0$ or $x^3 + 3x = 0$

29. $(x-3)[x - (1-i)][x - (1+i)] = 0$ or $x^3 - 5x^2 + 8x - 6 = 0$

31. $(x-1)(x-2)(x-3) = 0$ or $x^3 - 6x^2 + 11x - 6 = 0$

33. $(x-1)[x - (2-3i)][x - (2+3i)] = 0$ or $x^3 - 5x^2 + 17x - 13 = 0$

35. $(2x-1)(3x-1)(4x-1) = 0$ or $24x^3 - 26x^2 + 9x - 1 = 0$

37. $(x-i)(x+i)[x - (1+i)][x - (1-i)] = 0$ or $x^4 - 2x^3 + 3x^2 - 2x + 2 = 0$

39. $P(x) = x^3 + 5x^2 + 7x + 1$ has no sign change and $P(-x) = -x^3 + 5x^2 - 7x + 1$ has 3 sign changes. There are (a) 3 negative roots or (b) 1 negative root & 2 imaginary roots.

41. $P(x) = -x^3 - x^2 + 7x + 6$ has 1 sign change and $P(-x) = x^3 - x^2 - 7x + 6$ has 2 sign changes. There are (a) 1 positive root & 2 negative roots or (b) 1 positive root & 2 imginary roots.

43. $P(y) = y^4 + 5y^2 + 7 = P(-y)$ has no sign change. There are 4 imaginary roots.

45. $P(t) = t^4 - 3t^3 + 2t^2 - 5t + 7$ has 4 sign changes and $P(-t) = t^4 + 3t^3 + 2t^2 + 5t + 7$ has no sign change. There are (a) 4 positive roots or (b) 2 positive roots & 2 imaginary roots or (c) 4 imaginary roots.

47. $P(x) = x^5 + x^3 + 5x$ and $P(-x) = -x^5 - x^3 - 5x$ have no sign changes. 4 imaginary roots and 0.

49. Best integral bounds are $-1 < x < 3$. One checks that $1, 2$ are not upper bounds and that 3 is a bound.

3	2	-5	0	6
		6	3	9
	2	1	3	15

-1 is a lower bound since

-1	2	-5	0	6
		-2	7	-7
	2	-7	7	-1

3.4 THEORY OF EQUATIONS

51. Best integral bounds are $-3 < x < 2$. One checks that 1 is not an upper bound and that 2 is a bound.

```
2 | 4    8   -11  -15
  |      8   32   42
  |------------------------
    4   16   21   27
```

$-1, -2$ are not lower bounds but -3 is a bound since

```
-3 | 4    8   -11  -15
   |    -12   12   -3
   |------------------------
     4   -4    1   -18
```

53. Best integral bounds are $-1 < x < 5$. One checks that $1, 2, 3, 4$ are not upper bounds and that 5 is a bound

```
5 | 1   -5    3    2   -1
  |      5    0   15   85
  |---------------------------
    1    0    3   17   84
```

-1 is a lower bound since

```
-1 | 1   -5    3    2   -1
   |     -1    6   -9    7
   |---------------------------
     1   -6    9   -7    6
```

55. Best integral bounds are $-1 < x < 3$. Multiply equation by -1, this makes the leading coefficient positive. One checks that $1, 2$ are not upper bounds and that 3 is a bound.

```
3 | 2   -5    3   -9
  |      6    3   18
  |---------------------
    2    1    6    9
```

-1 is a lower bound since

```
-1 | 2   -5    3   -9
   |     -2    7  -10
   |---------------------
     2   -7   10  -19
```

57. Roots are $1, 5, -2$ since

```
1 | 1   -4   -7   10
  |      1   -3  -10
  |-------------------
    1   -3  -10    0
```

and $x^2 - 3x - 10 = (x-5)(x+2)$

59. Roots are $-3, \dfrac{3 \pm \sqrt{13}}{2}$ since

```
-3 | 1    0  -10   -3
   |     -3    9    3
   |-------------------
     1   -3   -1    0
```

and by the quadratic formula the roots of $x^2 - 3x - 1 = 0$ are $\dfrac{3 \pm \sqrt{13}}{2}$.

61. Roots are $2, -4, \pm i$ since

```
2 | 1    2   -7    2   -8
  |      2    8    2    8
  |---------------------------
    1    4    1    4    0
```

```
-4 | 1    4    1    4
   |     -4    0   -4
   |-------------------
     1    0    1    0
```

and the root of $x^2 + 1 = 0$ are $\pm i$.

63. Roots are $1/3, 1/2, -5$ since

```
1/3 | 6   25   -24    5
    |     2     9    -5
    |--------------------
      6   27   -15    0
```

and $6x^2 + 27x - 15 = 3(2x - 1)(x + 5)$.

65. Roots are $1, -2$ each with multiplicity 2 since

$$\begin{array}{r|rrrrr} 1 & 1 & 2 & -3 & -4 & 4 \\ & & 1 & 3 & 0 & -4 \\ \hline & 1 & 3 & 0 & -4 & 0 \end{array}$$

$$\begin{array}{r|rrrr} -2 & 1 & 3 & 0 & -4 \\ & & -2 & -2 & 4 \\ \hline & 1 & 1 & -2 & 0 \end{array}$$

and $x^2 + x - 2 = (x+2)(x-1)$.

67. Use synthetic division on the cubic factor in $x(x^3 - 6x^2 + 12x - 8) = 0$.

$$\begin{array}{r|rrrr} 2 & 1 & -6 & 12 & -8 \\ & & 2 & -8 & -8 \\ \hline & 1 & -4 & 4 & 0 \end{array}$$

Since $x^2 - 4x + 4 = (x-2)^2$, roots are $0, 2$ (with multiplicity 3).

69. Use synthetic division on the 5th degree factor in $x(x^5 - x^4 - x^3 + x^2 - 12x + 12) = 0$.

$$\begin{array}{r|rrrrrr} 1 & 1 & -1 & -1 & 1 & -12 & 12 \\ & & 1 & 0 & -1 & 0 & -12 \\ \hline & 1 & 0 & -1 & 0 & -12 & 0 \end{array}$$

$$\begin{array}{r|rrrrr} 2 & 1 & 0 & -1 & 0 & -12 \\ & & 2 & 4 & 6 & 12 \\ \hline & 1 & 2 & 3 & 6 & 0 \end{array}$$

$$\begin{array}{r|rrrr} -2 & 1 & 2 & 3 & 6 \\ & & -2 & 0 & -6 \\ \hline & 1 & 0 & 3 & 0 \end{array}$$

Since roots of $x^2 + 3 = 0$ are $\pm i\sqrt{3}$, all roots are $x = 0, 1, \pm 2, \pm i\sqrt{3}$.

71. By the Theorem of Bounds and the graph below

$$\begin{array}{r|rrrr} 6 & 2 & -3 & -50 & 18 \\ & & 12 & 54 & 24 \\ \hline & 2 & 9 & 4 & 42 \end{array}$$

$$\begin{array}{r|rrrr} -5 & 2 & -3 & -50 & 18 \\ & & -10 & 65 & -75 \\ \hline & 2 & -13 & 15 & -57 \end{array}$$

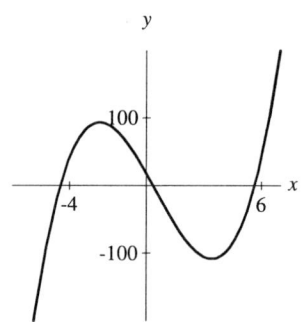

the intervals are $-5 < x < 6, -5 < x < 6$

73. By the Theorem of Bounds and the graph below

$$\begin{array}{r|rrrrr} 6 & 1 & 0 & -26 & 0 & 153 \\ & & 6 & 36 & 60 & 360 \\ \hline & 1 & 6 & 10 & 60 & 513 \end{array}$$

$$\begin{array}{r|rrrrr} -6 & 1 & 0 & -26 & 0 & 153 \\ & & -6 & 36 & -60 & 360 \\ \hline & 1 & -6 & 10 & -60 & 513 \end{array}$$

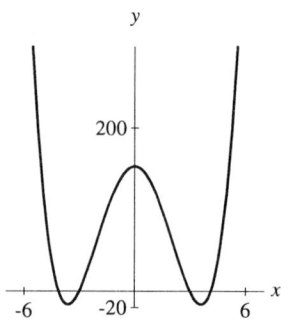

the intervals are $-6 < x < 6, -5 < x < 5$.

3.5 MISCELLANEOUS EQUATIONS

75. By the Theorem of Bounds and the graph below

```
 23 | 4   -90   -2    45
    |      92   46  1012
    |_____
      4    2    44  1057


 -1 | 4   -90   -2    45
    |     -4    94   -92
    |_____
      4   -94   92   -47
```

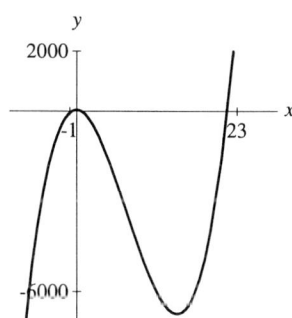

the intervals are $-1 < x < 23, -1 < x < 23$.

77. Multiplying by 100, $t^3 - 8t^2 + 11t + 20 = 0$.

```
  4 | 1   -8   11   20
    |      4  -16  -20
    |_____
      1   -4   -5    0
```

Factor, $t^2 - 4t - 5 = (t-5)(t+1) = 0$.
So $t = 4$ & 5 secs.

79. Let x be the radius of the cone.

The volume of the cone is $\frac{\pi}{3}x^2 \cdot 2$ and the volume of the cylinder with height $4x$ is $\pi x^2 \cdot (4x)$.

So $114\pi = \frac{\pi}{3}2x^2 + \pi x^2(4x)$.

Divide by π, multiply by 3, and simplify to obtain $12x^3 + 2x^2 - 342 = 0$.

```
  3 | 12   2    0   -342
    |     36  114   342
    |_____
     12   38  114    0
```

The radius is $x = 3$ in.

81.
$$\overline{(a+bi)+(c+di)} = \overline{(a+c)+i(b+d)}$$
$$= (a+c) - i(b+d)$$
$$\overline{a+bi} + \overline{c+di} = (a-bi) + (c-di)$$
$$= (a+c) - i(b+d)$$

So the conjugate of the sum of two complex numbers is equal to the sum of their conjugates.

83. If a is real then $\overline{a} = \overline{a+0i} = a - 0i = a$.

85. Let $f(x) = a(x-1)(x-2)(x-3)$.

Since $f(0) = -6a = 3$, $a = -\frac{1}{2}$.

Substitute and multiply out,

so $f(x) = -\frac{1}{2}x^3 + 3x^2 - \frac{11}{2}x + 3$.

For Thought

1. False, $(\sqrt{x-1}+\sqrt{x})^2 = (x-1) + 2\sqrt{x(x-1)} + x$.

2. False, since -1 is a solution of the first and not of the second equation.

3. False, since -27 is a solution of the first equation but not of the second.

4. False, rather let $u = x^{1/4}$ and $u^2 = x^{1/2}$.

5. True, since $x - 1 = \pm 4^{-3/2}$.

6. False, $\left(-\frac{1}{32}\right)^{-2/5} = (-32)^{2/5} = (-2)^2 = 4$.

7. False, $x = -2$ is not a solution.

8. True

9. True

10. False, $(x^3)^2 = x^6$.

3.5 Exercises

1. Factor: $x^2(x+3) - 4(x+3) = 0$
$(x^2 - 4)(x+3) = (x-2)(x+2)(x+3) = 0$
The solution set is $\{\pm 2, -3\}$.

3. Factor: $2x^2(x+500) - (x+500) = 0$
$(2x^2 - 1)(x+500) = 0$
The solution set is $\left\{\pm \dfrac{\sqrt{2}}{2}, -500\right\}$.

5. Set the right-hand side to 0 and factor
$$a(a^2 - 15a + 5) = 0$$
$a = \dfrac{15 \pm \sqrt{15^2 - 4(1)(5)}}{2}$ or $a = 0$
$a = \dfrac{15 \pm \sqrt{205}}{2}$ or $a = 0$
The solution set is $\left\{\dfrac{15 \pm \sqrt{205}}{2}, 0\right\}$.

7. Factor: $3y^2(y^2 - 4) = 3y^2(y-2)(y+2) = 0$
The solution set is $\{0, \pm 2\}$.

9. Factor: $(a^2 - 4)(a^2 + 4) = (a-2)(a+2)(a-2i)(a+2i) = 0$. The solution set is $\{\pm 2, \pm 2i\}$.

11. Squaring each side,
$x + 1 = x^2 - 10x + 25$
$0 = x^2 - 11x + 24 = (x-8)(x-3)$.
Checking $x = 3$, we get $2 \ne -2$. Then $x = 3$ is an extraneous root. The solution set is $\{8\}$.

13. Isolate the radical and then square each side.
$x = x^2 - 40x + 400$
$0 = x^2 - 41x + 400 = (x-25)(x-16)$
Checking $x = 16$, we get $2 \ne -6$. Then $x = 16$ is an extraneous root. The solution set is $\{25\}$.

15. Isolate the radical and then square each side.
$2w = \sqrt{1 - 3w}$
$4w^2 = 1 - 3w$
$4w^2 + 3w - 1 = (4w - 1)(w + 1) = 0$
$w = \dfrac{1}{4}, -1$
Checking $x = -1$, we get $-1 \ne 1$. Then $x = -1$ is an extraneous root. The solution set is $\left\{\dfrac{1}{4}\right\}$.

17. Multiply both sides by $z\sqrt{4z+1}$ and square each side,
$\sqrt{4z + 1} = 3z$
$4z + 1 = 9z^2$
$0 = 9z^2 - 4z - 1$
By the quadratic formula,
$z = \dfrac{4 \pm \sqrt{16 - 4(9)(-1)}}{18}$
$z = \dfrac{4 \pm \sqrt{52}}{18} = \dfrac{4 \pm 2\sqrt{13}}{18} = \dfrac{2 \pm \sqrt{13}}{9}$
Since $z = \dfrac{2 - \sqrt{13}}{9} < 0$ and the right-hand side of the original equation is nonnegative,
$z = \dfrac{2 - \sqrt{13}}{9}$ is an extraneous root.
The solution set is $\left\{\dfrac{2 + \sqrt{13}}{9}\right\}$.

19. Square each side and obtain
$x^2 - 2x - 15 = 9$
$x^2 - 2x - 24 = (x-6)(x+4) = 0$.
The solution set is $\{-4, 6\}$.

21. Isolate a radical and square each side.
$\sqrt{x + 40} = \sqrt{x} + 4$
$x + 40 = x + 8\sqrt{x} + 16$
$24 = 8\sqrt{x}$
$3 = \sqrt{x}$
$9 = x$
The solution set is $\{9\}$.

23. Isolate a radical and square each side.
$\sqrt{n + 4} = 5 - \sqrt{n - 1}$
$n + 4 = 25 - 10\sqrt{n-1} + (n-1)$
$-20 = -10\sqrt{n-1}$
$2 = \sqrt{n-1}$
$4 = n - 1$
The solution set is $\{5\}$.

25. Isolate a radical and square each side.
$$\sqrt{2x+5} = 9 - \sqrt{x+6}$$
$$2x+5 = 81 - 18\sqrt{x+6} + (x+6)$$
$$x - 82 = -18\sqrt{x+6}$$
$$x^2 - 164x + 6724 = 324(x+6)$$
$$x^2 - 488x + 4780 = 0$$
$$(x-10)(x-478) = 0$$

Checking $x = 478$ we get $53 \neq 9$ and $x = 478$ is an extraneous root. The solution set is $\{10\}$.

27. Raise each side to the power $3/2$ and obtain $x = \pm 2^{3/2} = \pm 8^{1/2} = \pm 2\sqrt{2}$. The solution set is $\{\pm 2\sqrt{2}\}$.

29. Raise each side to the power $-3/4$ and obtain $x = \pm(16)^{-3/4} = \pm(2)^{-3} = \pm\frac{1}{8}$. The solution set is $\{\pm\frac{1}{8}\}$.

31. Raise each side to the power -2 and obtain $t = (7)^{-2} = \frac{1}{49}$. The solution set is $\{\frac{1}{49}\}$.

33. Raise each side to the power -2 and obtain $s - 1 = (2)^{-2} = \frac{1}{4}$. Then $s = 1 + \frac{1}{4}$ and the solution set is $\{\frac{5}{4}\}$.

35. Since $(x^2 - 9)(x^2 - 3) = 0$, the solution set is $\{\pm 3, \pm\sqrt{3}\}$.

37. Let $u = \frac{2c-3}{5}$ and $u^2 = \left(\frac{2c-3}{5}\right)^2$. Then
$$u^2 + 2u - 8 = 0$$
$$(u+4)(u-2) = 0$$
$$u = -4, 2$$
$$\frac{2c-3}{5} = -4 \text{ or } \frac{2c-3}{5} = 2$$
$$2c - 3 = -20 \text{ or } 2c - 3 = 10$$
$$c = -\frac{17}{2} \text{ or } c = \frac{13}{2}.$$

The solution set is $\left\{-\frac{17}{2}, \frac{13}{2}\right\}$.

39. Let $u = \frac{1}{5x-1}$ and $u^2 = \left(\frac{1}{5x-1}\right)^2$. Then
$$u^2 + u - 12 = (u+4)(u-3) = 0$$
$$u = -4, 3$$
$$\frac{1}{5x-1} = -4 \text{ or } \frac{1}{5x-1} = 3$$
$$1 = -20x + 4 \text{ or } 1 = 15x - 3$$
$$x = \frac{3}{20} \text{ or } \frac{4}{15} = x.$$

The solution set is $\left\{\frac{3}{20}, \frac{4}{15}\right\}$.

41. Let $u = v^2 - 4v$ and $u^2 = (v^2 - 4v)^2$. Then
$$v^2 - 17v + 60 = (v-5)(v-12) = 0$$
$$v = 5, 12$$
$$v^2 - 4v = 5 \text{ or } v^2 - 4v = 12$$
$$v^2 - 4v - 5 = 0 \text{ or } v^2 - 4v - 12 = 0$$
$$(v-5)(v+1) = 0 \text{ or } (v-6)(v+2) = 0.$$

The solution set is $\{-1, -2, 5, 6\}$.

43. Factor the left-hand side as
$$(\sqrt{x} - 3)(\sqrt{x} - 1) = 0$$
$$\sqrt{x} = 3, 1$$
$$x = 9, 1.$$

The solution set is $\{1, 9\}$.

45. Factor the left-hand side as $(\sqrt{q} - 4)(\sqrt{q} - 3) = 0$. Then $\sqrt{q} = 4, 3$ and the solution set is $\{16, 9\}$.

47. Set the right-hand side to 0 and factor.
$$x^{2/3} - 7x^{1/3} + 10 = 0$$
$$\left(x^{1/3} - 5\right)\left(x^{1/3} - 2\right) = 0$$
$$x^{1/3} = 5, 2$$

The solution set is $\{8, 125\}$.

49. An equivalent statement is
$$w^2 - 4 = 3 \text{ or } w^2 - 4 = -3$$
$$w^2 = 7 \text{ or } w^2 = 1.$$
The solution set is $\left\{\pm\sqrt{7}, \pm 1\right\}$.

51. An equivalent statement assuming $5v \geq 0$ is
$$v^2 - 3v = 5v \text{ or } v^2 - 3v = -5v$$
$$v^2 - 8v = 0 \text{ or } v^2 + 2v = 0$$
$$v(v - 8) = 0 \text{ or } v(v + 2) = 0$$
$$v = 0, 8, -2.$$
Since $5v \geq 0$, $v = -2$ is an extraneous root and the solution set is $\{0, 8\}$.

53. An equivalent statement is
$$x^2 - x - 6 = 6 \text{ or } x^2 - x - 6 = -6$$
$$x^2 - x - 12 = 0 \text{ or } x^2 - x = 0$$
$$(x - 4)(x + 3) = 0 \text{ or } x(x - 1) = 0.$$
The solution set is $\{-3, 0, 1, 4\}$.

55. An equivalent statement is
$$x + 5 = 2x + 1 \text{ or } x + 5 = -(2x + 1)$$
$$4 = x \text{ or } x = -2.$$
The solution set is $\{-2, 4\}$.

57. Isolate a radical and square both sides.
$$\sqrt{16x + 1} = \sqrt{6x + 13} - 1$$
$$16x + 1 = (6x + 13) - 2\sqrt{6x + 13} + 1$$
$$10x - 13 = -2\sqrt{6x + 13}$$
$$100x^2 - 260x + 169 = 4(6x + 13)$$
$$100x^2 - 284x + 117 = 0$$
$$x = \frac{284 \pm \sqrt{284^2 - 4(100)(117)}}{200}$$
$$x = \frac{284 \pm 184}{200}$$
$$x = \frac{1}{2}, \frac{117}{50}$$
Checking $x = \frac{117}{50}$ we get $\sqrt{\frac{1922}{50}} - \sqrt{\frac{1352}{50}} > 0$ and so $x = \frac{117}{50}$ is an extraneous root.
The solution set is $\left\{\frac{1}{2}\right\}$.

59. Factor as a difference of two squares and then as a sum and difference of two cubes.
$$(v^3 - 8)(v^3 + 8) = 0$$
$$(v - 2)(v^2 + 2v + 4)(v + 2)(v^2 - 2v + 4) = 0$$
Then $v = \pm 2$ or
$$v = \frac{-2 \pm \sqrt{2^2 - 16}}{2} \text{ or } v = \frac{2 \pm \sqrt{2^2 - 16}}{2}$$
$$v = \frac{-2 \pm 2i\sqrt{3}}{2} \text{ or } v = \frac{2 \pm 2i\sqrt{3}}{2}$$
The solution set is $\left\{\pm 2, -1 \pm i\sqrt{3}, 1 \pm i\sqrt{3}\right\}$.

61. Raise both sides to the power 4. Then
$$7x^2 - 12 = x^4$$
$$0 = x^4 - 7x^2 + 12 = (x^2 - 4)(x^2 - 3)$$
$$x = \pm 2, \pm\sqrt{3}$$
Since the left-hand side of the given equation is nonnegative, $x = -2, -\sqrt{3}$ are extraneous roots. The solution set is $\left\{2, \sqrt{3}\right\}$.

63. Raise both sides to the power 3.
$$2 + x - 2x^2 = x^3$$
$$x^3 + 2x^2 - x - 2 = 0$$
$$x^2(x + 2) - (x + 2) = (x^2 - 1)(x + 2) = 0$$
$$x = \pm 1, -2$$
The solution set is $\{\pm 1, -2\}$.

65. Let $t = \dfrac{x - 2}{3}$ and $t^2 = \left(\dfrac{x - 2}{3}\right)^2$. Then
$$t^2 - 2t + 10 = 0$$
$$t^2 - 2t + 1 = -10 + 1$$
$$(t - 1)^2 = -9$$
$$t = 1 \pm 3i$$
$$\frac{x - 2}{3} = 1 \pm 3i$$
$$x - 2 = 3 \pm 9i$$
$$x = 5 \pm 9i.$$
The solution set is $\{5 \pm 9i\}$.

3.5 MISCELLANEOUS EQUATIONS

67. Raise both sides to the power 5/2. Then
$$3u - 1 = \pm 2^{5/2}$$
$$3u - 1 = \pm 32^{1/2}$$
$$3u = 1 \pm 4\sqrt{2}$$
The solution set is $\left\{\dfrac{1 \pm 4\sqrt{2}}{3}\right\}$.

69. Factor this quadratic type expression.
$$(x^2 + 1) - 11\sqrt{x^2 + 1} + 30 = 0$$
$$\left(\sqrt{x^2 + 1} - 5\right)\left(\sqrt{x^2 + 1} - 6\right) = 0$$
$$\sqrt{x^2 + 1} = 5 \quad \text{or} \quad \sqrt{x^2 + 1} = 6$$
$$x^2 = 24 \quad \text{or} \quad x^2 = 35$$
$$x = \pm 2\sqrt{6}, \pm\sqrt{35}$$
The solution set is $\left\{\pm\sqrt{35}, \pm 2\sqrt{6}\right\}$.

71. An equivalent statement is
$$x^2 - 2x = 3x - 6 \quad \text{or} \quad x^2 - 2x = -3x + 6$$
$$x^2 - 5x + 6 = 0 \quad \text{or} \quad x^2 + x - 6 = 0$$
$$(x - 3)(x - 2) = 0 \quad \text{or} \quad (x + 3)(x - 2) = 0$$
$$x = 2, \pm 3$$
The solution set is $\{2, \pm 3\}$.

73. Raise both sides to the power $-5/3$. Then
$$3m + 1 = \left(-\dfrac{1}{8}\right)^{-5/3}$$
$$3m + 1 = \left(-\dfrac{1}{2}\right)^{-5}$$
$$3m + 1 = -32.$$
The solution set is $\{-11\}$.

75. An equivalent statement assuming $x - 2 \geq 0$ is
$$x^2 - 4 = x - 2 \quad \text{or} \quad x^2 - 4 = -x + 2$$
$$x^2 - x - 2 = 0 \quad \text{or} \quad x^2 + x - 6 = 0$$
$$(x - 2)(x + 1) = 0 \quad \text{or} \quad (x + 3)(x - 2) = 0$$
$$x = 2, -1, -3.$$
Since $x - 2 \geq 0$, $x = -1, -3$ are extraneous roots and the solution set is $\{2\}$.

77. Solve for S.
$$21.24 + 1.25 S^{1/2} - 9.8(18.34)^{1/3} = 16.296$$
$$1.25 S^{1/2} - 25.84396 \approx -4.944$$
$$S^{1/2} \approx 16.72$$
$$S \approx 279.56$$
The maximum sailing area is 279.56 m^2.

79. Solve for x with $C = 83.50$.
$$0.5x + \sqrt{8x + 5000} = 83.50$$
$$\sqrt{8x + 5000} = 83.50 - 0.5x$$
$$8x + 5000 = 6972.25 - 83.50x + 0.25x^2$$
$$0 = 0.25x^2 - 91.50x + 1972.25$$
$$x = \dfrac{91.50 \pm \sqrt{(-91.50)^2 - 4(0.25)(1972.25)}}{0.5}$$
$$x = \dfrac{91.50 \pm 80}{0.5}$$
$$x = 23, 343$$
Checking $x = 343$, the value of the left-hand side of the first equation exceeds 83.50 and so $x = 343$ is an extraneous root. Then 23 loaves costs \$83.50.

81. Let x and $x + 6$ be two numbers. Then
$$\sqrt{x + 6} - \sqrt{x} = 1$$
$$\sqrt{x + 6} = \sqrt{x} + 1$$
$$x + 6 = x + 2\sqrt{x} + 1$$
$$5 = 2\sqrt{x}$$
$$25 = 4x$$
$$x = \dfrac{25}{4}$$
Since $\dfrac{25}{4} + 6 = \dfrac{49}{4}$, the numbers are $\dfrac{25}{4}$ and $\dfrac{49}{4}$.

83. Let x be the length of the short leg. Since $x+7$ is the other leg, by the Pythagorean Theorem the hypotenuse is $\sqrt{x^2+(x+7)^2}$. Then
$$x+(x+7)+\sqrt{x^2+(x+7)^2}=30$$
$$2x-23=-\sqrt{x^2+(x+7)^2}$$
$$4x^2-92x+529=2x^2+14x+49$$
$$2x^2-106x+480=0$$
$$x^2-53x+240=0$$
$$(x-5)(x-48)=0$$
$$x=5,48$$
Since the perimeter is 30 in., $x=48$ is an extraneous root. The short leg is $x=5$ in.

85. Let x be the length of one side of the original square foundation. From the 2100 ft^2 we have
$$(x-10)(x+30)=2100$$
$$x^2+20x-2400=0$$
$$(x+60)(x-40)=0.$$
Since x is nonnegative, $x=40$ and the area of the square foundation is $x^2=1600$ ft^2.

87. Solving for d, we find
$$598.9\left(\frac{d}{64}\right)^{-2/3}=14.26$$
$$\left(\frac{d}{64}\right)^{-2/3}=\frac{14.26}{598.9}$$
$$\frac{d}{64}=\left(\frac{14.26}{598.9}\right)^{-3/2}$$
$$d=64\left(\frac{14.26}{598.9}\right)^{-3/2}$$
$$d\approx 17,419.3\text{ lbs}.$$

89. Let x be the length of one side of the square base. The height of the box is $x+2$ and the volume of the box is $x^2(x+2)$. The space enclosed by the styrofoam has a square base whose side has length $(x-2)$ and whose height is x. Since this space has a volume one-half that of the box, we have
$$2x(x-2)^2=x^2(x+2)$$
$$2x(x^2-4x+4)=x^3+2x^2$$
$$x^3-10x^2+8x=0$$
$$x(x^2-10x+8)=0$$

Since $x\neq 0$,
$$x^2-10x+8=0$$
$$x^2-10x+25=-8+25$$
$$(x-5)^2=17$$
$$x=5\pm\sqrt{17}$$
$$x\approx 9.1231, 0.877$$
Exclude $x=0.877$ since this is too small for the box. So $x=9.1231$ and the volume of shrimp that can be shipped is
$x(x-2)^2\approx 9.1231(9.1231-2)^2\approx 462.89$ in.^{3}.

91. Let x be the number of hours after 10 : 00 a.m. so that the distance between Nancy & Edgar is 14 miles greater than the distance between Nancy & William. By the Pythagorean Theorem, $\sqrt{(5x)^2+(12x)^2}$ and $\sqrt{(5x)^2+[4(x+2)]^2}$ are the distances between Nancy & Edgar and Nancy & William, respectively. Since these distances are equal, we have
$$\sqrt{(5x)^2+(12x)^2}-14=\sqrt{(5x)^2+[4(x+2)]^2}$$
$$\sqrt{169x^2}-14=\sqrt{41x^2+64x+64}$$
$$13x-14=\sqrt{41x^2+64x+64}$$
$$169x^2-364x+196=41x^2+64x+64$$
$$128x^2-428x+132=0$$
$$32x^2-107x+33=0$$
$$x=\frac{107\pm\sqrt{(-107)^2-4(32)(33)}}{64}=3,\frac{11}{32}.$$
Since the left-hand side of the first equation is negative when $x=\frac{11}{32}$, $x=\frac{11}{32}$ is an extraneous root. So $x=3$ hours and the time is 1 : 00 p.m.

93. (a) $\sqrt[3]{(1.00)(1.40)(1.78)}\approx \$1.356$ billion

   (b) Let p be Compaq's profit in 1998. Then
$$\sqrt[4]{(1.00)(1.40)(1.78)p}=1.6$$
$$(1.00)(1.40)(1.78)p=1.6^4$$
$$p=\frac{1.6^4}{(1.40)(1.78)}$$
$$p\approx \$2.630\text{ billion}$$

3.6 GRAPHS OF POLYNOMIAL FUNCTIONS

95. The weight of a cylindrical tank is its volume times its density. So

$$\pi \left(\frac{d}{2}\right)^2 \cdot d \cdot 1600 = 25,850,060$$

$$400\pi d^3 = 25,850,060$$

$$d = \sqrt[3]{\frac{25,850,060}{400\pi}} \approx 27.4$$

The height of the tank is 27.4 meters.

For Thought

1. False, to be symmetric about the origin one must have $P(-x) = -P(x)$ for *all* x in the domain.

2. True 3. True

4. False, since $f(-x) = -f(x)$.

5. True 6. True 7. False

8. False, y-intercept is $(0,38)$. 9. True

10. False, only one x-intercept.

3.6 Exercises

1. Neither symmetry, crosses $(-2,0)$, does not cross $(1,0)$), $y \to \infty$ as $x \to \infty$, $y \to -\infty$ as $x \to -\infty$

3. Symmetric about y-axis, no x-intercepts, $y \to \infty$ as $x \to \infty$, $y \to \infty$ as $x \to -\infty$

5. y-axis, since $f(-x) = f(x)$

7. $x = 3/2$, since $\frac{3}{2}$ is the x-coordinate of the vertex of a parabola

9. None

11. Origin, since $f(-x) = -f(x)$

13. $x = 5$, since 5 is the x-coordinate of the vertex of a parabola

15. Origin, since $f(-x) = -f(x)$

17. It does not cross $(4,0)$ since $x-4$ is raised to an even power.

19. It crosses $(1/2,0)$.

21. It crosses $(1/4,0)$

23. No x-intercepts since $x^2 - 3x + 10 = 0$ has no real root.

25. It crosses $(3,0)$ and not $(0,0)$ since $x^3 - 3x^2 = x^2(x-3)$.

27. It crosses $(1/2,0)$ and not $(1,0)$ since

$$\begin{array}{r|rrrr} 1 & 2 & -5 & 4 & -1 \\ & & 2 & -3 & 1 \\ \hline & 2 & -3 & 1 & 0 \end{array}$$

and $2x^3 - 5x^2 + 4x - 1 = (x-1)(2x^2 - 3x + 1) = (x-1)^2(2x-1)$.

29. It crosses $(2,0)$ and not $(-3,0)$ since

$$\begin{array}{r|rrrr} -3 & -2 & -8 & 6 & 36 \\ & & 6 & 6 & -36 \\ \hline & -2 & -2 & 12 & 0 \end{array}$$

and $(x+3)(-2x^2 - 2x + 12) = -2(x+3)(x^2 + x - 6) = -2(x+3)(x+3)(x-2) = -2(x+3)^2(x-2)$.

31. $y \to \infty$ 33. $y \to -\infty$

35. $y \to -\infty$ 37. $y \to \infty$

39. $y \to \infty$ 41. e 43. g

45. b 47. c

49. $f(x) = x - 30$ has x-intercept $(30,0)$ and y-intercept $(0,-30)$

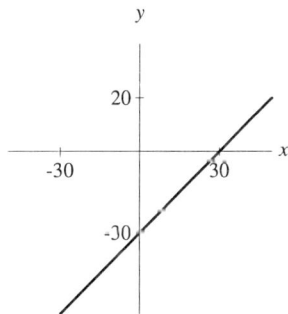

51. $f(x) = (x-30)^2$ does not cross $(30,0)$, y-intercept $(0,900)$, $y \to \infty$ as $x \to \infty$ and as $x \to -\infty$

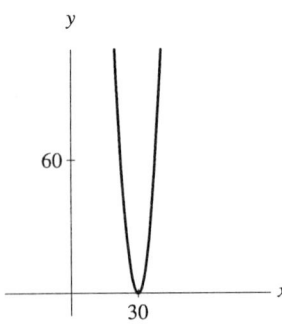

53. $f(x) = x^2(x-40)$ crosses $(40,0)$ but does not cross $(0,0)$, $y \to \infty$ as $x \to \infty$, $y \to -\infty$ as $x \to -\infty$

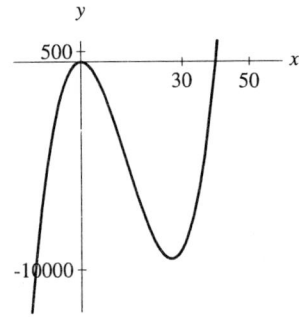

55. $f(x) = (x-20)^2(x+20)^2$ does not cross $(20,0), (-20,0)$, y-intercept $(0, 160000)$, $y \to \infty$ as $x \to \infty$ and as $x \to -\infty$

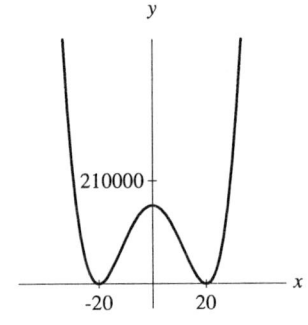

57. Since $f(x) = -x^3 - x^2 + 5x - 3$

```
1 | -1   -1    5   -3
  |      -1   -2    3
  |_____
    -1   -2    3    0
```

$f(x) = (x-1)(-x^2 - 2x + 3) = -(x-1)(x+3)(x-1)$ then graph crosses $(-3,0)$ but not $(1,0)$, y-intercept $(0,-3)$, $y \to -\infty$ as $x \to \infty$, and $y \to \infty$ as $x \to -\infty$

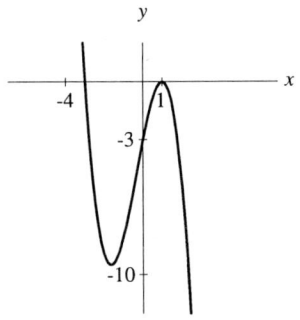

59. Since $x^3 - 10x^2 - 600x = x(x-30)(x+20)$, graph crosses $(0,0), (30,0), (-20,0)$, $y \to \infty$ as $x \to \infty$, $y \to -\infty$ as $x \to -\infty$

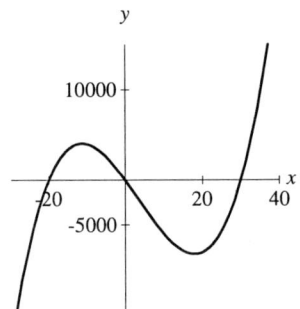

61. $f(x) = x^3 + 18x^2 - 37x + 60$ has only one x-intercept since

```
-20 | 1   18   -37   60
    |     -20   40  -60
    ---------------------
      1   -2    3    0
```

and $x^2 - 2x + 3$ has no real root. Graph crosses $(-20, 0)$, y-intercept $(0, 60)$, $y \to \infty$ as $x \to \infty$, $y \to -\infty$ as $x \to -\infty$

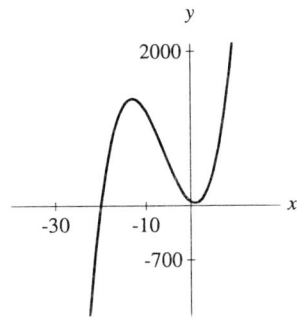

63. Since $-x^2(x^2 - 196) = -x^2(x - 14)(x + 14)$, graph crosses $(\pm 14, 0)$ and does not cross $(0, 0)$, $y \to -\infty$ as $x \to \infty$ and as $x \to -\infty$

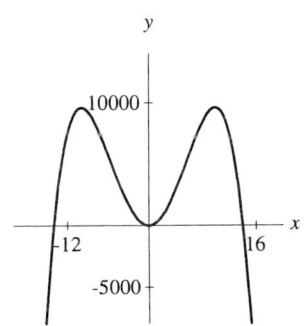

65. Use synthetic division on $f(x) = x^3 + 3x^2 + 3x + 1$ with $c = 1$.

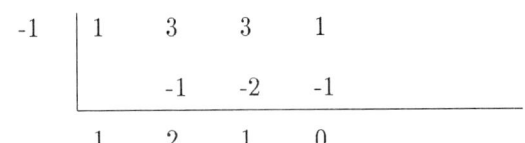

Since $x^2 + 2x + 1 = (x+1)^2$ then $f(x) = (x+1)^3$. Graph crosses $(-1, 0)$, y-intercept $(0, 1)$, $y \to \infty$ as $x \to \infty$, $y \to -\infty$ as $x \to -\infty$.

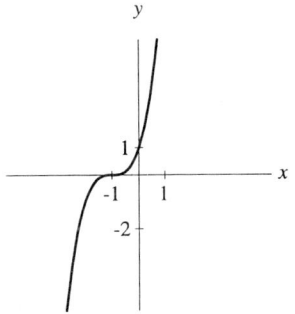

67. Since $x^3 - 3x = x(x^2 - 3)$, the x-intercepts are $(0, 0), (\pm\sqrt{3}, 0)$. From the graph of $y = x^3 - x$, we conclude the solution set is $(-\sqrt{3}, 0) \cup (\sqrt{3}, \infty)$.

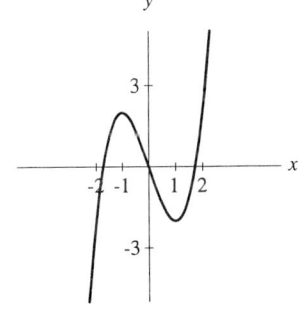

69. The x-intercepts of $y = x^2(2-x^2)$ are $(0,0)$, $(\pm\sqrt{2},0)$. From the graph of $y = x^2(2-x^2)$, it follows that the solution set is $(-\infty, -\sqrt{2}] \cup [\sqrt{2}, \infty) \cup \{0\}$.

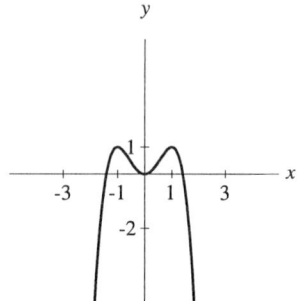

71. Use synthetic division on $f(x) = x^3 + 4x^2 - x - 4$.

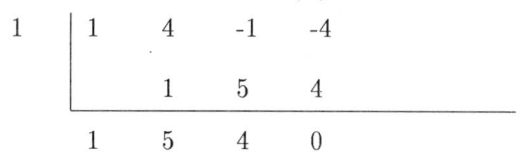

Since $x^2 + 5x + 4 = (x+1)(x+4)$, the x-intercepts of $f(x) = x^3 + 4x^2 - x - 4$, are $(\pm 1, 0)$ and $(-4, 0)$. The solution set to $x^3 + 4x^2 - x - 4 > 0$ is $(-4, -1) \cup (1, \infty)$.

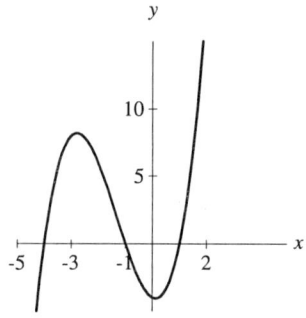

73. Local max. ≈ 3.11, local min. $\approx .37$

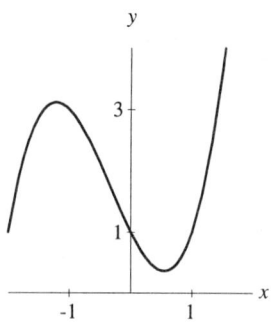

75. Local max. ≈ 23.7, local min. ≈ -163.7

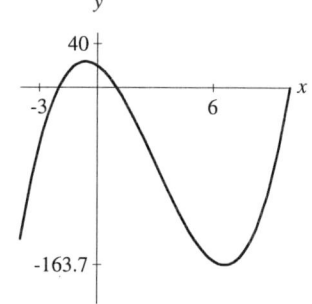

77. Local max. ≈ 21.01, local min. ≈ 13.99

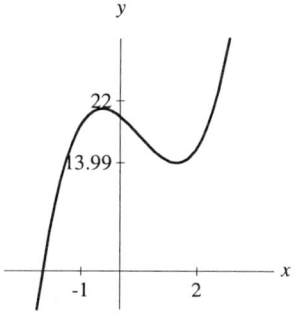

79. From the graph of the profit function, we find a local maximum of $3,400$ and a local minimum of $2,600$. To get a profit higher than $3,400$, the company must spend more than $2,200$ on advertising.

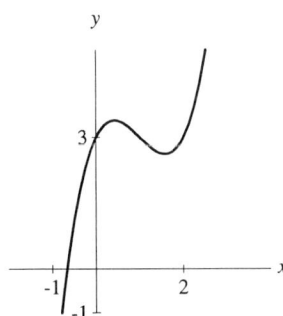

81. Since $P = \dfrac{x}{10}(x^2 - 60x + 900) = \dfrac{x}{10}(x-30)^2$, graph does not cross x-intercept $(30, 0)$. Profit increases on $(0, 10)$ and on $(30, \infty)$ since $P \to \infty$ as $x \to \infty$.

83. Since $3x + 2y = 12$, $y = \dfrac{12-3x}{2}$. The volume V of the block is given by

$$V = xy\frac{4y}{3}$$
$$= \frac{4x}{3}y^2$$
$$= \frac{4x}{3}\left(\frac{12-3x}{2}\right)^2$$
$$= 3x^3 - 24x^2 + 48x.$$

One finds from the graph of $V = \dfrac{4x}{3}\left(\dfrac{12-3x}{2}\right)^2$ that the optimal dimensions of the block are $x = \dfrac{4}{3}$ in. and $y = \dfrac{12-4}{2} = 4$ in. by $\dfrac{16}{3}$ in. long.

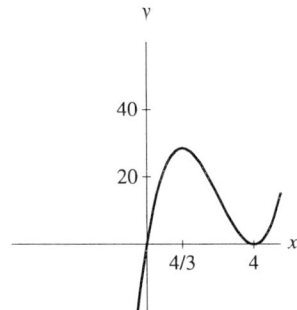

85. If the surface area is 50 square feet, then no paint can be used (since all the paint is used to coat the inside of the can).

Since amount of paint in a 1-pint can is $\dfrac{1}{8}\dfrac{1}{7.5}$ or $\dfrac{1}{60}$ feet^3, then $\dfrac{1}{60} = \pi r^2 h$. Solving for h, one obtains $h = \dfrac{1}{60\pi r^2}$. Then

$$2\pi r^2 + 2\pi rh = 50$$
$$2\pi r^2 + 2\pi r\left(\frac{1}{60\pi r^2}\right) = 50$$
$$2\pi r^2 + \frac{1}{30r} - 50 = 0.$$

With a graphing calculator, one finds $r \approx \pm 2.82$ or $r \approx 6.67 \times 10^{-4}$.
If $r \approx 2.82$, then $h \approx 6.67 \times 10^{-4}$ feet since $h = \dfrac{1}{60\pi r^2}$. If $r \approx 6.67 \times 10^{-4}$, then $h \approx 11,936.62$ feet.

Thus, either the radius and height of the can are $r \approx 2.82$ and $h \approx 6.67 \times 10^{-4}$ feet, respectively, or $r \approx 6.67 \times 10^{-4}$ and $h \approx 11,924.69$ feet.

For Thought

1. False, $\sqrt{x} - 3$ is not a polynomial.

2. False, domain is $(-\infty, 2) \cup (2, \infty)$.

3. False

4. False, it has three vertical asymptotes.

5. True

6. False, $y = 5$ is the horizontal asymptote.

7. True   8. False

9. True, it is an even function.

10. True, since $x = -3$ is not a vertical asymptote.

3.7 Exercises

1. $(-\infty, -2) \cup (-2, \infty)$

3. $(-\infty, -2) \cup (-2, 2) \cup (2, \infty)$

5. $(-\infty, 3) \cup (3, \infty)$

7. $(-\infty, 0) \cup (0, \infty)$

9. $(-\infty, -1) \cup (-1, 0) \cup (0, 1) \cup (1, \infty)$ since
$$f(x) = \frac{3x^2 - 1}{x(x^2 - 1)}$$

11. $(-\infty, -3) \cup (-3, -2) \cup (-2, \infty)$ since
$$f(x) = \frac{-x^2 + x}{(x+3)(x+2)}$$

13. Domain $(-\infty, 2) \cup (2, \infty)$, asymptotes $y = 0, x = 2$

15. Domain $(-\infty, 0) \cup (0, \infty)$, asymptotes $y = x$, $x = 0$

17. Asymptotes $x = 2$, $y = 0$

19. Asymptotes $x = \pm 3$, $y = 0$

21. Asymptotes $x = 1$, $y = 2$

23. Asymptotes $x = 0$, $y = x - 2$ since $f(x) = x - 2 + \frac{1}{x}$

25. Asymptotes $x = -1$, $y = 3x - 3$ since

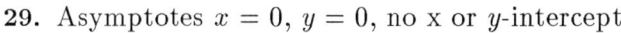

and $f(x) = 3x - 3 + \dfrac{7}{x+1}$

27. Asymptotes $x = -2$, $y = -x + 6$ since

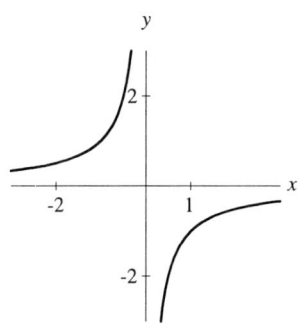

and $f(x) = -x + 6 + \dfrac{-12}{x+2}$

29. Asymptotes $x = 0$, $y = 0$, no x or y-intercept

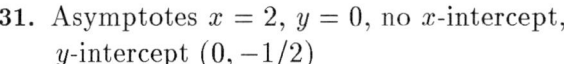

31. Asymptotes $x = 2$, $y = 0$, no x-intercept, y-intercept $(0, -1/2)$

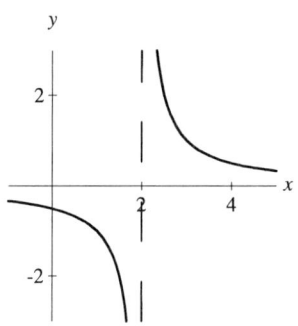

33. Asymptotes $x = \pm 2$, $y = 0$, no x-intercept, y-intercept $(0, -1/4)$

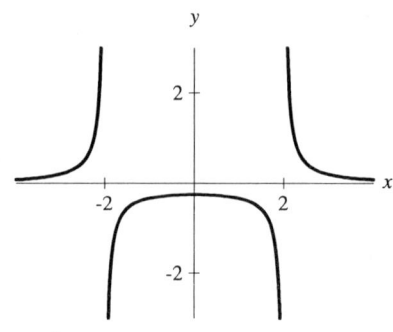

35. Asymptotes $x = -1$, $y = 0$, no x-intercept, y-intercept $(0, -1)$

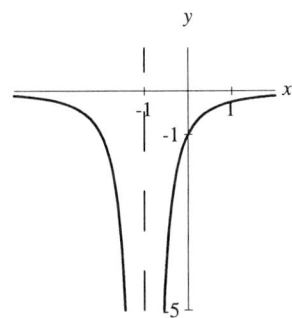

37. Asymptotes $x = 1$, $y = 2$, x-intercept $(-1/2, 0)$, y-intercept $(0, -1)$

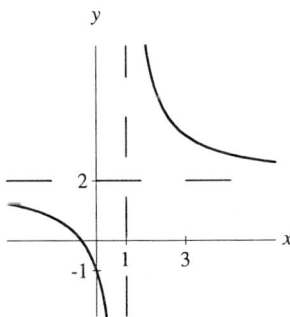

39. Asymptotes $x = -2$, $y = 1$, x-intercept $(3, 0)$, y-intercept $(0, -3/2)$

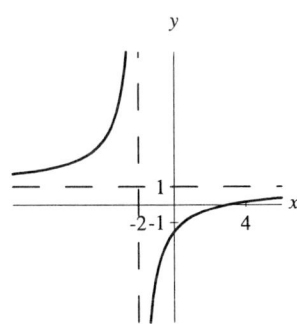

41. Asymptotes $x = \pm 1$, $y = 0$, x-intercept $(0, 0)$,

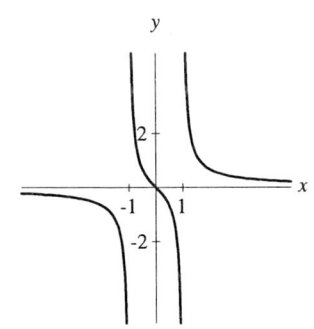

43. Since $f(x) = \dfrac{4x}{(x-1)^2}$, asymptotes are $x = 1$, $y = 0$, x-intercept $(0, 0)$

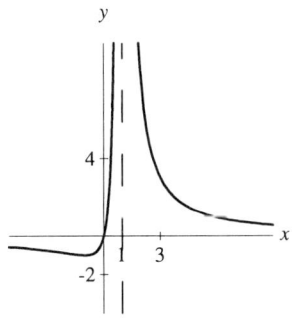

45. Asymptotes are $x = \pm 3$, $y = -1$, x-intercept $(\pm 2\sqrt{2}, 0)$, y-intercept $(0, -8/9)$

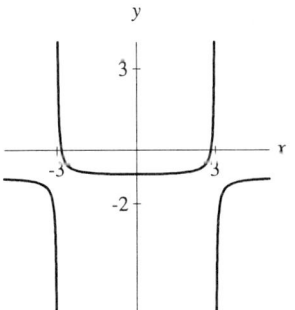

47. Since $f(x) = \dfrac{2x^2 + 8x + 2}{(x+1)^2}$, asymptotes are $x = -1$, $y = 2$, by solving $2x^2 + 8x + 2 = 0$ one gets the x-intercepts $\left(-2 \pm \sqrt{3}, 0\right)$, y-intercept $(0, 2)$

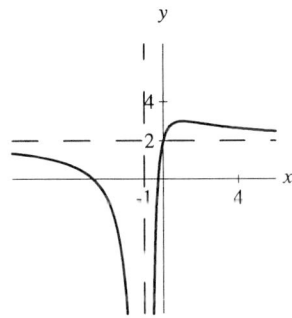

49. Since $f(x) = x + \dfrac{1}{x}$, oblique asymptote is $y = x$, asymptote $x = 0$, no x-intercept, no y-intercept, graph goes through $(1,2), (-1,-2)$

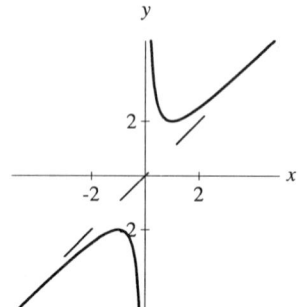

51. Since $f(x) = x - \dfrac{1}{x^2}$, oblique asymptote is $y = x$, asymptote $x = 0$, x-intercept $(1,0)$, no y-intercept, graph goes through $(-1,-2)$

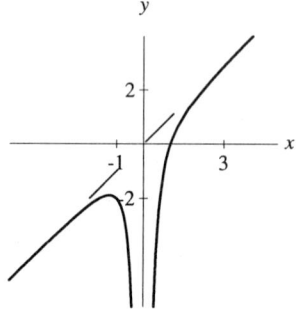

53. $f(x) = x - 1 + \dfrac{1}{x+1}$ since

$$
\begin{array}{r}
x - 1 \\
x+1 \,\overline{)\,x^2 + 0x } \\
\underline{x^2 + x } \\
-x + 0 \\
\underline{-x - 1} \\
1
\end{array}
$$

Oblique asymptote $y = x - 1$, asymptote $x = -1$, x-intercept $(0,0)$, graph goes through $(-2,-4)$

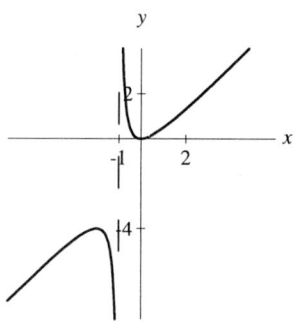

55. $f(x) = 2x + 1 + \dfrac{1}{x-1}$ since

$$
\begin{array}{r}
2x + 1 \\
x-1 \,\overline{)\,2x^2 - x + 0} \\
\underline{2x^2 - 2x } \\
x + 0 \\
\underline{x - 1} \\
1
\end{array}
$$

Oblique asymptote $y = 2x+1$, asymptote $x = 1$,

x-intercepts $(0,0), (1/2, 0)$, graph goes through $(2,6)$

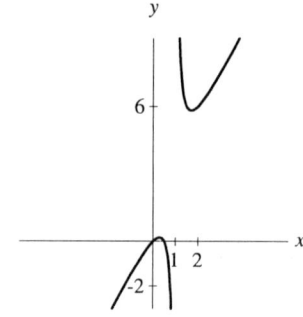

57. e   59. a

61. b   63. c

65. $f(x) = \dfrac{1}{x-1}$ where $x \neq -1$, 'hole' at $(-1, -1/2)$, asymptotes $x = 1$, $y = 0$, no x-intercept, y-intercept $(0, -1)$

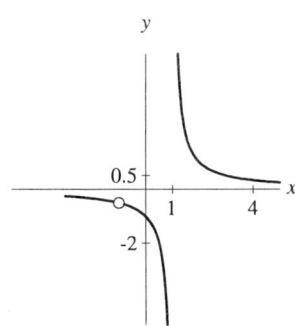

67. $f(x) = x + 1$ where $x \neq 1$, a line with a 'hole' at $(1, 2)$

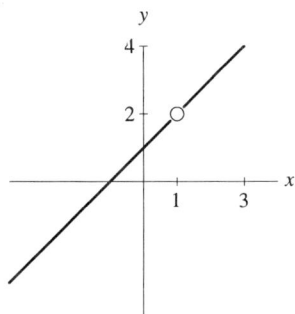

69. Symmetric about y-axis, asymptote $x = 0$, goes through $(0, 2), (1, 1)$

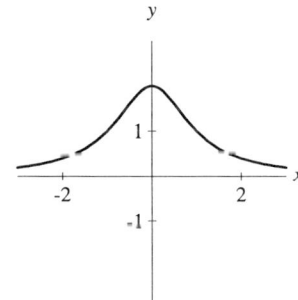

71. Since $f(x) = \dfrac{x-1}{x(x^2-9)}$, asymptotes are $x = \pm 3$, $x = 0$, $y = 0$, crosses $(1, 0)$,

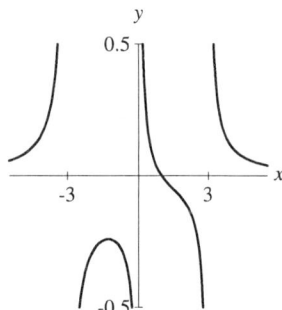

73. Asymptotes $x = 0, y = 0$, x-intercept $(-1, 0)$, no y-intercept, graph goes through $(1, 2)$

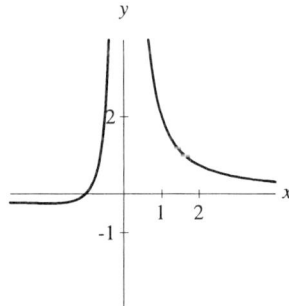

75. $f(5000) = \dfrac{14,999}{9991} \approx 1.5$, $f(5000) - \dfrac{3}{2} \approx .0013$.

77. $f(5000) = \dfrac{4}{15,040} \approx 0.0$, $f(5000) - 0 \approx 2.7 \times 10^{-4}$

79. $f(-5000) = \dfrac{224,999,999}{74,999,998} \approx 3.0$, $f(-5000) - 3 \approx 6.7 \times 10^{-8}$

81. $f(-5000) \approx \dfrac{-14,996}{2.49775 \times 10^{11}} \approx -0.0$, $f(-5000) - 0 \approx -0.0 \times 10^{-8}$

83. No vertical asymptote since $x^2 + 1 = 0$ has no real root, domain $(-\infty, \infty)$, range $(0, 100]$, horizontal asymptote $y = 0$

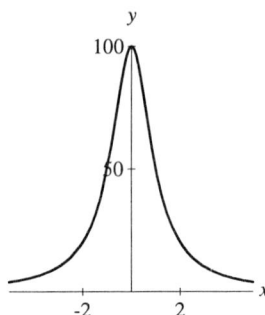

85. No vertical asymptote since $x^2 + 2x + 6 = 0$ has no real root, domain $(-\infty, \infty)$, range $(0, 1]$, horizontal asymptote $y = 0$

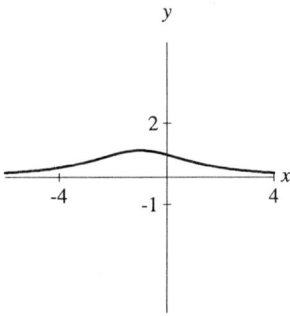

87. No vertical asymptote since $x^2 + x + 1 = 0$ has no real root, domain $(-\infty, \infty)$, range $[-5.55, 0.24]$, horizontal asymptote $y = 0$

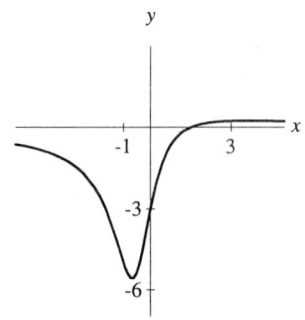

89. Since $\dfrac{3}{9x + 2} < 0$ is equivalent to $9x + 2 < 0$, the solution set is $\left(-\infty, -\dfrac{2}{9}\right)$.

91. From the graph, the x-intercept is $\left(\dfrac{5}{6}, 0\right)$, vertical asymptotes are $x = \pm\sqrt{5}$, and the solution set is $\left(-\sqrt{5}, \dfrac{5}{6}\right) \cup (\sqrt{5}, \infty)$.

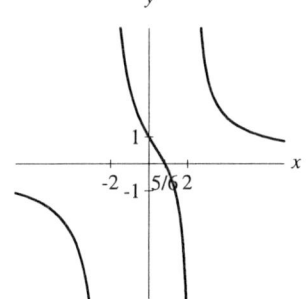

93. Re-writing the inequality, we have
$$\dfrac{2x}{x + 2} - \dfrac{6}{x - 3} \leq 0$$
$$\dfrac{2x(x - 3) - 6(x + 2)}{(x + 2)(x - 3)} \leq 0$$
$$\dfrac{2x^2 - 12x - 12}{(x + 2)(x - 3)} \leq 0.$$

the x-intercepts of $y = \dfrac{2x^2 - 12x - 12}{(x + 2)(x - 3)}$ as shown in its graph are $(3 \pm \sqrt{15}, 0)$, vertical asymptotes are $x = 3$, $x = -2$, and the solution set is $(-2, 3 - \sqrt{15}] \cup (3, 3 + \sqrt{15}]$.

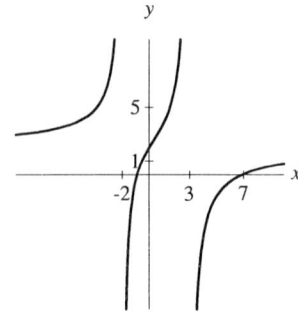

95. Average cost is $C = \dfrac{100 + x}{x}$. If she visits the zoo 100 times then her average cost per visit is $C = \dfrac{200}{100} = \$2$. Since $C \to 1$ as $x \to \infty$, over a long period her average cost per visit is \$1.

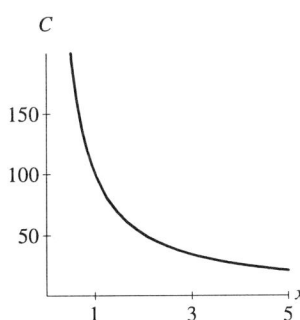

97. Since the second half is 100 miles long and it must be completed in $4 - x$ hours, the average speed for the second half is $S(x) = \dfrac{100}{4 - x}$. The asymptote $x = 4$ implies the average speed in the second half goes to ∞ as the completion time in the first half nears four hours.

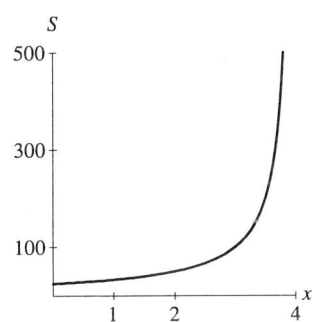

99. (a) 25 minutes, (b) less than 10 minutes, (c) coma and permanent brain damage,

(d) Let t be the number of minutes and let L be the level of carbon monoxide. Approximately, the vertical asymptote is $t = 0$; if the level of carbon dioxide is very high, death can occur in a short time.

Approximately, the horizontal asymptote is $L = 0$; long exposures to carbon dioxide with a low level can cause death.

101.

(a) $h = \dfrac{500}{\pi r^2}$ since $500 = \pi r^2 h$

(b) $S = 2\pi r^2 + 2\pi r h = 2\pi r^2 + 2\pi r \dfrac{500}{\pi r^2}$

or equivalently $S = 2\pi r^2 + \dfrac{1000}{r}$

(c) As can be seen from the graph of $S = 2\pi r^2 + \dfrac{1000}{r}$, S is minimized when $r \approx 4.3$ feet.

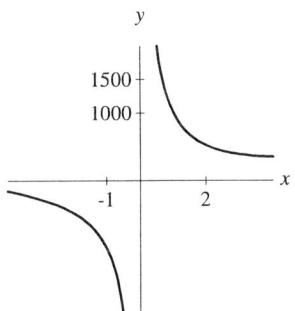

(d) If it costs \$8 per square foot to construct the tank in part (c), the tank would cost $8\left(2\pi (4.3)^2 + \dfrac{1000}{4.3}\right) \approx \$2{,}789.87$.

Chapter 3 Review Exercises

1. Since $y = \dfrac{3}{5}x - \dfrac{8}{5}$, slope is $\dfrac{3}{5}$

3. The slope of $2x - 3y = 6$ is $m = 2/3$.
$$y + 2 = \dfrac{-3}{2}(x - 1)$$
$$y = \dfrac{-3}{2}x + \dfrac{3}{2} - 2$$
So $y = \dfrac{-3}{2}x - \dfrac{1}{2}$.

5. Complete the square.
$$f(x) = 3\left(x^2 - \dfrac{2}{3}x + \dfrac{1}{9}\right) - \dfrac{1}{3} + 1$$
$$= 3\left(x - \dfrac{1}{3}\right)^2 + \dfrac{2}{3}$$

7. Since $y = 2(x^2 - 2x + 1) - 2 - 1 = 2(x - 1)^2 - 3$, vertex is $(1, -3)$, axis of symmetry $x = 1$.

Set y to 0, so $x - 1 = \pm\dfrac{\sqrt{6}}{2}$ and the x-intercepts are $x = \left(\dfrac{2 \pm \sqrt{6}}{2}, 0\right)$.

y-intercept is $(0, -1)$.

9. From the x-intercepts, $y = a(x + 1)(x - 3)$. Substitute $(0, 6)$, so $6 = a(-3)$ or $a = -2$. The equation of the parabola is $y = -2(x^2 - 2x - 3)$ or $y = -2x^2 + 4x + 6$.

11. $y = 2x^2 + 5x - 3 = (2x - 1)(x + 3)$ is a parabola opening down with x-intercepts $(1/2, 0), (-3, 0)$. The set of x's for which y is negative is the interval $(-3, 1/2)$.

13. $1/3$ 15. $\pm 2\sqrt{2}$

17. Factor, $m(x) = (2x - 1)(4x^2 + 2x + 1)$.
Apply the quadratic formula to the second factor,
$$x = \frac{-2 \pm \sqrt{-12}}{8} = \frac{-2 \pm 2i\sqrt{3}}{8}$$
Zeros are $\frac{1}{2}, \frac{-1 \pm i\sqrt{3}}{4}$.

19. Since $P(t) = (t^2 - 10)(t^2 + 10)$, zeros are $\pm\sqrt{10}, \pm i\sqrt{10}$

21. Factor, $R(s) = 4s^2(2s - 1) - (2s - 1) = (4s^2 - 1)(2s - 1)$. Zeros are $\pm\frac{1}{2}$.

23. Find the roots of the second factor of $f(x) = x(x^2 + 2x - 6)$.
Since $x^2 + 2x - 6 = (x^2 + 2x + 1) - 6 - 1 = (x + 1)^2 - 7$, the zeros are $0, -1 \pm \sqrt{7}$.

25. $P(3) = 108 - 27 + 3 - 1 = 83$. By synthetic division,

```
3 |  4   -3    1    -1
  |      12   27    84
  ---------------------
     4    9   28    83
```

So remainder is $P(3) = 83$.

27. $P(-1/2) = \frac{1}{4} - \frac{1}{4} + 3 + 2 = 5$. By synthetic division,

```
-1/2| -8   0    2    0   -6    2
    |      4   -2    0    0    3
    ------------------------------
      -8   4    0    0   -6    5
```

So remainder is $P(-1/2) = 5$.

29. $\pm\left\{1, \frac{1}{3}, 2, \frac{2}{3}\right\}$

31. $\pm\left\{1, \frac{1}{2}, \frac{1}{3}, \frac{1}{6}, 3, \frac{3}{2}\right\}$

33.
$$2\left(x + \frac{1}{2}\right)(x - 3) = (2x + 1)(x - 3)$$
$$= 2x^2 - 5x - 3$$
So an equation is $2x^2 - 5x - 3 = 0$.

35. $(x - (3 - 2i))(x - (3 + 2i)) =$
$= ((x - 3) + 2i)((x - 3) - 2i)$
$= (x - 3)^2 + 4$
$= x^2 - 6x + 13$
So an equation is $x^2 - 6x + 13 = 0$.

37. $(x - 2)(x - (1 - 2i))(x - (1 + 2i)) =$
$= (x - 2)((x - 1) + 2i)((x - 1) - 2i)$
$= (x - 2)\left((x - 1)^2 + 4\right)$
$= (x - 2)(x^2 - 2x + 5)$
$= x^3 - 4x^2 + 9x - 10$
So an equation is $x^3 - 4x^2 + 9x - 10 = 0$.

39. $\left(x - (2 - \sqrt{3})\right)\left(x - (2 + \sqrt{3})\right) =$
$= \left((x - 2) + \sqrt{3}\right)\left((x - 2) - \sqrt{3}\right)$
$= \left((x - 2)^2 - 3\right)$
$= x^2 - 4x + 1$
So an equation is $x^2 - 4x + 1 = 0$.

41. $P(x) = P(-x) = x^8 + x^6 + 2x^2$ has no sign variation. There are 6 imaginary roots and 0 has multiplicity 2.

43. $P(x) = 4x^3 - 3x^2 + 2x - 9$ has 3 sign variations and $P(-x) = -4x^3 - 3x^2 - 2x - 9$ has no sign variation. There are
(a) 3 positive roots or
(b) 1 positive & 2 imaginary roots.

45. $P(x) = x^3 + 2x^2 + 2x + 1$ has no sign variation and $P(-x) = -x^3 + 2x^2 - 2x + 1$ has 3 sign variations. There are
(a) 3 negative roots or
(b) 1 negative root & 2 imaginary roots.

REVIEW EXERCISES

47. Best integral bounds: $-4 < x < 3$ One checks that $1, 2$ are not upper bounds and that 3 is a bound

$$
\begin{array}{r|rrr}
3 & 6 & 5 & -50 \\
 &   & 18 & 69 \\
\hline
 & 6 & 23 & 19
\end{array}
$$

One checks that $-1, -2, -3$ are not lower bounds and that -4 is a bound since

$$
\begin{array}{r|rrr}
-4 & 6 & 5 & -50 \\
 &   & -24 & 76 \\
\hline
 & 6 & -19 & 26
\end{array}
$$

49. Best integral bounds: $-1 < x < 8$ One checks that $1, 2, 3, 4, 5, 6, 7$ are not upper bounds and that 8 is a bound.

$$
\begin{array}{r|rrrr}
8 & 2 & -15 & 31 & -12 \\
 &   & 16 & 8 & 312 \\
\hline
 & 2 & 1 & 39 & 300
\end{array}
$$

-1 is a lower bound since

$$
\begin{array}{r|rrrr}
-1 & 2 & -15 & 31 & -12 \\
 &   & -2 & 17 & -48 \\
\hline
 & 2 & -17 & 48 & -60
\end{array}
$$

51. Best integral bounds: $-1 < x < 1$

$$
\begin{array}{r|rrrr}
1 & 12 & -4 & -3 & 1 \\
 &   & 12 & 8 & 5 \\
\hline
 & 12 & 8 & 5 & 6
\end{array}
$$

$$
\begin{array}{r|rrrr}
-1 & 12 & -4 & -3 & 1 \\
 &   & -12 & 16 & -13 \\
\hline
 & 12 & -16 & 13 & -12
\end{array}
$$

53. Roots are $1, 2, 3$ since

$$
\begin{array}{r|rrrr}
1 & 1 & -6 & 11 & -6 \\
 &   & 1 & -5 & 6 \\
\hline
 & 1 & -5 & 6 & 0
\end{array}
$$

and $x^2 - 5x + 6 = (x - 3)(x - 2)$.

55. Roots are $\pm i, 1/3, 1/2$ since

$$
\begin{array}{r|rrrrr}
1/2 & 6 & -5 & 7 & -5 & 1 \\
 &   & 3 & -1 & 3 & -1 \\
\hline
 & 6 & -2 & 6 & -2 & 0
\end{array}
$$

$$
\begin{array}{r|rrrr}
1/3 & 6 & -2 & 6 & -2 \\
 &   & 2 & 0 & 2 \\
\hline
 & 6 & 0 & 6 & 0
\end{array}
$$

and the zeros of $6x^2 + 6 = 6(x^2 + 1) = 0$ are $\pm i$.

57. Roots are $3, 3 \pm i$ since

$$
\begin{array}{r|rrrr}
3 & 1 & -9 & 28 & -30 \\
 &   & 3 & -18 & 30 \\
\hline
 & 1 & -6 & 10 & 0
\end{array}
$$

and the zeros of
$x^2 - 6x + 10 = (x - 3)^2 + 1 = 0$ are $3 \pm i$.

59. Roots are $2, 1 \pm i\sqrt{2}$ since

$$
\begin{array}{r|rrrr}
2 & 1 & -4 & 7 & -6 \\
 &   & 2 & -4 & 6 \\
\hline
 & 1 & -2 & 3 & 0
\end{array}
$$

and the zeros of
$x^2 - 2x + 3 = (x - 1)^2 + 2 = 0$ are $1 \pm i\sqrt{2}$.

61. Apply synthetic division to the second factor in $x(2x^3 - 5x^2 - 2x + 2) = 0$.

$$\begin{array}{c|cccc} 1/2 & 2 & -5 & -2 & 2 \\ & & 1 & -2 & -2 \\ \hline & 2 & -4 & -4 & 0 \end{array}$$

By completing the square, the zeros of $2(x^2 - 2x) - 4 = 2(x-1)^2 - 6 = 0$ are $1 \pm \sqrt{3}$. All the roots are $x = 0, 1/2, 1 \pm \sqrt{3}$

63. Solve an equivalent statement assuming $3v \geq 0$.
$$2v - 1 = 3v \quad \text{or} \quad 2v - 1 = -3v$$
$$-1 = v \quad \text{or} \quad 5v = 1$$
Since $3v \geq 0$, $v = -1$ is an extraneous root.
The solution set is $\{1/5\}$.

65. Let $w = x^2$ and $w^2 = x^4$. Then
$$w^2 + 7w = 18$$
$$(w+9)(w-2) = 0$$
$$w = -9, 2$$
$$x^2 = -9 \quad \text{or} \quad x^2 = 2.$$
Since $x^2 = -9$ has no real solution, the solution set is $\{\pm\sqrt{2}\}$.

67. Isolate a radical and square both sides
$$\sqrt{x+6} = \sqrt{x-5} + 1$$
$$x + 6 = x - 5 + 2\sqrt{x-5} + 1$$
$$5 = \sqrt{x-5}$$
$$25 = x - 5$$
The solution set is $\{30\}$.

69. Let $w = \sqrt[4]{y}$ and $w^2 = \sqrt{y}$. Then
$$w^2 + w - 6 = 0$$
$$(w+3)(w-2) = 0$$
$$w = -3 \quad \text{or} \quad w = 2$$
$$y^{1/4} = -3 \quad \text{or} \quad y^{1/4} = 2$$
Since $y^{1/4} = -3$ has no real solution, the solution set is $\{16\}$.

71. Let $w = x^2$ and $w^2 = x^4$. Then
$$w^2 - 3w - 4 = 0$$
$$(w+1)(w-4) = 0$$
$$x^2 = -1 \quad \text{or} \quad x^2 = 4.$$
Since $x^2 = -1$ has no real solution, the solution set is $\{\pm 2\}$.

73. Raise to the power $3/2$ and get $x - 1 = \pm(4^{1/2})^3$. So $x = 1 \pm 8$. The solution set is $\{-7, 9\}$.

75. No solution since $(x+3)^{1/4}$ is nonnegative.

77. Since $3x - 7 = 4 - x$ then $4x = 11$. The solution set is $\{11/4\}$.

79. Symmetric about $x = 3/4$ since $\dfrac{-b}{2a} = \dfrac{3}{4}$.

81. Symmetric about y-axis since $f(-x) = f(x)$.

83. Symmetric about the origin for $f(-x) = -f(x)$.

85. $(-\infty, -5/2) \cup (-5/2, \infty)$

87. $(-\infty, \infty)$

89. $f(x) = -\dfrac{1}{2}x + 4$ has x-intercept $(8, 0)$, y-intercept $(0, 4)$

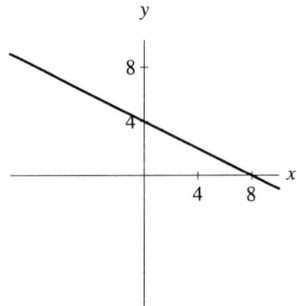

91. Since $f(x) = (x-2)(x+1)$, x-intercepts are $(2,0), (-1,0)$, y-intercept is $(0,-2)$.
Since $\dfrac{-b}{2a} = \dfrac{1}{2}$, vertex is $(1/2, -9/4)$.

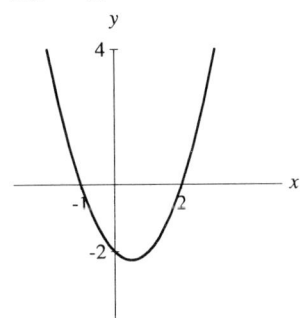

93. Use synthetic division on $f(x) = x^3 - 3x - 2$.

$$
\begin{array}{r|rrrr}
-1 & 1 & 0 & -3 & -2 \\
   &   & -1 & 1 & 2 \\
\hline
   & 1 & -1 & -2 & 0
\end{array}
$$

Since $x^2 - x - 2 = (x-2)(x+1)$, $f(x) = (x+1)^2(x-2)$. Graph crosses $(2,0)$ but does not cross $(-1,0)$. y-intercept is $(0,-2)$.

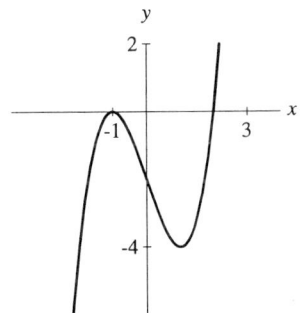

95. Use synthetic division on
$$f(x) = \dfrac{1}{2}x^3 - \dfrac{1}{2}x^2 - 2x + 2.$$

$$
\begin{array}{r|rrrr}
1 & 1/2 & -1/2 & -2 & 2 \\
  &   & 1/2 & 0 & -2 \\
\hline
  & 1/2 & 0   & -2 & 0
\end{array}
$$

Since the roots of $\dfrac{1}{2}x^2 - 2 = 0$ are ± 2, the graph crosses x-intercepts $(\pm 2, 0), (1, 0)$, y-intercept is $(0, 2)$.

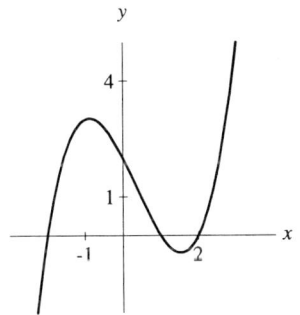

97. Factor, $f(x) = \dfrac{1}{4}(x^4 - 8x^2 + 16) = \dfrac{1}{4}(x^2 - 4)^2$. Graph does not cross x-intercepts $(\pm 2, 0)$. y-intercept is $(0, 4)$.

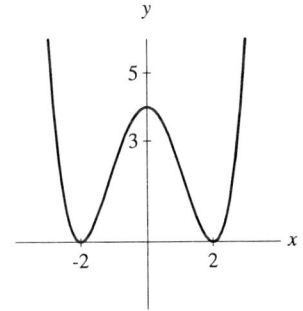

99. $f(x) = \dfrac{2}{x+3}$ has no x-intercept, y-intercept is $(0, 2/3)$, asymptotes are $x = -3$, $y = 0$

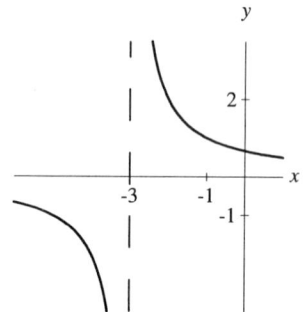

101. $f(x) = \dfrac{2x}{x^2 - 4}$ has x-intercept $(0, 0)$, asymptotes are $x = \pm 2$, $y = 0$. Symmetric about the origin. Graph goes through $(1, -2/3)$, $(3, 6/5)$, and $(-3, -6/5)$.

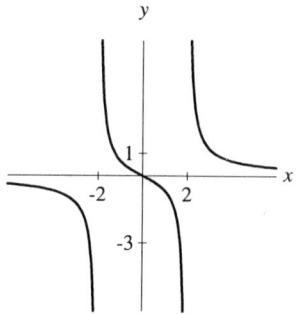

103. $f(x) = \dfrac{(x-1)^2}{x-2}$ has x-intercept $(1, 0)$, y-intercept is $(0, -1/2)$, asymptote $x = 2$, and oblique asymptote $y = x$ since

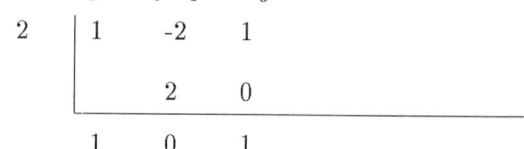

and $f(x) = x + \dfrac{1}{x-2}$.

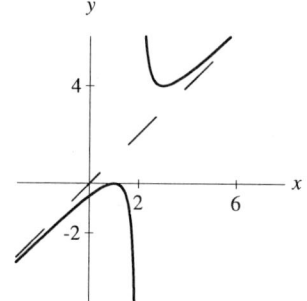

105. $f(x) = \dfrac{2x-1}{2-x}$ has x-intercept $(1/2, 0)$, y-intercept is $(0, -1/2)$, asymptotes $x = 2$, $y = -2$, graph goes through $(1, 1), (3, -5)$.

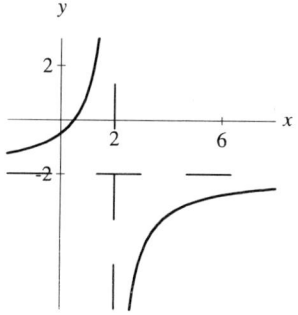

107. Since $f(x) = \dfrac{x^2 - 4}{x - 2} = x + 2$ if $x \neq 2$, x-intercept is $(-2, 0)$, y-intercept is $(0, 2)$, no asymptotes

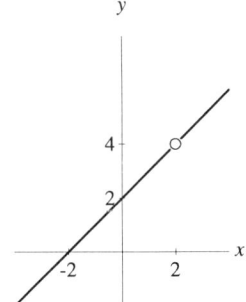

109. Using the Rational Zero Theorem and synthetic division, we obtain

$$\begin{array}{r|rrrr} 1/2 & 4 & -400 & -1 & 100 \\ & & 2 & -199 & -100 \\ \hline & 4 & -398 & -200 & 0 \end{array}$$

Note, $4x^2 - 398x - 200 = (4x + 2)(x - 100)$. Using the zeros, we derive the sign graph of $4x^3 - 400x^2 - x + 100 \geq 0$

```
- - - - - - - - - - - - - - - - 0 + + +
- - - - - - - - - - - 0 + + + + + +
- - - - - 0 + + + + + + + + +
←————————————————————→
     -1/2      1/2         100
```

The solution set is $\left[-\dfrac{1}{2}, \dfrac{1}{2}\right] \cup [100, \infty)$.

111. Let $R(x) = \dfrac{x + 10}{x + 2} - 5 = \dfrac{-4x}{x + 2} < 0$. Since $R(-3) < 0$, $R(-1) > 0$, and $R(1) < 0$ then from the graph

```
      test         test        test
       pt     U     pt    0    pt
←————————————————————————→
    -3    -2    -1    0    1
```

the solution is $(-\infty, -2) \cup (0, \infty)$.

113. Let $R(x) = \dfrac{12 - 7x}{x^2} + 1 = \dfrac{(x - 3)(x - 4)}{x^2} > 0$.

Since $R(-1) > 0, R(1) > 0, R(3.5) < 0$, and $R(5) > 0$,

```
   test        test        test        test
    pt    U    pt    0     pt    0     pt
←————————————————————————————→
  -1   0    1      3   3.5   4      5
```

the solution is $(-\infty, 0) \cup (0, 3) \cup (4, \infty)$.

115. Let $R(x) = \dfrac{(x - 1)(x - 2)}{(x - 3)(x - 4)}$.

Since $R(0) > 0, R(1.5) < 0, R(2.5) > 0$, $R(3.5) < 0, R(5) > 0$ and the graph below

```
  test       test       test       test       test
   pt   0    pt    0    pt    U    pt    U    pt
←————————————————————————————————→
  0    1   1.5   2   2.5   3   3.5   4     5
```

the solution is $(-\infty, 1] \cup [2, 3) \cup (4, \infty)$.

The graph is ⇐══]——[——)——(⇒
                1   2   3   4

117. No solution since $\dfrac{x + 3}{x + 3} = 1$ for $x \neq -3$.

119. Since $(x - 1)^2 \geq 0$, the solution set is $(-\infty, 1) \cup (1, \infty)$.

121. Quotient $x^2 - 3x$, remainder -15

$$\begin{array}{r|rrrr} 3 & 1 & -6 & 9 & -15 \\ & & 3 & -9 & 0 \\ \hline & 1 & -3 & 0 & -15 \end{array}$$

123. Let x be the year with $x = 0$ corresponding to 1985 and let y be the enrollment in millions. Since the slope is given by $m = \dfrac{13.9 - 12.2}{5 - 0} = .34$ and the y-intercept is $(0, 12.2)$, an equation of the line is $y = 0.34x + 12.2$.

Using this equation, in 1988 the enrollment was 13.22 million $(= 0.34(3) + 12.2)$.

125. Let x be a woman's age and let y be the percentage of body fat. An equation of the line is given by
$$y - 23 = \frac{47 - 23}{50 - 20}(x - 20)$$
$$y - 23 = \frac{4}{5}(x - 20)$$
$$y = \frac{4}{5}x + 7$$
Using this equation, the average percentage of body fat in 50-year old women is $y(65) \approx \frac{4}{5}(65) + 7 = 59\%$.

127. Since $\frac{-b}{2a} = \frac{-156}{-32} = 4.875$, the maximum height is $-16(4.875)^2 + 156(4.875) = 380.25$ ft.

129. Let x and y be the maximum time allowed and the pulse rate, respectively. The slope is
$$\frac{19.1 - 16.2}{110 - 170} = -\frac{2.9}{60}$$ and a point-slope form is $y - 16.2 = -\frac{2.9}{60}(x - 170)$.
Substitute $x = 145$, so $y \approx 17.41$.
Tina is not in the high fitness category.

131. Let w be the height and let l be the length of the given cross-section of the room. Since the ratios of corresponding sides of similar triangles are equal, we obtain
$$\frac{5}{12} = \frac{w}{(48 - l)/2}$$
$$w = \frac{5}{24}(48 - l)$$
and the area, A, of the cross-section of the room is $A = wl = \frac{5}{24}(48 - l)l$. Since $l = 24$ maximizes A, the optimal dimensions are $l = 24$ feet and $w = \frac{5}{24}(48 - 24) = 5$ feet.

133.
  (a) $V(10) = \frac{10,000}{58} \approx 172.4$ feet
  (b) $V = 200$
  (c) 200 feet per second

Chapter 3 Test

1. $\frac{2 + 6}{-1 - 3} = -2$

2. Since $y = \frac{3}{4}x - \frac{9}{4}$, slope is 3/4.

3. Since the slope of $y = -3x - 5$ is -3, $y + 4 = \frac{1}{3}(x - 2)$. So $y = \frac{1}{3}x - \frac{14}{3}$.

4. Complete the square: $y = 3(x^2 - 4x + 4) + 1 - 12 = 3(x - 2)^2 - 11$

5. $y = 3(x - 2)^2 - 11$ has vertex $(2, -11)$, axis of symmetry $x = 2$, y-intercept $(0, 1)$, x-intercepts $\left(\frac{6 \pm \sqrt{33}}{3}, 0\right)$, range $[-11, \infty)$

6. Minimum value is -11 since vertex is $(2, -11)$

7. Quotient $2x^2 - 6x + 14$, remainder -37

   $$\begin{array}{c|cccc} -3 & 2 & 0 & -4 & 5 \\ & & -6 & 18 & -42 \\ \hline & 2 & -6 & 14 & -37 \end{array}$$

8. By the Remainder Theorem, remainder is -14.

9. $\pm\left\{1, \frac{1}{3}, 2, \frac{2}{3}, 3, 6\right\}$

10.
$$(x + 3)(x - 4i)(x + 4i) = (x + 3)(x^2 + 16)$$
$$= x^3 + 3x^2 + 16x + 48$$
So an equation is $x^3 + 3x^2 + 16x + 48 = 0$.

11. $P(x) = x^3 - 3x^2 + 5x + 7$ has 2 sign variations and $P(-x) = -x^3 - 3x^2 - 5x + 7$ has 1 sign variation. There are
    (a) 2 positive roots & 1 negative root or
    (b) 1 negative root & 2 imaginary roots.

12. Since $\frac{-b}{2a} = \frac{-128}{-32} = 4$, the maximum height is $S(4) = 256$ feet.

13. Let x be the year with $x = 0$ corresponding to 1994 and let y be the median price of a new home in Springfield. Since the slope is given by $m = \dfrac{92,800 - 88,000}{3 - 0} = 1600$ and the y-intercept is $(0, 88000)$, an equation of the line is $y = 1,600x + 88,000$.

Using this equation, the median price in the year 2003 is $y = 1600(9) + 88,000 = \$102,400$.

14. ± 3 15. $\pm 2, \pm 2i$

16.

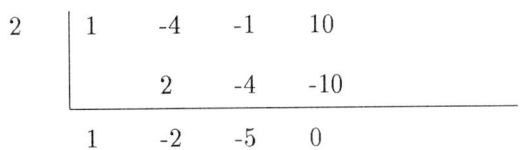

Since $x^2 - 2x - 5 = (x-1)^2 - 6$, all zeros are $2, 1 \pm \sqrt{6}$.

17. Zeros are $x = \pm i$, each with multiplicity 2 since $f(x) = (x^2 + 1)^2 = (x - i)^2(x + i)^2$

18. $2, 0$ with multiplicity 2, $-3/2$ with mulitiplicity 3

19.

```
1/2 | 2   -9   14   -5
    |      1   -4    5
    | 2   -8   10    0
```

By completing the square, the roots of $2(x^2 - 4x) + 10 = 2(x-2)^2 + 2 = 0$ are $2 \pm i$. All zeros are $x = 2 \pm i, 1/2$.

20. $y = -\dfrac{1}{2}x + 3$ with y-intercept $(0, 3)$

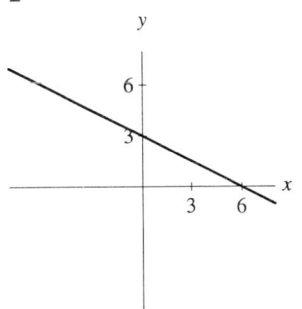

21. Parabola $y = 2(x-3)^2 + 1$ with vertex $(3, 1)$

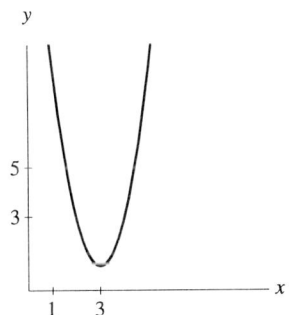

22. $y = (x-2)^2(x+1)$ crosses $(-1, 0)$ but does not cross $(2, 0)$. y-intercept $(0, 4)$.

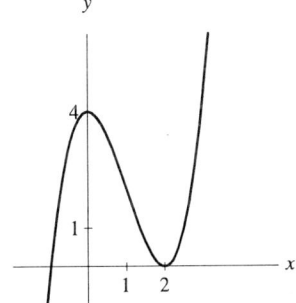

23. $y = x(x-2)(x+2)$ crosses $(0,0), (\pm 2, 0)$, and goes through $(1, -3)$

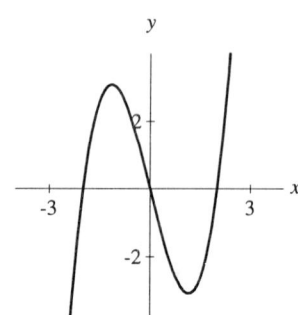

24. $y = \dfrac{1}{x-2}$ has asymptotes $x = 2$, $y = 0$, and goes through $(1, -1), (3, 1), (4, 1/2)$

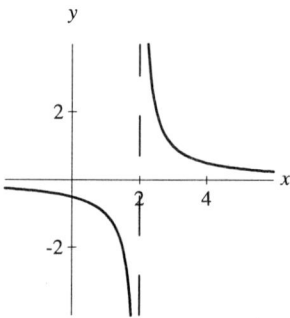

25. $y = \dfrac{2x-3}{x-2}$ has asymptotes $x = 2, y = 2$. x-intercept $(3/2, 0)$, y-intercept $(0, 3/2)$

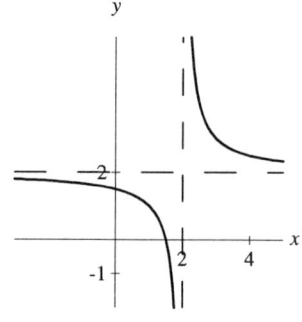

26. $f(x) = x + \dfrac{1}{x}$ has oblique asymptote $y = x$, asymptote $x = 0$, goes through $(1, 2), (-1, -2)$

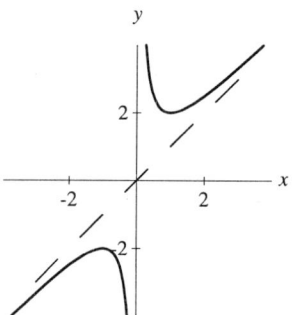

27. $y = \dfrac{4}{x^2 - 4}$ has asymptotes $x = \pm 2, y = 0$, y-intercept $(0, -1)$

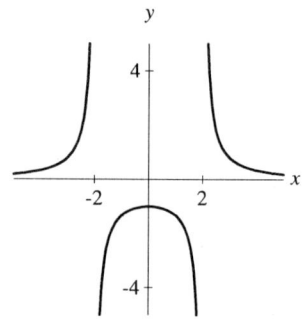

28. Observe, $\dfrac{x^2 - 2x + 1}{x - 1} = x - 1$ provided $x \neq 1$; no x-intercept, y-intercept $(0, -1)$

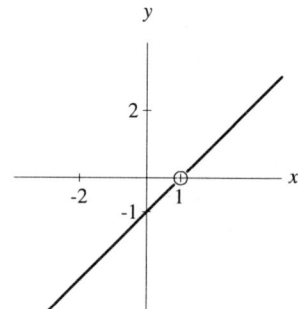

29. The sign graph of $(x-4)(x+2) < 0$ is

```
- - - - - - - - 0 + + + +
- - - - 0 + + + + + + +
←————————————————————→
     -2          4
```

The solution set is the interval $(-2, 4)$.

30. Set the right-hand side to zero to obtain $\dfrac{2x-1}{x-3} > 0$.
The sign chart is

```
- - - - - - - - 0 + + + +
- - - - 0 + + + + + + +
←————————————————————→
     1/2         3
```

The solution set is $(-\infty, 1/2) \cup (3, \infty)$.

31. The sign chart of $\dfrac{x+3}{(4-x)(x+1)} \geq 0$ is

```
+++++++++++++++++0 - - -
- - - - - - - - - - 0 +++++++
- - - - - 0 +++++++++++
←————————————————————→
   -3       -1        4
```

The solution set is $(-\infty, -3] \cup (-1, 4)$.

32. Raise both sides to the power $-3/2$. Then $x - 3 = \pm 27^{1/2}$ and the solution set is $\{3 \pm 3\sqrt{3}\}$.

33. Isolate a radical and square both sides
$$\sqrt{x} = \sqrt{x-7} + 1$$
$$x = x - 7 + 2\sqrt{x-7} + 1$$
$$3 = \sqrt{x-7}$$
$$9 = x - 7$$

The solution set is $\{16\}$.

Tying It All Together

1. 2 **2.** $-1/2, 1/3$ **3.** $-1/2$

4. Since $3x - 1 = 2x + 1$, solution set is $\{2\}$.

5. Multiply by $6(3x - 1)(2x + 1)$.
$$6(2x+1) - 6(3x-1) = (3x-1)(2x+1)$$
$$-6x + 12 = 6x^2 + x - 1$$
$$0 = 6x^2 + 7x - 13$$
$$0 = (x-1)(6x+13)$$
Solution set is $\left\{1, -\dfrac{13}{6}\right\}$.

6.

1/3	6	1	5	1	-1
		2	1	2	1
	6	3	6	3	0

-1/2	6	3	6	3
		-3	0	-3
	6	0	6	0

Since second quotient is $6(x^2 + 1)$, solution set is $x = \dfrac{1}{3}, -\dfrac{1}{2}, \pm i$.

7. Square both sides, so $2x + 1 = 3x - 1$. Solution set is $\{2\}$.

8. Since $x^2 = -3$, solution set is $\{\pm i\sqrt{3}\}$.

9.
$$\sqrt{3x+4} = \sqrt{2x+1} + 1$$
$$3x + 4 = 2x + 1 + 2\sqrt{2x+1} + 1$$
$$x + 2 = 2\sqrt{2x+1}$$
$$x^2 + 4x + 4 = 4(2x+1)$$
$$x^2 - 4x = 0$$
$$x(x-4) = 0$$
$$x = 0, 4$$

10. An equivalent statement is $2x + 1 = 3$ or $2x + 1 = -3$. Solutions are $1, -2$.

11. Raise both sides to the power 3/2.
$$2x + 1 = \pm 9^{3/2} = \pm 3^3$$
$$2x + 1 = 27 \quad \text{or} \quad 2x + 1 = -27$$
$$x = 13 \quad \text{or} \quad x = -14$$
Solution set is $\{13, -14\}$.

12. An equivalent statement is
$$2x + 1 = 3x - 1 \quad \text{or} \quad 2x + 1 = -3x + 1$$
$$2 = x \quad \text{or} \quad 5x = 0$$
Solution set is $\{0, 2\}$.

13. $y = 3 - x$ goes through $(3, 0), (0, 3)$, domain $(-\infty, \infty)$, range $(-\infty, \infty)$

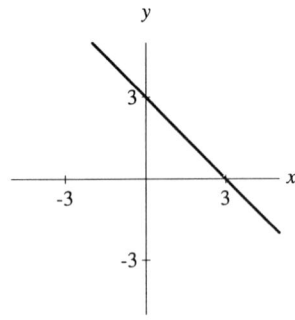

14. $y = |3 - x|$ goes through $(4, 1), (2, 1), (3, 0)$, domain $(-\infty, \infty)$, range $[0, \infty)$,

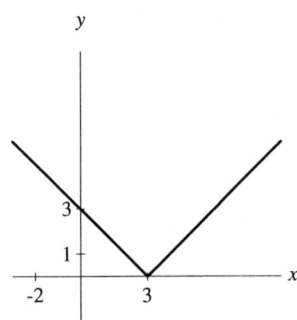

15. Parabola $y = 3 - x^2$ with vertex $(0, 3)$, domain $(-\infty, \infty)$, range $(-\infty, 3]$

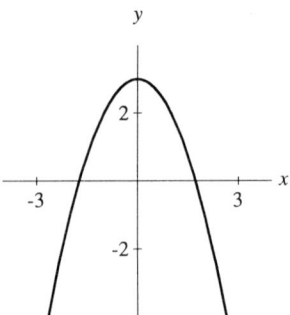

16. $y = \dfrac{x - 3}{|x - 3|}$ has domain $(-\infty, 3) \cup (3, \infty)$, range $\{\pm 1\}$

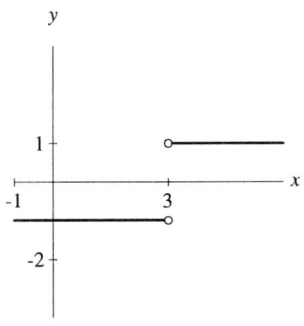

17. $y = 3 - \sqrt{x}$ goes through $(0, 3), (1, 2), (4, 1)$, domain $[0, \infty)$, range $(-\infty, 3]$

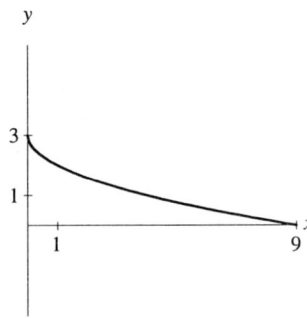

18. $y = \dfrac{1}{3-x}$ has asymptotes $x = 3, y = 0$, and goes through $(4,-1), (2,1), (5,-1/2)$, domain $(-\infty, 3) \cup (3, \infty)$, and range $(-\infty, 0) \cup (0, \infty)$

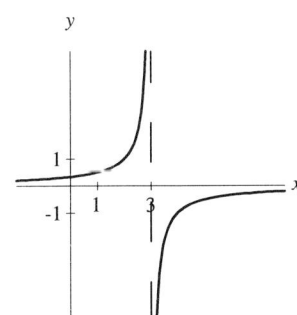

19. $y = (x-3)^2$ has vertex $(3,0)$, domain $(-\infty, \infty)$, range $[0, \infty)$

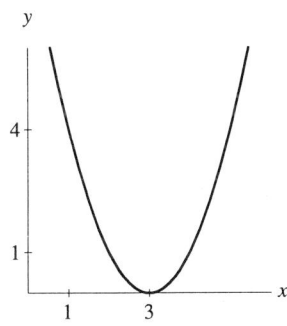

20. $y = \dfrac{1}{(x-3)^2}$ has asymptotes $x = 3, y = 0$, and goes through $(4,1), (2,1), (5, 1/4)$, domain $(-\infty, 3) \cup (3, \infty)$, range $(0, \infty)$

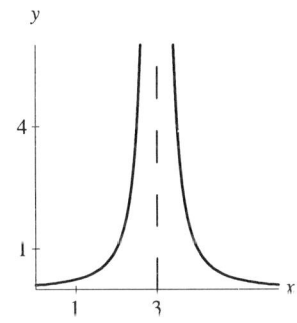

21. $y = x - \dfrac{3}{x}$ has asymptotes $y = x, x = 0$ and goes through $(1,-2), (-1, 2)$, domain $(-\infty, 0) \cup (0, \infty)$, range $(-\infty, \infty)$

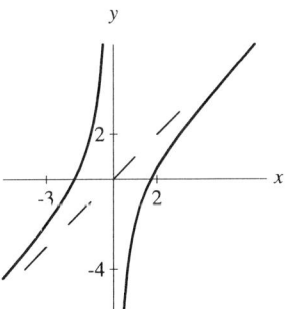

22. Semicircle $y = \sqrt{3 - x^2}$ with radius $\sqrt{3}$, domain $[-\sqrt{3}, \sqrt{3}]$, range $[0, \sqrt{3}]$

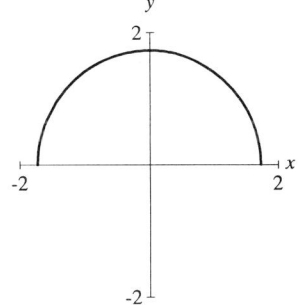

23. $y = \sqrt{3-x}$ goes through $(3,0), (2,1), (-1, 2)$, domain $(-\infty, 3]$, range $[0, \infty)$

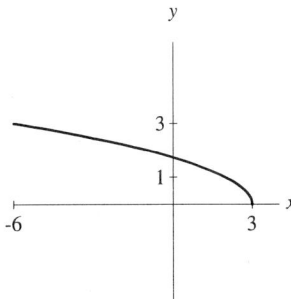

24. $y = x^2(x-3)$ crosses $(3,0)$ but does not cross $(0,0)$, and goes through $(1,-2)$, domain $(-\infty, \infty)$, range $(-\infty, \infty)$

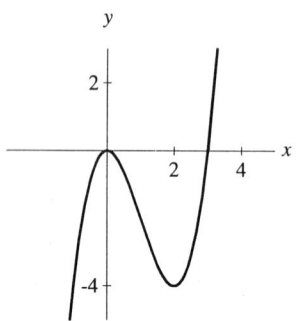

25. 1000

26. 1

27. $\dfrac{1}{2}$

28. 0.001

29. $\dfrac{1}{9}$

30. 3

31. $3^{-2} = \dfrac{1}{9}$

32. $2^3 = 8$

4.1 EXPONENTIAL FUNCTIONS

For Thought

1. False, the base of an exponential function is positive.   2. True

3. False, since there is no solution.

4. True   5. True   6. True   7. True

8. False, since it is decreasing.

9. True, since $0.25 = 4^{-1}$.

10. True, since $\sqrt[100]{2^{173}} = \left(2^{173}\right)^{1/100}$.

4.1 Exercises

1. $3^2 = 9$ 3. $3^{-2} = 1/9$

5. $2^{-1} = 1/2$

7. $2^3 = 8$

9. $(1/4)^{-1} = 4$

11. $4^{1/2} = 2$

13. $f(x) = 5^x$ goes through $(-1, 1/5), (0, 1), (1, 5)$, domain is R, range is $(0, \infty)$, increasing

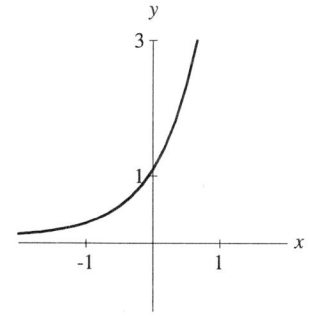

15. $f(x) = 10^{-x}$ goes through $(-1, 10), (0, 1), (1, 1/10)$, domain is R, range is $(0, \infty)$, decreasing

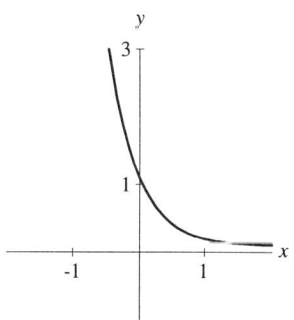

17. $f(x) = (1/4)^x$ goes through $(-1, 4), (0, 1), (1, 1/4)$, domain is R, range is $(0, \infty)$, decreasing

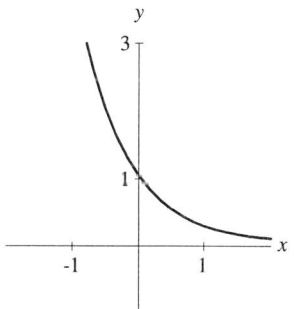

19. Shift $y = 2^x$ down by 3 units; $f(x) = 2^x - 3$ goes through $(-1, -2.5), (0, -2), (2, 1)$, asymptote $y = -3$

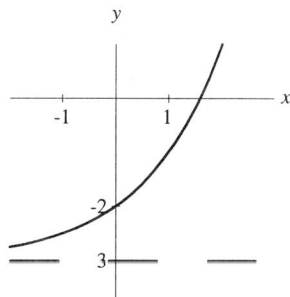

21. Reflect $y = 2^{-x}$ about x-axis, $f(x) = -2^{-x}$ goes through $(-1, -2), (0, -1), (1, -1/2)$, asymptote $y = 0$

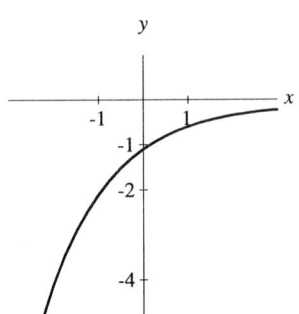

23. Reflect $y = 2^x$ about x-axis and shift up by 1 unit, $f(x) = 1 - 2^x$ goes through $(-1, 0.5), (0, 0), (1, -1)$, asymptote $y = 1$

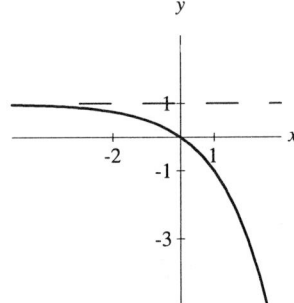

25. Shift $y = 3^x$ to right by 2 and shrink by a factor of 0.5, $f(x) = 0.5 \cdot 3^{x-2}$ goes through $(0, 1/18), (2, 0.5), (3, 1.5)$, asymptote $y = 0$

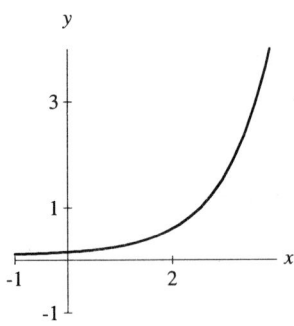

27. Stretch $y = (0.5)^x$ by a factor of 500, $f(x) = 500 \cdot (0.5)^x$ goes through $(0, 500), (1, 250)$, asymptote $y = 0$

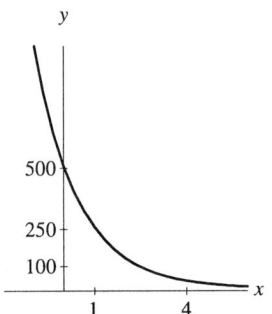

29. Shift $y = 2^x$ to left by 3 units and down by 5 units; $f(x) = 2^{x+3} - 5$ goes through $(-4, -4.5), (-3, -4), (0, 3)$, asymptote $y = -5$

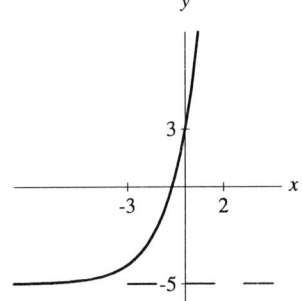

31. $y = 2^{x-5} - 2$

33. $y = -\left(\dfrac{1}{4}\right)^{x-1} - 2$

35. Since $2^x = 2^6$, solution set is $\{6\}$.

37. $\{-1\}$

39. Multiplying the equation by -1, $3^x = 3^3$ and the solution set is $\{3\}$.

41. $\{-2\}$

43. Since $(2^3)^x = 2^{3x} = 2$, $3x = 1$. The solution set is $\left\{\dfrac{1}{3}\right\}$.

45. $\{-2\}$

47. Since $(2^{-1})^x = 2^{-x} = 2^3$, $-x = 3$. The solution set is $\{-3\}$.

49. Since $10^{x-1} = 10^{-2}$, $x - 1 = -2$. The solution set is $\{-1\}$.

51. Since $2^x = 4$, the solution set is $\{2\}$.

53. Since $2^x = \dfrac{1}{2}$, the solution set is $\{-1\}$.

55. Since $\left(\dfrac{1}{3}\right)^x = 1$, the solution set is $\{0\}$.

57. Since $\left(\dfrac{1}{3}\right)^x = 3^3$, the solution set is $\{-3\}$.

59. Since $10^x = 1000$, the solution set is $\{3\}$.

61. Since $10^x = 0.1 = 10^{-1}$, the solution set is $\{-1\}$.

63. 1 **65.** -1

67. $(2, 9), (1, 3), (-1, 1/3), (-2, 1/9)$

69. $(0, 1), (-2, 25), (-1, 5), (1, 1/5)$

71. $(4, -16), (-2, -1/4), (-1, -1/2), (5, -32)$

73. From the x-intercept of $y = 3^x - 5$ the solution set is $\{1.46\}$.

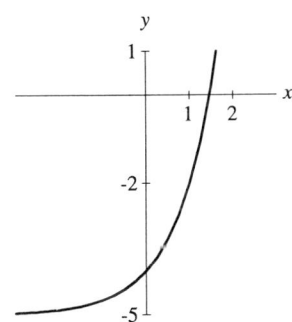

75. From the x-intercept of $y = 2^x - 3^{x+1}$ the solution set is $\{-2.71\}$.

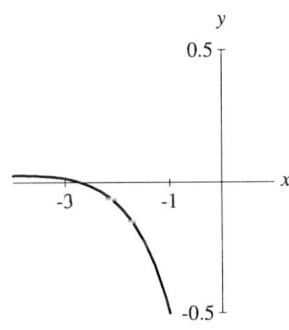

77. From the x-intercept of $y = e^{x+1} - 9$ the solution set is $\{1.20\}$.

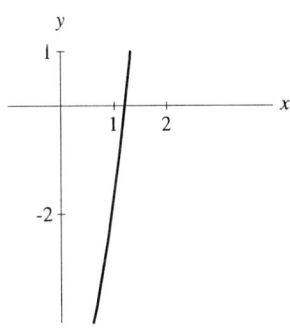

79. From the x-intercept of $y = 200e^{0.06x} - 400$, the solution set is $\{11.55\}$.

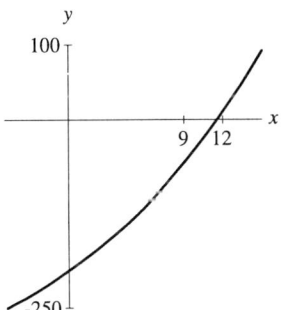

81. $f(x) = 2^{|x|}$ is an even function that goes through $(0, 1), (\pm 1, 2)$, domain $(-\infty, \infty)$, range $[1, \infty)$

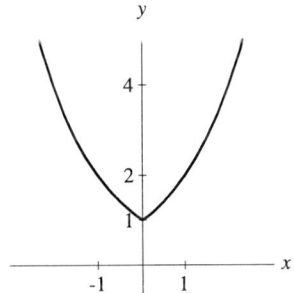

83. $f(x) = 2^{-x} + 2^x$ is an even function that goes through $(0, 2), (\pm 1, 2.5)$, domain $(-\infty, \infty)$, range $[2, \infty)$

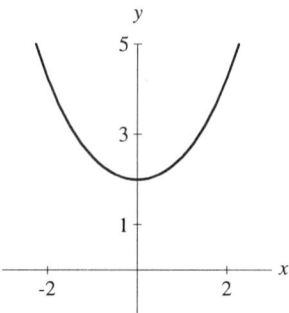

85. For $y = 2^{-x^2}$, the domain is $(-\infty, \infty)$ and the range is $(0, 1]$.

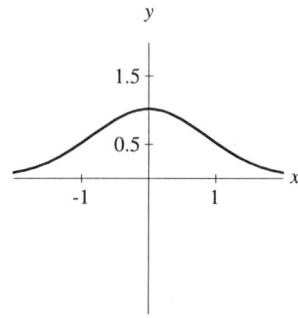

87. The amount at the end of 6 years is
$$4500\left(1 + \frac{0.08}{4}\right)^{24} = \$7,237.97 \ .$$ Interest earned in 6 years is $7,237.97 - 4,500 = \$2,737.97$.

89. Assume there are 365 days in a year and 30 days in a month. The deposit is worth
$$2,300\left(1 + \frac{0.065}{365}\right)^{5(365)+120} = \$3,251.93 \ .$$

91. At 5% compounded continuously, it will amount to $100,000 \cdot e^{0.05 \cdot (5+45/365)} = \$129,196.51$. Compounded daily it will amount to
$$100,000\left(1 + \frac{0.05}{365}\right)^{(365)(5+45/365)} = \$129,194.24 \ .$$

93. In 1990 where $t = 0$, the population is $2,400,000$. In 2001, $t = 11$, so the population is $2,400,000 \cdot e^{0.03(11)} = 3,338,324$.

95. (a) $1000e^{.24(10)} - 1000e^{.05(10)} = \$9,374.46$
(b) $20,000e^{.17(13)} - 20,000e^{.05(13)} = \$144,003.51$

97. (a) decreasing
(b) $89.7e^{-0.0058(0)} = 89.7\%$,
$89.7e^{-0.0058(40)} \approx 71.1\%$
$89.7e^{-0.0058(48)} \approx 67.9\%$
(c) $89.7e^{-0.0058(70)} \approx 59.8\%$

99. $P = \dfrac{10}{2^n}$

101. Tabulated values of $\left(1 + \dfrac{1}{n}\right)^n$ are given below.

n	$\left(1 + \dfrac{1}{n}\right)^n$
20	2.653297704
200	2.711517124
2000	2.717602569
20,000	2.718213874

The difference between e and the last entry above is 6.8×10^{-5}.

103. When $t = 31$ the number of O-rings damaged is $n = 644e^{-0.15(31)} \approx 6$.

105. $f^{-1}(100) = 2$ since $f(2) = 100$

For Thought

1. True   2. False, since $log_{100}(10) = 1/2$.
3. True   4. True
5. False, the domain is $(0, \infty)$.   6. True
7. True   8. False, since $log_a(a) = 1$.
9. True   10. True

4.2 Exercises

1. $f(1) = 5$, $f^{-1}(5) = 1$.
3. $f(-1) = \dfrac{1}{5}$, $f^{-1}\left(\dfrac{1}{5}\right) = -1$.
5. Since $f(3) = 125$, $f^{-1}(125) = 3$.
7. Since $2^6 = 64$, $log_2(64) = 6$.
9. Since $3^{-4} = \dfrac{1}{81}$, $log_3\left(\dfrac{1}{81}\right) = -4$.
11. Since $16^{1/4} = 2$, $log_{16}(2) = \dfrac{1}{4}$.
13. Since $\left(\dfrac{1}{5}\right)^{-3} = 125$, $log_{1/5}(125) = -3$.
15. Since $10^{-1} = 0.1$, $log(0.1) = -1$
17. Since $10^0 = 1$, $log(1) = 0$.
19. Since $e^1 = e$, $ln(e) = 1$.
21. -5

23. $y = log_3(x)$ goes through $(1/3, -1), (1, 0), (3, 1)$, domain $(0, \infty)$, range R

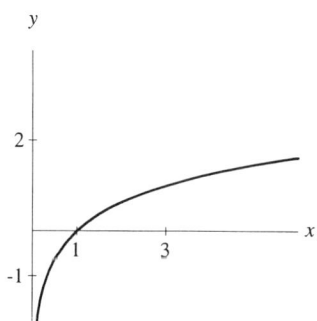

25. $f(x) = log_5(x)$ goes through $(1/5, -1), (1, 0), (5, 1)$, domain $(0, \infty)$, range R

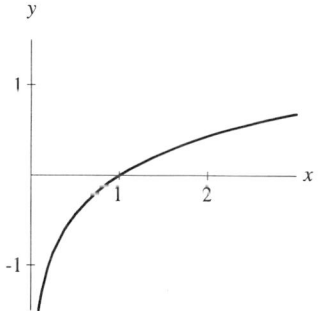

27. $y = log_{1/2}(x)$ goes through $(2, -1), (1, 0), (1/2, 1)$, domain $(0, \infty)$, range R

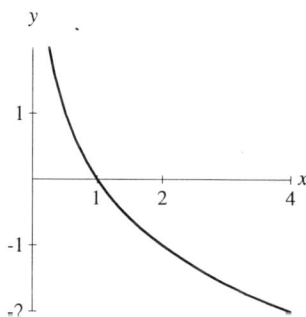

29. $h(x) = log_{1/5}(x)$ goes through $(5,-1),(1,0),(1/5,1)$, domain $(0,\infty)$, range R

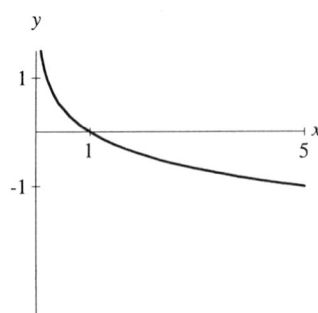

31. $f(x) = ln(x-1)$ goes through $\left(1+\dfrac{1}{e},-1\right)$, $(2,0),(1+e,1)$, domain $(1,\infty)$, range R

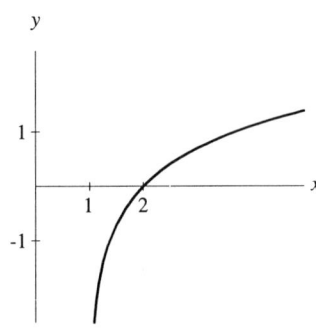

33. $f(x) = -3 + log(x+2)$ goes through $(-1.9,-4)$, $(-1,-3),(8,-2)$, domain $(-2,\infty)$, range R

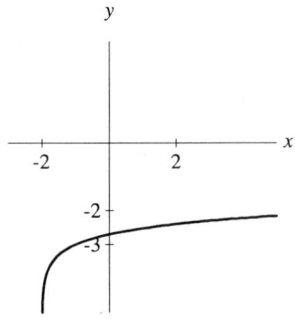

35. $f(x) = -\dfrac{1}{2}log(x-1)$ goes through $(1.1, 0.5)$, $(2,0),(11,-0.5)$, domain $(1,\infty)$, range R

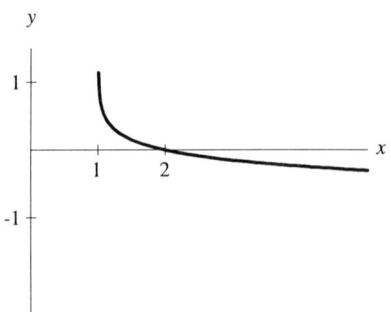

37. $y = \ln(x-3) - 4$

39. $y = -log_2(x-5) - 1$

41. $10^y = 30$ 43. $log_5(7) = y$

45. $2^0 = 1$ 47. $ln(2) = 0.09t$

49. $b^3 = N/M$

51. $log_b(y) = 3x$

53. Since $2^8 = x$, the solution set is $\{256\}$.

55. Since $3^{1/2} = x$, the solution set is $\{\sqrt{3}\}$.

57. Since $x^2 = 16$ and the base of a logarithm is positive, $x = 4$.

59. By definition of logarithm, the solution set is $\{log_3(77)\}$.

61. Since $y = ln(x)$ is one-to-one, $x - 3 = 2x - 9$. The solution set is $\{6\}$.

63. Since $x^2 = 18$, $x = \sqrt{18} = 3\sqrt{2}$. The base of a logarithm is positive.

65. Since $x + 1 = log_3(7)$, $x = log_3(7) - 1$.

67. Since $y = log(x)$ is one-to-one,
$$x = 6 - x^2$$
$$x^2 + x - 6 = 0$$
$$(x+3)(x-2) = 0$$
$$x = -3, 2$$
But $log(-3)$ is undefined, so the solution is $\{2\}$.

69. $x = \left[(100)^{1/2}\right]^{-3} = 10^{-3} = 0.001$

4.2 LOGARITHMIC FUNCTIONS

71. Since $x^{-2/3} = \dfrac{1}{9}$ then $x = \pm\left(\dfrac{1}{9}\right)^{-3/2} = \pm\left(\dfrac{1}{3}\right)^{-3} = \pm 27$. But the base must be positive, so the solution set is $\{27\}$.

73. Since $(2^2)^{2x-1} = 2^{4x-2} = 2^{-1}$, $4x - 2 = -1$. The solution set is $\left\{\dfrac{1}{4}\right\}$.

75. $x = \log(25) \approx 1.3979$

77. $x = \dfrac{\ln(7)}{\ln(3)} \approx \dfrac{1.94591}{1.09861} \approx 1.7712$

79.
$$x \cdot \ln(8) = 5 + \ln(20)$$
$$x = \dfrac{5 + \ln(20)}{\ln(8)}$$
$$x \approx 3.8451$$

81. Factor x,
$$x[\ln(2) - \ln(3)] = 5$$
$$x = \dfrac{5}{\ln(2) - \ln(3)}$$
$$x \approx -12.3315$$

83. Multiply and group the $x's$ on one side,
$$x \cdot \log(5) - \log(5) = x \cdot \log(9) - 2\log(9)$$
$$x[\log(5) - \log(9)] = \log(5) - 2\log(9)$$
$$x = \dfrac{\log(5) - 2\log(9)}{\log(5) - \log(9)}$$
$$x \approx 4.7381$$

85. The x-intercept of $y = \ln(x-2) - 3.2$ is the solution which is $\{26.53\}$

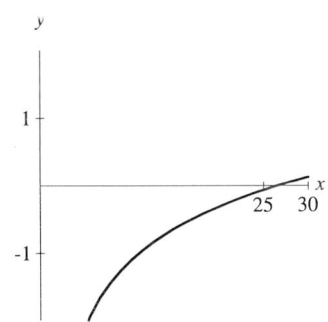

87. The x-intercept of $y = \log(x+1) + \ln(x+2)$ is the solution which is $\{-0.56\}$

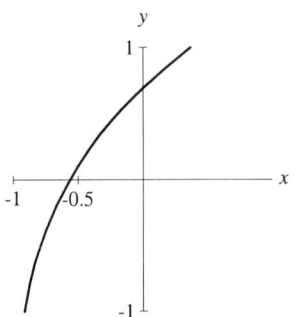

89. $f^{-1}(x) = \log_2(x)$

91. $f^{-1}(x) = 7^x$

93. Replace $f(x)$ by y, interchange x and y, solve for y, and replace y by $f^{-1}(x)$.
$$y = \ln(x - 1)$$
$$x = \ln(y - 1)$$
$$e^x = y - 1$$
$$y = f^{-1}(x) = e^x + 1$$

95. Replace $f(x)$ by y, interchange x and y, solve for y, and replace y by $f^{-1}(x)$.
$$y = 3^{x+2}$$
$$x = 3^{y+2}$$
$$y + 2 = \log_3(x)$$
$$y = f^{-1}(x) = \log_3(x) - 2$$

97. domain $(-\infty, 0) \cup (0, \infty)$, range $(-\infty, \infty)$

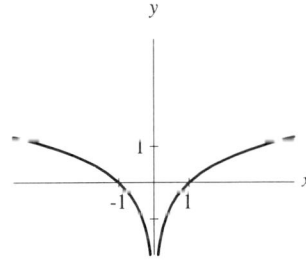

99. domain $(0, \infty)$, range $(-\infty, \infty)$

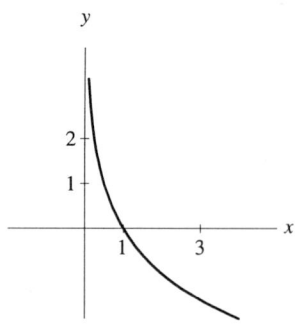

101. Let t be the number of years.
$$1000 \cdot e^{0.14t} = 10^6$$
$$e^{0.14t} = 1000$$
$$0.14t = ln(1000)$$
$$t \approx 49.341 \text{ years}$$
Note, $0.341(365) \approx 125$
It takes 49 years and 125 days.

103. Since $e^{rt} = A/P$, $rt = ln(A/P)$ and $r = \dfrac{ln(A/P)}{t}$. So $\$1,000$ will double in 3 years if the the rate is $r = \dfrac{ln(2000/1000)}{3} = \dfrac{ln(2)}{3} = 0.231$ or 23.1%.

105.

(a) Let t be the number of years. Then
$$P \cdot e^{0.1t} = 2P$$
$$e^{0.1t} = 2$$
$$0.1t = ln(2)$$
$$t = \dfrac{ln(2)}{0.1} \approx 6.9$$
An investment at 10% doubles every 6.9 years.

(b) If t is the number of years it takes before an investment doubles, then
$$P \cdot e^{rt} = 2P$$
$$e^{rt} = 2$$
$$rt = ln(2)$$
$$t = \dfrac{ln(2)}{r}$$
$$t \approx \dfrac{0.70}{r}$$
$$t \approx \dfrac{70}{100r}$$

Thus, the number of years it takes an investment to double is approximately 70 divided by the interest rate (given in percentage).

107. Let r be the interest rate.
$$4,000 \cdot e^{200r} = 4,500,000$$
$$e^{200r} = 1,125$$
$$200r = ln(1,125)$$
$$r = \dfrac{ln(1,125)}{200} \approx 0.035$$
The rate is 3.5%.

109. Let t be the number of years.
$$F_o \cdot e^{-0.052t} = 0.6 F_o$$
$$e^{-0.052t} = 0.6$$
$$-0.052t = ln(0.6)$$
$$t = \dfrac{ln(0.6)}{-0.052} \approx 9.8$$
Only 60% of the present forest will remain after 9.8 yrs.

111. Let r be the annual rate from 1950 to 1987.
$$2.5 \cdot e^{37r} = 5$$
$$e^{37r} = 2$$
$$37r = ln(2)$$
$$r = \dfrac{ln(2)}{37} \approx 0.0187$$
The annual rate is 1.87%. If the annual rate is 1.63% and the initial population is 5 billion in 1987, the world population in year 2000 is $5 \cdot e^{0.163(13)} \approx 6.2$ billion.

113.

(a) $0.1 e^{0.46(7)} \approx 2.5$ billion gigabits

(b) Solving for t, we find
$$.1 e^{.46t} = 14$$
$$e^{.46t} = 140$$
$$.46t = ln(140)$$
$$t = \dfrac{ln(140)}{.46}$$
$$t \approx 11.$$
In the year 2005, the long distance data transmission is expected to be 14 billion gigabits.

115.

(a) The function is given by
$$p - 100 = \frac{100 - 10}{2 - 4}(x - 2)$$
$$p - 100 = -45(x - 2)$$
$$p = -45x + 90 + 100$$
$$p = -45x + 190$$
$$p = -45\log(I) + 190.$$

(b) A graph of $p = -45\log(I) + 190$ is given

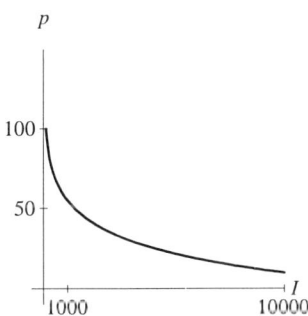

A log scale is better for this function since the graph appears like a straight line and is easier to read than a non-linear graph.

(c) $p = -45\log(100,000) + 190 = -35\%$ or 0% of the population is expected to be without safe drinking water.

117. $pH = -\log\left(10^{-4.1}\right) = 4.1$

119. $pH = -\log\left(10^{-3.7}\right) = 3.7$

121. By substituting $x = 1$, we find a formula for c
$$a \cdot b^x = a \cdot e^{cx}$$
$$b^x = e^{cx}$$
$$b = e^c$$
$$c = \ln(b).$$

By using this formula, we find $y = 50(1.036)^x = 500e^{\ln(1.036)x}$ and the continuous growth rate is $\ln(1.036) \cdot 100 \approx 3.54\%$.

For Thought

1. False, since $\log(8) - \log(3) = \log(8/3) \neq \dfrac{\log(8)}{\log(3)}$.

2. True, since $\ln(3^{1/2}) = \dfrac{1}{2} \cdot \ln(3) = \dfrac{\ln(3)}{2}$.

3. True, since $\dfrac{\log_{19}(8)}{\log_{19}(2)} = \log_2(8) = 3 = \log_3(27)$.

4. True, because of the base-change formula.

5. False

6. False, since $\log(x) - \log(2) = \log(x/2)$.

7. False, since the solution of the first equation is $x = -2$ and the second equation is not defined when $x = -2$.

8. True   9. False, since x can be negative.

10. False, since a can be negative and $\ln(a)$ is undefined.

4.3 Exercises

1. $\log(15)$

3. $\log_2((x-1)x) = \log_2(x^2 - x)$

5. $\log_4(6)$ 7. $\ln\left(\dfrac{x^8}{x^3}\right) = \ln(x^5)$

9. $\log(5^2) = 2\log(5)$

11. $\log(5^{-3}) = -3\log(5)$

13. $\log(5^{1/3}) = (1/3)\log(5)$

15. $\sqrt{y}$

17. $y + 1$

19. 999

21. $\log(5) + \log(2)$

23. $\log(5/2) = \log(5) - \log(2)$

25. $\log(\sqrt{2^2 \cdot 5}) = \dfrac{1}{2}(2\log(2) + \log(5)) = \log(2) + \dfrac{1}{2}\log(5)$

27. $\log(4) - \log(25) = \log(2^2) - \log(5^2) = 2\log(2) - 2\log(5)$

29. $log_3(5) + log_3(x)$

31. $log_2(5) - log_2(2y) = log(5) - log_2(2) - log_2(y)$

33. $log(3) + \frac{1}{2}log(x)$

35. $log(3) + (x-1)log(2)$

37. $\frac{1}{3} \cdot ln(xy) - \frac{4}{3}ln(t) = \frac{1}{3} \cdot ln(x) + \frac{1}{3} \cdot ln(y) - \frac{4}{3}ln(t)$

39. $ln(6\sqrt{x-1}) - ln(5x^3) =$
    $ln(6) + \frac{1}{2} \cdot ln(x-1) - ln(5) - 3 \cdot ln(x)$

41. $log_2(5x^3)$

43. $log_7(x^5) - log_7(x^8) = log(x^5/x^8) = log_7(x^{-3})$

45. $log(2xy/z)$

47. $log\left(\frac{\sqrt{x}}{y}\right) + log\left(\frac{z}{\sqrt[3]{w}}\right) = log\left(\frac{z\sqrt{x}}{y\sqrt[3]{w}}\right)$

49. $log_4(x^6) + log_4(x^{12}) + log_4(x^2) = log_4(x^{20})$

51. $x = log_2(9) = \frac{ln(9)}{ln(2)} \approx \frac{2.197224}{0.693147} \approx 3.1699$

53. $x = log_{0.56}(8) = \frac{ln(8)}{ln(0.56)} \approx \frac{2.0794415}{-0.579819}$
    ≈ -3.5864

55. $x = log_{1.06}(2) = \frac{ln(2)}{ln(1.06)} \approx \frac{0.6931472}{0.0582689}$
    ≈ 11.8957

57. $x = log_{0.73}(0.5) = \frac{ln(0.5)}{ln(0.73)} \approx \frac{-0.6931472}{-0.3147107}$
    ≈ 2.2025

59. $\frac{ln(9)}{ln(4)} \approx \frac{2.1972246}{1.3862944} \approx 1.5850$

61. $\frac{ln(2.3)}{ln(9.1)} \approx \frac{0.8329091}{2.2082744} \approx 0.3772$

63. $4t = log_{1.02}(3) = \frac{ln(3)}{ln(1.02)}$
    $t = \frac{ln(3)}{4 \cdot ln(1.02)} \approx 13.8695$

65. $365t = log_{1.0001}(3.5) = \frac{ln(3.5)}{ln(1.0001)}$
    $t = \frac{ln(3.5)}{365 \cdot ln(1.0001)} \approx 34.3240$

67. $1 + r = \sqrt[3]{2.3}$, so $r = \sqrt[3]{2.3} - 1 \approx 0.3200$

69.
$$\left(1 + \frac{r}{12}\right)^{360} = 4.2$$
$$1 + \frac{r}{12} = \sqrt[360]{4.2}$$
$$r = 12\left(\sqrt[360]{4.2} - 1\right)$$
$$r \approx 0.0479$$

71. $x^5 = 33.4$, so $x = \sqrt[5]{33.4} \approx 2.0172$.

73. $x^{-1.3} = 0.546$, so $x = 0.546^{1/(-1.3)} \approx 1.5928$.

75. True, since $log_3(81) = 4$ and $2 = log_3(9)$.

77. False, since $ln(3^2) = 2 \cdot ln(3) \neq (ln(3))^2$.

79. True, since $4 \cdot log_2(8) = 4 \cdot 3 = 12$.

81. False, since $3 + log(6) = log(10^3) + log(6) = log(6000)$.

83. False, since $log_2(16) = 4$, $log_2(8) = 3$, and $\frac{3}{4} \neq 3 - 4$.

85. False, since $log_2(25) = 2log_2(5) = 2 \cdot \frac{log(5)}{log(2)} \neq 2 \cdot log(5)$.

87. True, because of the base-changing formula.

89. Let t be the number of years.
$$800\left(1 + \frac{0.08}{365}\right)^{365t} = 2000$$
$$\left(1 + \frac{0.08}{365}\right)^{365t} = 2.5$$
$$(1.0002192)^{365t} \approx 2.5$$
$$365t \approx log_{1.0002192}(2.5)$$
$$t \approx \frac{1}{365} \cdot \frac{ln(2.5)}{ln(1.0002192)}$$
$$t \approx 11.453753$$
$$t \approx 11 \text{ years}, 166 \text{ days}.$$

91. Let t be the number of years.
$$W\left(1+\frac{0.1}{4}\right)^{4t} = 3W$$
$$(1.025)^{4t} = 3$$
$$4t \approx \log_{1.025}(3)$$
$$t \approx \frac{1}{4} \cdot \frac{\ln(3)}{\ln(1.025)}$$
$$t \approx 11.122 \text{ years}$$
$$t \approx 11.122(4) \approx 44 \text{ quarters}$$

93. Let t be the number of years.
$$500(1+r)^{25} = 2000$$
$$(1+r)^{25} = 4$$
$$1+r \approx \sqrt[25]{4}$$
$$r = \sqrt[25]{4} - 1$$
$$r \approx 0.057 \text{ or } 5.7\%$$

95. Let t be the number of years.
$$4000(1+r)^{200} = 4.5 \times 10^6$$
$$(1+r)^{200} = 1125$$
$$r = \sqrt[200]{1125} - 1$$
$$r \approx 0.035752 \text{ or } 3.58\%$$

97. The Richter scale rating is
$$\log(I) - \log(I_o) = \log\left(\frac{I}{I_o}\right).$$
When $I = 1000 \cdot I_o$, the Richter scale rating is
$$\log\left(\frac{1000 \cdot I_o}{I_o}\right) = \log(1000) = 3.$$

99. $t = \frac{1}{r}\ln(P/P_o) = \frac{1}{r}\ln(P) - \frac{1}{r}\ln(P_o)$

101.

(a) p decreases as n increases

(b) Solving for n, one can take the logarithm of both sides and to note that $y = \log(x)$ is an increasing function.

$$\left(\frac{7,059,051}{7,059,052}\right)^n > \frac{1}{2}$$
$$\log\left(\left(\frac{7,059,051}{7,059,052}\right)^n\right) > \log\left(\frac{1}{2}\right)$$
$$n\log\left(\frac{7,059,051}{7,059,052}\right) > \log\left(\frac{1}{2}\right)$$
$$n < \frac{\log(1/2)}{\log(7,059,051/7,059,052)}$$
$$n < 4,892,962$$

The probability of rollover is greater than 50% if less than 4,892,962 tickets are purchsed.

103. $MR(x) = MR(x+1) - MR(x) =$
$500 \cdot \log(x+2) - 500 \cdot \log(x+1) =$
$$500 \cdot \log\left(\frac{x+2}{x+1}\right) = \log\left(\left(\frac{x+2}{x+1}\right)^{500}\right).$$
As $x \to \infty$, then $\frac{x+2}{x+1} \to 1$ and $MR(x) \to 0$.

105.

(a) Let x be the number of years since 1990 and let y be the number of internet hosts in millions. An exponential model is
$y = (0.201)(1.876815073)^x$.

(b) Since $\ln(1.876815073) \approx 0.6295762301$, then equivalently
$y \approx (0.201)e^{0.6295762301x}$.

(c) annual rate of 62.96%, growing continuously

(d) In the year 2002, the number of internet hosts is predicted to be $y = (0.201)(1.876815073)^{12}$
≈ 383.9 million.

(e) Solving $(0.201)(1.876815073)^x = 1,000$, one finds $x \approx 13.5$, i.e., in the year 2003 the number of internet hosts is expected to reach 1 billion.

Suppose the earth's population was 6 billion in 1990 and grows at an annual rate of 2% compounded continuously. Solving
$$(0.201)e^{0.6295762301x} = 6,000e^{.02x}$$
one finds $x \approx 16.9$. Thus, the number of internet hosts is expected to surpass the earth's population in the year 2007. Of course, the number of internet hosts will never surpass the population of the earth, we hope.

107. Let $a > 0$, $a \neq 1$. Using the definition of a logarithm and properties of exponents, we have
$$a^{\log_a(M^p)} = M^p = \left(a^{\log_a(M)}\right)^p = a^{p \cdot \log_a(M)}.$$
Then $a^{\log_a(M^p)} = a^{p \cdot \log_a(M)}$. By recalling that $a^s = a^t$ implies $s = t$, we have $\log_a(M^p) = p \cdot \log_a(M)$.

For Thought

1. True, since $(1.02)^x = 7$ is equivalent to $x = \log_{1.02}(7)$.

2. True. Since $x(1 - \ln(3)) = 8$, $x = \dfrac{8}{1 - \ln(3)}$.

3. False, $\ln(1 - \sqrt{6})$ is undefined.

4. True, by definition of a logarithm.

5. False, the exact solution is $x = \log_3(17) \neq 2.5789$

6. False, since $x = -2$ is not a solution of the first equation but is a solution of the second one.

7. True, since $4^x = 2^{2x} \neq 2^{x-1}$. 8. True

9. True, since $\dfrac{\ln(2)}{\ln(7)} = \dfrac{\log(2)}{\log(7)}$.

10. True, since $\log(e) \cdot \ln(10) = \ln\left(10^{\log(e)}\right) = \ln(e) = 1$.

4.4 Exercises

1. Since $10^2 = x + 20$, $x = 80$.

3. Since $x > 0$ and $x^2 = 9$, $x = 3$.

5.
$$\log_2(x^2 - 4) = 5$$
$$x^2 - 4 = 2^5$$
$$x^2 = 36$$
$$x = \pm 6$$
Checking $x = -6$, one gets $\log_2(-6 + 2)$ which is undefined. The solution is $x = 6$.

7.
$$\log(5) + \log(x) = 2$$
$$\log(5x) = 2$$
$$5x = 10^2$$
$$x = 20$$

9. Since $\ln(x(x+2)) = \ln(8)$ and $y = \ln(x)$ is one-to-one,
$$x^2 + 2x = 8$$
$$x^2 + 2x - 8 = 0$$
$$(x + 4)(x - 2) = 0$$
$$x = -4, 2$$
Since $\ln(-4)$ is undefined, solution is $x = 2$.

11. Since $\log(4x) = \log\left(\dfrac{5}{x}\right)$ and $y = \log(x)$ is one-to-one,
$$4x = \dfrac{5}{x}$$
$$4x^2 = 5$$
$$x^2 = \dfrac{5}{4}$$
$$x = \pm\dfrac{\sqrt{5}}{2}.$$
But $\log\left(-\dfrac{\sqrt{5}}{2}\right)$ is undefined, so $x = \dfrac{\sqrt{5}}{2}$.

13. Since $\log_2\left(\dfrac{x}{3x - 1}\right) = 0$,
$$\dfrac{x}{3x - 1} = 1$$
$$x = 3x - 1$$
$$1 = 2x$$
$$x = \dfrac{1}{2}$$

15.
$$x \cdot \ln(3) + x = 2$$
$$x(\ln(3) + 1) = 2$$
$$x = \dfrac{2}{\ln(3) + 1}$$

17. $x - 1 = \log_2(7)$
$$x = \dfrac{\ln(7)}{\ln(2)} + 1 \approx 3.8074.$$

19. $4x = \log_{1.09}(3.4)$
$$x = \dfrac{1}{4} \cdot \dfrac{\ln(3.4)}{\ln(1.09)} \approx 3.5502.$$

21. $-x = \log_3(30)$
$$x = -\dfrac{\ln(30)}{\ln(3)} \approx -3.0959.$$

23. Since $-3x^2 = \ln(9)$, no solution since the left-hand side is non-negative.

25.
$$ln(6^x) = ln(3^{x+1})$$
$$x \cdot ln(6) = (x+1) \cdot ln(3)$$
$$x \cdot ln(6) = x \cdot ln(3) + ln(3)$$
$$x(ln(6) - ln(3)) = ln(3)$$
$$x = \frac{ln(3)}{ln(6) - ln(3)}$$
$$x \approx 1.5850$$

27.
$$ln(e^{x+1}) = ln(10^x)$$
$$(x+1) \cdot ln(e) = x \cdot ln(10)$$
$$x + 1 = x \cdot ln(10)$$
$$1 = x(ln(10) - 1)$$
$$x = \frac{1}{ln(10) - 1}$$
$$x \approx 0.7677$$

29.
$$2^{x-1} = (2^2)^{3x}$$
$$2^{x-1} = 2^{6x}$$
$$x - 1 = 6x$$
$$-1 = 5x$$
$$x = -0.2$$

31.
$$ln(6^{x+1}) = ln(12^x)$$
$$(x+1) \cdot ln(6) = x \cdot ln(12)$$
$$x \cdot ln(6) + ln(6) = x \cdot ln(12)$$
$$ln(6) = x(ln(12) - ln(6))$$
$$x = \frac{ln(6)}{ln(12) - ln(6)}$$
$$x \approx 2.5850$$

33. Since $e^{-ln(w)} = e^{ln(1/w)} = 1/w$, $\frac{1}{w} = 3$.
So $w = \frac{1}{3}$.

35.
$$(log(z))^2 = 2 \cdot log(z)$$
$$(log(z))^2 - 2 \cdot log(z) = 0$$
$$log(z) \cdot (log(z) - 2) = 0$$
$$log(z) = 0 \text{ or } log(z) = 2$$
$$z = 10^0 \text{ or } z = 10^2$$
$$z = 1 \text{ or } z = 100$$

37. Divide the equation by $4(1.03)^x$,
$$\left(\frac{1.02}{1.03}\right)^x = \frac{3}{4}$$
$$ln\left(\left(\frac{1.02}{1.03}\right)^x\right) = ln\left(\frac{3}{4}\right)$$
$$x \cdot ln\left(\frac{1.02}{1.03}\right) = ln\left(\frac{3}{4}\right)$$
$$x = \frac{ln\left(\frac{3}{4}\right)}{ln\left(\frac{1.02}{1.03}\right)}$$
$$x \approx 29.4872$$

39. Note that $e^{ln((x^2)^3) - ln(x^2)} = e^{ln(x^6) - ln(x^2)} = e^{ln(x^6/x^2)} = e^{ln(x^4)} = x^4$.
So $x^4 = 16$ and $x = \pm 2$.
But $ln(-2)$ is undefined, so $x = 2$.

41. Since $\left(\frac{1}{2}\right)^2 = \frac{1}{4}$,
$$\left(\frac{1}{2}\right)^{2x-1} = \left(\frac{1}{2}\right)^{6x+4}$$
$$2x - 1 = 6x + 4$$
$$-5 = 4x$$
$$x = -\frac{5}{4}$$

43. Graph $y = 2^x - 3^{x-1} - 5^{-x}$.
Solutions are $x \approx 0.194, \; 2.70$

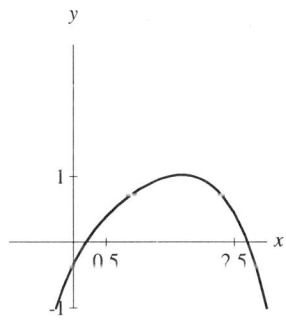

45. Graph $y = ln(x+51) - log(-48-x)$.
    Solution is $x \approx -49.73$

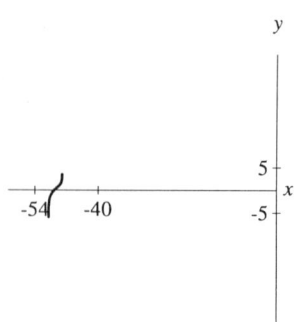

47. Graph $y = x^2 - 2^x$. Solutions are
    $x \approx -0.767, 2, 4$

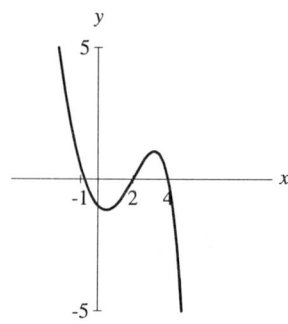

49. From $A = A_o e^{rt}$ with $A_o = 1$, $\dfrac{1}{2} = e^{5730r}$.

    By definition of a logarithm, $5730r = ln\left(\dfrac{1}{2}\right)$ and $r \approx -0.000120968$. When $A = 0.1$,
    $$0.1 = e^{-0.000120968t}$$
    $$ln(0.1) = -0.000120968t$$
    $$t = \dfrac{ln(0.1)}{-0.000120968} \approx 19,035 \text{ years}$$

51. From Number 43, $r \approx -0.000120968$.
    When $A = 10$ and $A_o = 12$,
    $$10 = 12 \cdot e^{-0.000120968t}$$
    $$ln(10/12) = -0.000120968t$$
    $$t = \dfrac{ln(5/6)}{-0.000120968} \approx 1,507 \text{ years}$$

53. From $A = A_o e^{rt}$ where $A_o = 25$, $A = 20$ and $t = 8000$ one obtains
    $$20 = 25 \cdot e^{8000r}$$
    $$ln(0.8) = 8000r$$
    $$r \approx -0.000027892$$

    For the half-life, let $A = 12.5$. Thus,
    $$12.5 = 25 \cdot e^{-0.000027892t}$$
    $$ln(0.5) = -0.000027892t$$
    $$t \approx 24,850 \text{ years.}$$

55. $\dfrac{2.5(0.5)^{24/14}}{2.5} \times 100 \approx 30.5\%$, the percentage of the last dosage that remains before the next dosage is taken

57. Since half-life is $t = 5730$ years and $A_o = 1$ and $A = 0.5$,
    $$0.5 = e^{(5730) \cdot r}$$
    $$ln(0.5) = 5730 \cdot r$$
    $$r \approx -0.000121$$

    If 79.3% of the carbon is still present then
    $$0.793 = e^{(-0.000121) \cdot t}$$
    $$ln(0.793) = -0.000121 \cdot t$$
    $$t \approx 1917$$
    The scrolls were made in the year $1951 - 1917 = 34$ AD.

59. The initial difference in temperature is $325 - 35 = 290$ and $t = 3$ hours after the difference is $325 - 140 = 185$. Then
    $$185 = 290 \cdot e^{3k}$$
    $$ln\left(\dfrac{185}{290}\right) = 3k$$
    $$k \approx -0.1498417$$

    The difference when the roast is well-done is $325 - 170 = 155$. Thus,
    $$155 = 290 \cdot e^{(-0.1498417) \cdot t}$$
    $$ln\left(\dfrac{155}{290}\right) = (-0.1498417) \cdot t$$
    $$t \approx 4.18 \text{ hrs.}$$
    $$t \approx 4 \text{ hrs. and } 11 \text{ min.}$$

    James must wait 1 hour and 11 minutes longer. If the oven temperature is set at $170°$ then the initial and final differences are 135 and 0, respectively. Since $0 = 135 \cdot e^{(-0.1498417) \cdot t}$ has no solution then James has to wait forever.

4.4 MORE EQUATIONS AND APPLICATIONS

61. At $7:00$ a.m., the difference in temperature is $80 - 40 = 40$ and $t = 1$ hour later the difference is $72 - 40 = 32$. Then
$$32 = 40 \cdot e^{1 \cdot k}$$
$$ln\left(\frac{32}{40}\right) = k$$
$$k \approx -0.2231436$$
Let n be the number of hours before $7:00$ a.m. when death occured. At death the difference in temperature is $98 - 40 = 58$. So
$$40 = 58 \cdot e^{-0.2231436 \cdot n}$$
$$ln\left(\frac{40}{58}\right) = -0.2231436 \cdot n$$
$$n \approx 1.665 \text{ hrs.}$$
$$n \approx 1 \text{ hr. and } 40 \text{ min.}$$
Death occured at $5:20$ a.m.

63. Set $A = B$ and solve for t.
$$30 \cdot e^{-0.05t} = 20 \cdot e^{0.07t}$$
$$1.5 = e^{0.12t}$$
$$ln(1.5) = 0.12t$$
$$t \approx 3.4 \text{ years}$$
The number of people below the poverty level will be equal to the number people above the poverty level after 3.4 years.

65. The future values of the $1000 and $1100 investments are equal. Then
$$1,000 \cdot e^{0.06t} = 1100\left(1 + \frac{0.06}{365}\right)^{365t}$$
$$e^{0.06t} = 1.1\left(1 + \frac{0.06}{365}\right)^{365t}$$
$$.06t = ln(1.1) + 365t \ln\left(1 + \frac{0.06}{365}\right)$$
$$.06t - 365t \ln\left(1 + \frac{0.06}{365}\right) = ln(1.1)$$
$$t = \frac{ln(1.1)}{.06 - 365 \ln\left(1 + \frac{0.06}{365}\right)}$$
$$t \approx 19,328.84173 \text{ years}$$
They will be equal after 19,328 yrs. & 307 days.

67. The present number of rabbits is $P = 12,300 + 1000 \cdot ln(1) = 12,300 + 0 = 12,300$. The number of years before there will be 15,000 rabbits is given by
$$12,300 + 1000 \cdot ln(t+1) = 15,000$$
$$1000 \cdot ln(t+1) = 2,700$$
$$ln(t+1) = 2.7$$
$$t + 1 = e^{2.7}$$
$$t \approx 13.9 \text{ years.}$$

69. Since $m = 0$ and $M_v = 4.39$, the distance to α Centauri is given by
$$4.39 - 5 + 5 \cdot log(d) = 0$$
$$5 \cdot log(d) = 0.61$$
$$log(d) = 0.122$$
$$d = 10^{0.122} \approx 1.32 \text{ parsecs.}$$

71.

(a) Let $y = log(P)$. A formula for P is
$$y - 4.5 = \frac{4.5 - 1.5}{0 - 18}(x - 0)$$
$$y = -\frac{1}{6}x + 4.5$$
$$log(P) = -\frac{1}{6}x + 4.5$$
$$P = 10^{(-\frac{1}{6}x + 4.5)}$$

(b) In 1991, $n = 9$ and $P = 10^{(-\frac{1}{6}9 + 4.5)} = 10^3 = \$1,000$ per gigabit.

(c) Solving for x, we derive
$$10^1 = 10^{(-\frac{1}{6}x + 4.5)}$$
$$1 = -\frac{1}{6}x + 4.5$$
$$\frac{1}{6}x = 3.5$$
$$x = 21$$
In the year 2003, according to this model a hard drive with one gigabit will cost $10.

73. If the sound level is 90 db then the intensity of the sound is given by
$$10 \cdot log(I \times 10^{12}) = 90$$
$$log(I) + log(10^{12}) = 9$$
$$log(I) + 12 = 9$$
$$log(I) = -3$$
$$I = 10^{-3} \text{ watts per } m^2$$

75. Since $P \cdot e^{(0.06)18} = 20,000$, the investment will grow to $P = \dfrac{20,000}{e^{(0.06)18}} \approx \$6,791.91$

77. If $R = 1 \times 10^5$, then $T \approx 24.999938°$ C.

One finds $T = 35°$ C if $R \approx 63,491$ as shown in the graph

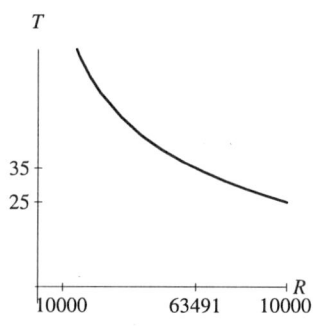

79. If $x = .4$, then $\ln(1.4) \approx$

$.4 - \dfrac{.4^2}{2} + \dfrac{.4^3}{3} - \dfrac{.4^4}{4} + \dfrac{.4^5}{5} \approx 0.33698.$

With a calculator, $\ln(1.4) \approx 0.33647$.

Review Exercises

1. 64   3. 6   5. 0

7. 17   9. $2^5 = 32$

11. $log(10^3) = 3$ 13. $log_2(2^9) = 9$

15. $log(1000) = 3$ 17. $log_2(1) - log_2(8) = 0 - 3 = -3$

19. $log_2(8) = 3$

21. $log((x-3)x) = log(x^2 - 3x)$

23. $ln(x^2) + ln(3y) = ln(3x^2y)$

25. $log(3) + log(x^4) = log(3) + 4 \cdot log(x)$

27. $log_3(5) + log_3(x^{1/2}) - log_3(y^4) = log_3(5) + \dfrac{1}{2} \cdot log_3(x) - 4 \cdot log_3(y)$

29. $ln(5 \cdot 2) = ln(5) + ln(2)$

31. $ln(2^5 \cdot 2) = ln(2^5) + ln(2) = 2 \cdot ln(5) + ln(2)$

33. $log_{10}(x) = 10$, so $x = 10^{10}$.

35. Since $x^4 = 81$ and $x > 0$, $x = 3$.

37. $log_{1/3}(27) = -3 = x + 2$, so $x = -5$.

39. Since $3^{x+2} = 3^{-2}$, $x + 2 = -2$. So $x = -4$.

41. $x - 2 = ln(9)$, so $x = ln(9) + 2$.

43. Since $(2^2)^{x+3} = 2^{2x+6} = 2^{-x}$, $2x + 6 = -x$, $6 = -3x$, so $x = -2$.

45.
$$log(2x^2) = 5$$
$$2x^2 = 10^5$$
$$x^2 = 50,000$$
$$x = \pm 100\sqrt{5}$$
But $log(-100\sqrt{5})$ is undefined, so $x = 100\sqrt{5}$.

47.
$$log_2\left(x^2 - 4x\right) = log_2(x + 24)$$
$$x^2 - 4x = x + 24$$
$$x^2 - 5x - 24 = 0$$
$$(x - 8)(x + 3) = 0$$
$$x = 8, -3$$
But $log(-3)$ is undefined, so $x = 8$.

49. Since $ln((x+2)^2) = ln(4^3)$,
$$(x+2)^2 = 64$$
$$x+2 = \pm 8$$
$$x = -2 \pm 8$$
$$x = 6, -10$$
Checking $x = -10$ one gets $2ln(-8)$ which is undefined. So $x = 6$.

51.
$$x \cdot log(4) + x \cdot log(25) = 6$$
$$x(log(4) + log(25)) = 6$$
$$x \cdot log(100) = 6$$
$$x \cdot 2 = 6$$
$$x = 3$$

53. The missing coordinates are

(i) 3 since $\left(\dfrac{1}{3}\right)^{-1} = 3$,

(ii) -3 since $\left(\dfrac{1}{3}\right)^{-3} = 27$,

(iii) $\sqrt{3}$ since $\left(\dfrac{1}{3}\right)^{-1/2} = \sqrt{3}$, and

(iv) 0 since $\left(\dfrac{1}{3}\right)^{0} = 1$.

55. c   57. b

59. d   61. e

63. Domain $(-\infty, \infty)$, range $(0, \infty)$, increasing, asymptote $y = 0$

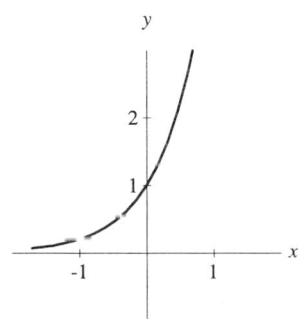

65. Domain $(-\infty, \infty)$, range $(0, \infty)$, decreasing, asymptote $y = 0$

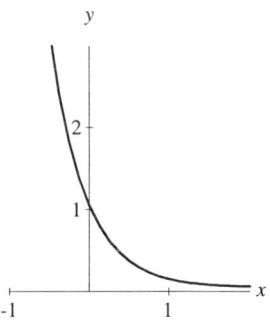

67. Domain $(0, \infty)$, range $(-\infty, \infty)$, increasing, asymptote $x = 0$

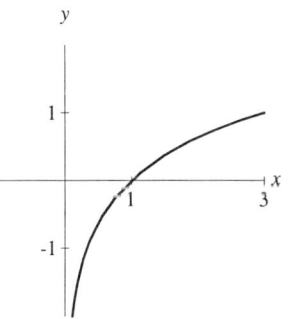

69. Domain $(-3, \infty)$, range $(-\infty, \infty)$, increasing, asymptote $x = -3$

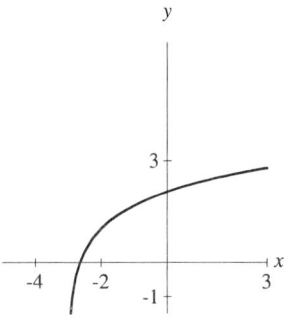

71. Domain $(-\infty, \infty)$, range $(1, \infty)$, increasing, asymptote $y = 1$

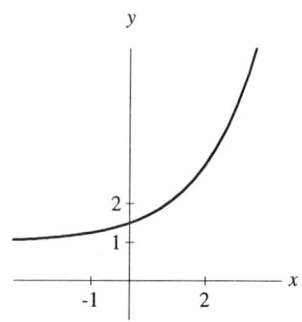

73. Domain $(-\infty, 2)$, range $(-\infty, \infty)$, decreasing, asymptote $x = 2$

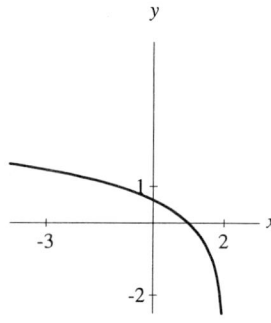

75. $f^{-1}(x) = log_7(x)$ 77. $f^{-1}(x) = 5^x$

79. Since $3^x = 10$, $x = log_3(10) = \dfrac{ln(10)}{ln(3)} \approx 2.0959$.

81. Since $log_3(x) = 1.876$, $x = 3^{1.876} \approx 7.8538$.

83. After taking the natural logarithm of both sides, we have
$$ln(5^x) = ln(8^{x+1})$$
$$x \cdot ln(5) = (x+1) \cdot ln(8)$$
$$x \cdot ln(5) = x \cdot ln(8) + ln(8)$$
$$x \cdot (ln(5) - ln(8)) = ln(8)$$
$$x = \dfrac{ln(8)}{ln(5) - ln(8)}$$
$$x \approx -4.4243.$$

85. If H_B^+ is the hydrogen ion concentration of liquid B then $10 \cdot H_B^+$ is the hydrogen ion concentration of liquid A. The pH of A is
$$-log(10 \cdot H_B^+) = -1 - log(H_B^+)$$
i.e. the pH of A is one less than the pH of B.

87. The value at the end of 18 yrs. is
$$50,000 \left(1 + \dfrac{0.05}{4}\right)^{18 \cdot 4} \approx \$122,296.01 \ .$$

89. Let t be the number of years.
$$50,000 \left(1 + \dfrac{0.05}{4}\right)^{4t} = 100,000$$
$$(1.0125)^{4t} = 2$$
$$4t = log_{1.0125}(2)$$
$$t = \dfrac{1}{4} \cdot \dfrac{ln(2)}{ln(1.0125)}$$
$$t \approx 13.9 \ \text{years}$$

It doubles in $4 \cdot 13.9 \approx 56$ quarters.

91. The number of grams present is $A = 25 \cdot e^0 = 25$. After $t = 1000$ yrs., the amount present is $A = 25 \cdot e^{-0.32} \approx 18.15$ gms.

For the half-life, let $A = 12.5$. Then
$$25 \cdot e^{-0.00032t} = 12.5$$
$$e^{-0.00032t} = 0.5$$
$$-0.00032t = ln(0.5)$$
$$t \approx 2166$$

The half-life is 2166 years.

93. Let $f(t) = 10,000$.
$$40,000 \cdot (1 - e^{-0.0001t}) = 10,000$$
$$1 - e^{-0.0001t} = 0.25$$
$$0.75 = e^{-0.0001t}$$
$$ln(0.75) = -0.0001t$$
$$t \approx 2,877 \ \text{hrs.}$$

It takes $2,877$ hrs. to learn $10,000$ words.

95. (a) If on the nth bet a gambler gets her/his first win, then the gambler has a net gain of $2.

(b) If the nth bet is more than $1 million, then
$$2^n > 10^6$$
$$n \ln(2) > 6 \ln(10)$$
$$n > \dfrac{6 \ln(10)}{\ln(2)}$$
$$n \approx 19.9.$$

The first bet that exceeds $1,000,000 is the 20th bet.

97. In the following equations, one finds y.

$$\frac{25005}{2240}\left(\frac{35+\frac{1}{12}}{100}\right)^y = 258.51$$

$$\left(\frac{35+\frac{1}{12}}{100}\right)^y = \frac{258.51(2240)}{25005}$$

$$y = \frac{\ln\left(\frac{258.51(2240)}{25005}\right)}{\ln\left(\frac{35+\frac{1}{12}}{100}\right)}$$

$$y \approx -3$$

Chapter 4 Test

1. 3   **2.** -2 **3.** 6.47   **4.** $\sqrt{2}$

5. $f^{-1}(x) = e^x$ **6.** $f^{-1}(x) = \log_8(x)$

7. $\log(x) + \log(y^3) = \log(xy^3)$

8. $\ln(\sqrt{x-1})$ $\ln(33) - \ln\left(\frac{\sqrt{x-1}}{33}\right)$

9. $\log_a(2^2 \cdot 7) = \log_a(2^2) + \log_a(7) = 2 \cdot \log_a(2) + \log_a(7)$

10. $\log_a\left(\frac{7}{2}\right) = \log_a(7) - \log_a(2)$

11.
$$\log_2(x^2 - 2x) = 3$$
$$x^2 - 2x = 2^3$$
$$x^2 - 2x - 8 = 0$$
$$(x-4)(x+2) = 0$$
$$x = 4, -2$$

Since $\log_2(-2)$ is undefined, $x = 4$.

12.
$$\log\left(\frac{10x}{x+2}\right) = \log(3^2)$$
$$\frac{10x}{x+2} = 9$$
$$10x = 9x + 18$$
$$x = 18$$

13.
$$\ln(3^x) = \ln(5^{x-1})$$
$$x \cdot \ln(3) = (x-1) \cdot \ln(5)$$
$$x \cdot \ln(3) = x \cdot \ln(5) - \ln(5)$$
$$x(\ln(3) - \ln(5)) = -\ln(5)$$
$$x(\ln(5) - \ln(3)) = \ln(5)$$
$$x = \frac{\ln(5)}{\ln(5) - \ln(3)}$$
$$x \approx 3.1507$$

14. By definition of a logarithm $x - 1 = 3^{5.46}$. So $x = 1 + 3^{5.46} \approx 403.7931$.

15. Domain $(-\infty, \infty)$, range $(1, \infty)$, increasing, $y = 1$

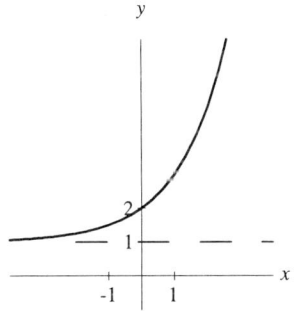

16. Domain $(1, \infty)$, range $(-\infty, \infty)$, decreasing, $x = 1$

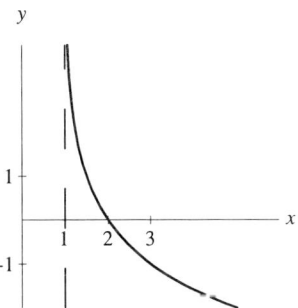

17. (1,0)

18. Compounded quarterly, the investment is worth
$$2000\left(1+\frac{0.08}{4}\right)^{80} \approx \$9,750.88\ .$$
Compounded continuously, the investment is worth
$2000 \cdot e^{0.08(20)} \approx \$9,906.06$

19. The amount of power at the end of $t = 200$ days is
$P = 50 \cdot e^{-200/250} \approx 22.5$ watts.
For the half-life, let $P = 25$. Then
$$50 \cdot e^{-t/250} = 25$$
$$e^{-t/250} = 0.5$$
$$-\frac{t}{250} = ln(0.5)$$
$$t \approx 173.3$$

The half-life is 173.3 days. The operational power of $P = 9$ watts is given by
$$50 \cdot e^{-t/250} = 9$$
$$e^{-t/250} = \frac{9}{50}$$
$$-\frac{t}{250} = ln\left(\frac{9}{50}\right)$$
$$t \approx 428.7 \text{ days}$$

20.
$$4,000 \cdot \left(1+\frac{0.06}{4}\right)^{4t} = 10,000$$
$$(1.015)^{4t} = 2.5$$
$$4t = log_{1.015}(2.5)$$
$$t \approx 15.38576 \text{ years}$$
$$t \approx 61.5 \text{ quarters}$$

21. Substituting $t = 100$,
$$-50 \cdot ln(1-p) = 100$$
$$ln(1-p) = -2$$
$$1-p = e^{-2}$$
$$p = 1 - e^{-2}$$
$$p \approx 0.86$$

The level reached after 100 hrs. is $p = 0.86$. When $p = 1$, then $t = -50 ln(0)$ which is undefined, so it is impossible to master MGM.

Tying It All Together

1. $x - 3 = \pm 2$, so $x = 3 \pm 2 = 1, 5$.

2. Since $log((x-3)^2) = log(4)$,
$$(x-3)^2 = 4$$
$$x - 3 = \pm 2$$
$$x = 3 \pm 2$$
$$x = 5, 1.$$

Checking $x = 1$, one gets $2 log(-2)$ which is undefined, so $x = 5$.

3. By definition of a logarithm, $x - 3 = 2^4$. Solution is $x = 16 + 3 = 19$.

4. Since $2^{x-3} = 2^2$, $x - 3 = 2$. So $x = 5$.

5. Squaring both sides of the equation, $x - 3 = 16$. Solution is $x = 19$.

6. An equivalent statement is $x - 3 = \pm 4$. So $x = 3 \pm 4 = -1, 7$.

7. Completing the square,
$$x^2 - 4x + 4 = -2 + 4$$
$$(x-2)^2 = 2$$
$$x - 2 = \pm\sqrt{2}$$
$$x = 2 \pm \sqrt{2}$$

8. Since $2^{x-3} = 2^{2x}$, $x - 3 = 2x$. So $x = -3$.

9. Isolate a radical in one side and square.
$$\sqrt{x-5} = 5 - \sqrt{x}$$
$$x - 5 = 25 - 10\sqrt{x} + x$$
$$-30 = -10\sqrt{x}$$
$$3 = \sqrt{x}$$
$$x = 9$$

10. By definition of a logarithm, $x = log_2(3)$.

11.
$$log(4x - 12) = log(x)$$
$$4x - 12 = x$$
$$3x = 12$$
$$x = 4$$

12. Use synthetic division with $c = -1$.

$$\begin{array}{r|rrrr} -1 & 1 & -4 & 1 & 6 \\ & & -1 & 5 & -6 \\ \hline & 1 & -5 & 6 & 0 \end{array}$$

The quotient factors as $x^2 - 5x + 6 = (x-3)(x-2)$. The solutions are $x = -1, 2, 3$.

13. Parabola $y = x^2$ goes through $(0,0), (\pm 1, 1)$

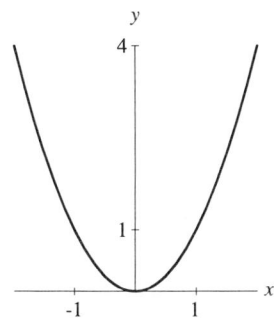

14. Parabola $y = (x-2)^2$ goes through $(0, 4), (1, 1), (3, 1)$

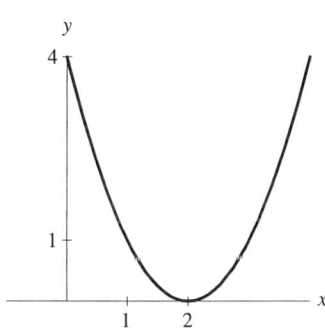

15. $y = 2^x$ goes through $\left(-1, \dfrac{1}{2}\right), (0, 1), (1, 2)$

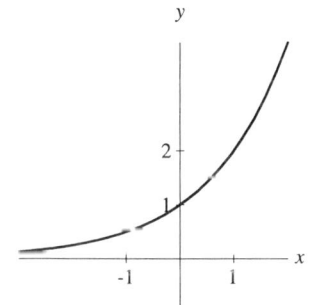

16. $y = x^{-2}$ goes through $(\pm 1, 1)$, $\left(\pm 2, \dfrac{1}{4}\right), \left(\pm \dfrac{1}{2}, 4\right)$

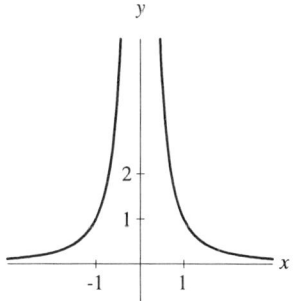

17. $y = \log_2(x-2)$ goes through $(3, 0), (4, 1), (2.5, -1)$

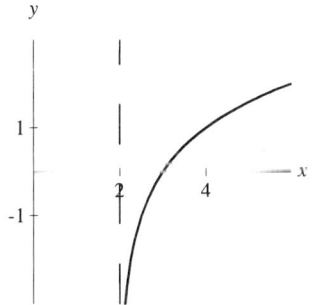

18. Line $y = x - 2$ goes through $(0, -2), (2, 0)$

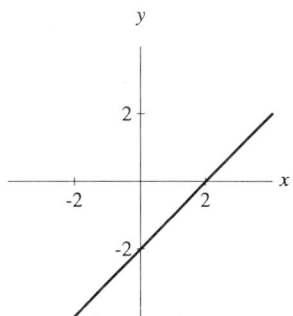

19. Line $y = 2x$ goes through $(0,0)$, $(1,2)$

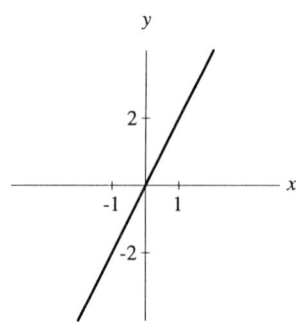

20. Line $y = x \cdot log(2)$ goes through $(0,0)$, $(1, log(2))$

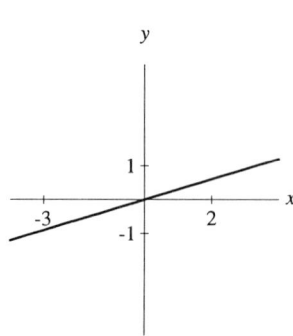

21. Horizontal line $y = e^2$ goes through $(0, e^2)$

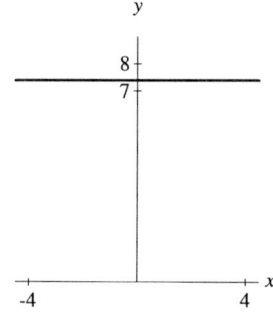

22. Parabola $y = 2 - x^2$ goes through $(0,2)$, $(\pm 1, 1)$

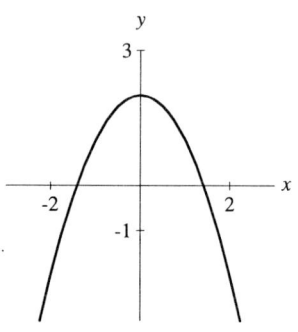

23. $y = \dfrac{2}{x}$ is symmetric about the origin and goes through $(1,2)$, $(2,1)$, $\left(\dfrac{1}{2}, 4\right)$, $\left(4, \dfrac{1}{2}\right)$

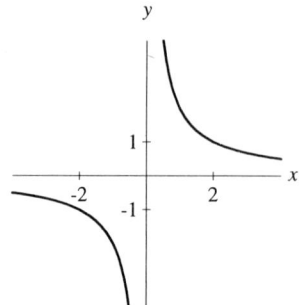

24. Graph of $y = \dfrac{1}{x-2}$ is obtained by shifting $y = \dfrac{1}{x}$ to the right by 2 units.

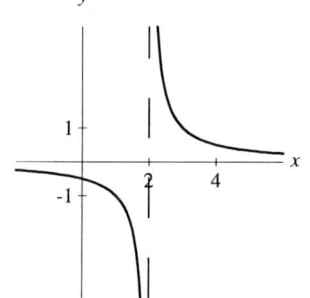

25. $f^{-1}(x) = 3x$

26. $f^{-1}(x) = -log_3(x)$ since $f(x) = 3^{-x}$

27. $f^{-1}(x) = x^2 + 2$ for $x \geq 0$

28. $f^{-1}(x) = \sqrt[3]{x-2} + 5$

29. $f^{-1}(x) = (10^x + 3)^2$

30. $\{(1,3), (4,5)\}$

31. $f^{-1}(x) = 5 + \dfrac{1}{x-3}$

32. $f^{-1}(x) = (ln(3-x))^2$, $x \leq 2$

33. $(p \circ m)(x) = e^{x+5}$, domain $(-\infty, \infty)$, range $(0, \infty)$

34. $(p \circ q)(x) = e^{\sqrt{x}}$, domain $[0, \infty)$, range $[1, \infty)$

35. $(q \circ p \circ m)(x) = q\left(e^{x+5}\right) = \sqrt{e^{x+5}}$, domain $(-\infty, \infty)$, range $(0, \infty)$

36. $(m \circ r \circ q)(x) = m\left(ln(\sqrt{x})\right) = ln(\sqrt{x}) + 5$, domain $(0, \infty)$, range $(-\infty, \infty)$

Tying It All Together

37. $(p \circ r \circ m)(x) = p(\ln(x+5)) = e^{\ln(x+5)}$,
    domain $(-5, \infty)$, range $(0, \infty)$

38. $(r \circ q \circ p)(x) = r\left(\sqrt{e^x}\right) = \ln\left(\sqrt{e^x}\right) = \dfrac{1}{2}x$,
    domain $(-\infty, \infty)$, range $(-\infty, \infty)$

39. $F(x) = (f \circ g \circ h)(x)$

40. $H(x) = (g \circ h \circ f)(x)$

41. $G(x) = (h \circ g \circ f)(x)$

42. $M(x) = (h \circ f \circ g)(x)$

For Thought

1. True

2. False, since $(2,3)$ does not satisfy $x - y = 1$.

3. False, it is independent since the two lines are perpendicular and intersect at only one point.

4. True, multiplying $x - 2y = 4$ by -3 and adding to the second equation,

$$-3x + 6y = -12$$
$$3x - 6y = 8$$
$$\overline{}$$
$$0 = -4$$

A contradiction and so there is no solution.

5. True   6. True

7. False, it is dependent because substituting $x = 5 + 3y$ results in,

$$9y - 3(5 + 3y) = -15$$
$$9y - 15 - 9y = -15$$
$$-15 = -15$$

8. False, it is dependent and the solution set is $\{(5 + 3y, y) | y \text{ is any real number}\}$.

9. False, it is dependent, there are an infinite number of solutions.

10. True, since both lines have slope $1/2$.

5.1 Exercises

1. $\{(1,2)\}$

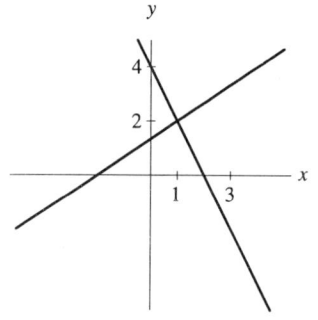

3. No solution since the lines are parallel.

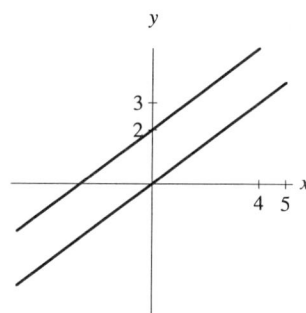

5. $\{(3,2)\}$

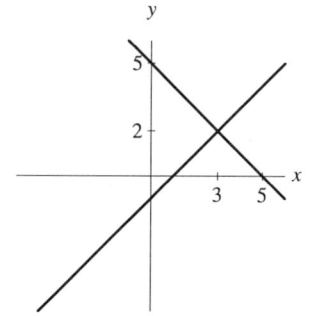

7. $\{(3,1)\}$

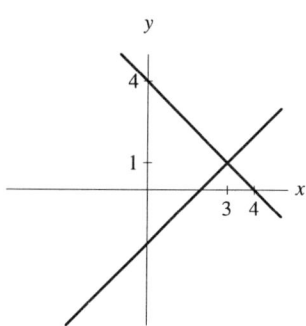

9. No solution since lines are parallel.

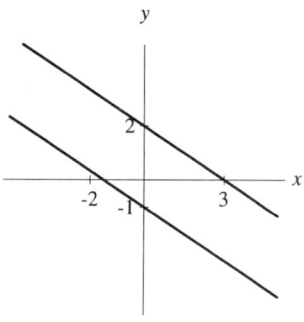

11. Since lines are identical, the solution set is $\{(x,y): x - 2y = 6\}$.

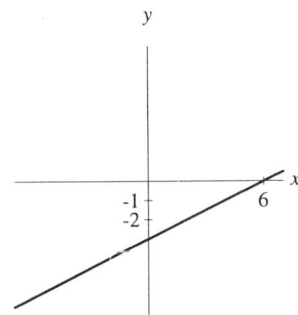

13. Substitute $y = 2x + 1$ into $3x - 4y = 1$. So
$$3x - 4(2x+1) = 1$$
$$3x - 8x - 4 = 1$$
$$-5x = 5$$
$$x = -1$$
From $y = 2x + 1$, $y = 2(-1) + 1 = -1$.
Independent and solution set is $\{(-1,-1)\}$.

15. Substitute $y = 1 - x$ into $2x - 3y = 8$. So
$$2x - 3(1-x) = 8$$
$$2x - 3 + 3x = 8$$
$$5x = 11$$
$$x = 11/5$$
From $y = 1 - x$, $y = 1 - 11/5 = -6/5$.
Independent and solution set is $\{(11/5, -6/5)\}$.

17. Substitute $y = 3x + 5$ into $3(x+1) = y - 2$. So
$$3x + 3 = (3x+5) - 2$$
$$3x + 3 = 3x + 3$$
$$3 = 3$$
Dependent and solution set is $\{(x,y) | y = 3x + 5\}$.

19. Multiplying $\frac{1}{2}x + \frac{1}{3}y = 3$ by 6,
$3x + 2y = 18$. Then substitute $2y = 6 - 3x$ into $3x + 2y = 18$. So
$$3x + (6 - 3x) = 18$$
$$6 = 18$$
Inconsistent and no solution.

21. Multiplying $0.05x + 0.06y = 10.50$ by 100. $5x + 6y = 1050$. Then substitute $y = 200 - x$ into $5x + 6y = 1050$. So
$$5x + 6(200 - x) = 1050$$
$$5x + 1200 - 6x = 1050$$
$$-x = -150$$
$$x = 150$$
From $y = 200 - x$, $y = 200 - 150 = 50$.
Independent and solution set is $\{(150, 50)\}$.

23. Since $3x + 1 = 3x - 7$, $1 = -7$.
Inconsistent and no solution.

25. Multiplying the first and second equations by 6 and 4, respectively, we have $3x - 2y = 72$ and $x - 2y = 4$.
Substitute $2y = x - 4$ into $3x - 2y = 72$.
$$3x - (x - 4) = 72$$
$$2x + 4 = 72$$
$$2x = 68$$
$$x = 34$$
From $2y = x - 4$, $y = \frac{34 - 4}{2} = 15$.
Independent and solution set is $\{(34, 15)\}$.

27. Adding the two equations, $2x = 26$.
So $x = 13$ and from $x + y = 20$, $13 + y = 20$ or $y = 7$.
Independent and solution set is $\{(13, 7)\}$.

29. Multiplying $x - y = 5$ by 2 and adding to the second equation,
$$2x - 2y = 10$$
$$3x + 2y = 10$$
$$\overline{}$$
$$5x = 20$$
$$x = 4$$
From $x - y = 5$, $4 - y = 5$ or $y = -1$.
Independent and solution set is $\{(4, -1)\}$.

31. Adding the two equations, $0 = 12$.
Inconsistent and no solution.

33. Multiply $2x + 3y = 1$ by -3 and $3x - 5y = -8$ by 2. Then add the equations.

$$-6x - 9y = -3$$
$$6x - 10y = -16$$
$$\overline{}$$
$$-19y = -19$$
$$y = 1$$

From $2x + 3y = 1$, $2x + 3 = 1$ or $x = -1$. Independent and solution set is $\{(-1, 1)\}$.

35. Multiply $0.05x + 0.1y = 0.6$ by 100 and $x + 2y = 12$ by 5. Then add the equations.

$$-5x - 10y = -60$$
$$5x + 10y = 60$$
$$\overline{}$$
$$0 = 0$$

Dependent and solution set is $\{(x, y) | x + 2y = 12\}$.

37. Multiplying $\frac{x}{2} + \frac{y}{2} = 5$ by 2 and $\frac{3x}{2} - \frac{2y}{3} = 2$ by 6, we have $x + y = 10$ and $9x - 4y = 12$, respectively. Then multiply $x + y = 10$ by 4 and add to the second equation. So

$$4x + 4y = 40$$
$$9x - 4y = 12$$
$$\overline{}$$
$$13x = 52$$
$$x = 4$$

From $x + y = 10$, $4 + y = 10$ and $y = 6$. Independent and solution set is $\{(4, 6)\}$.

39. Multiply $3x - 2.5y = -4.2$ by -4 and $0.12x + 0.09y = 0.4932$ by 100. Then add the equations. So

$$-12x + 10y = 16.8$$
$$12x + 9y = 49.32$$
$$\overline{}$$
$$19y = 66.12$$
$$y = 3.48$$

From $3x - 2.5y = -4.2$,

$$3x - 2.5(3.48) = -4.2$$
$$3x - 8.7 = -4.2$$
$$3x = 4.5$$
$$x = 1.5$$

Independent and solution set is $\{(1.5, 3.48)\}$.

41. Independent 43. Dependent

45. Point of intersection is $(-1000, -497)$

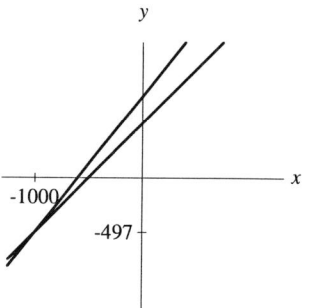

47. Point of intersection is $(6.18, -0.54)$

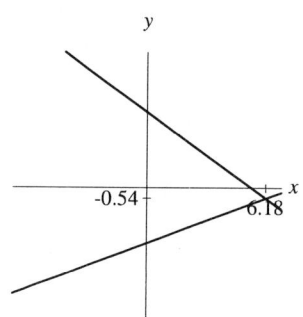

49. Let x and y be the amounts invested at 10% and 8%, respectively. So

$$x + y = 25,000$$
$$0.1x + 0.08y = 2,200$$

Multiply the second equation by -10 and add to the first.

$$x + y = 25,000$$
$$-x - 0.8y = -22,000$$
$$\overline{}$$
$$0.2y = 3,000$$
$$y = 15,000$$

Carmen invested $15,000 at 8% and $10,000 at 10%.

5.1 SYSTEMS OF LINEAR EQUATIONS IN TWO VARIABLES

51. Let x and y be the prices of an adult ticket and child ticket, respectively. So
$$2x + 5y = 33$$
$$x + 3y = 18.50$$
Multiply the second equation by -2 and add to the first equation.
$$2x + 5y = 33$$
$$-2x - 6y = -37$$
$$\overline{}$$
$$-y = -4$$
$$y = 4$$
A child ticket costs $\$4$. From $x + 3y = 18.50$, $x + 12 = 18.50$ and so an adult ticket costs $x = \$6.50$.

53. If m and f are the number of male and female memberships, respectively, then a system of equations is
$$m + f = 12$$
$$500m + 500f = 6,000.$$
Dividing the second equation by 500, one finds $m + f = 12$; i.e. the system of equations is redundant. Therefore, one cannot conclude the number of female memberships and male memberships. All we know is that there are 12 memberships, in total.

55. If c and m are the prices of a coffee and a muffin, respectively, then a system of equations is
$$3c + 7m = 7.77$$
$$6c + 14m = 14.80.$$
But if one multiplies the second equation by two, then one obtains $6c + 14m = 15.54$; this last equation contradicts the second equation in the system. Therefore, the system of equations is inconsistent.

57. Let x and y be the number of male and female students, respectively. So
$$0.5x + 0.3y = 230$$
$$0.2x + 0.6y = 260$$
Multiply first equation by -2 and the second by 5. Then add resulting equations.
$$-x - 0.6y = -460$$
$$x + 3y = 1300$$
$$\overline{}$$
$$2.4y = 840$$
$$y = 350$$

From $0.2x + 0.6y = 260$, $0.2x + 210 = 260$ and so $x = 250$. There are $250 + 350 = 600$ students at CHS.

59. Let x and y be the number of nickels and pennies, respectively. So
$$x + y = 87$$
$$0.05x + 0.01y = 1.75$$
Multiply second equation by -100 and then add it to the second equation.
$$x + y = 87$$
$$-5x - y = -175$$
$$\overline{}$$
$$-4x = -88$$
$$x = 22$$
From $x + y = 87$, $22 + y = 87$ and $y = 65$. Isabelle has 22 nickels and 65 pennies.

61. The weights x and y must satisfy
$$5x = 3y$$
$$3(4 + x + y) = 6y.$$
The second equation can be written as $12 + 3x = 3y$. Substituting the first equation one finds
$$12 + 3x = 5x$$
$$12 = 2x$$
$$6 = x.$$
Then $x = 6$ oz. and $y = 10$ oz. since
$$y = \frac{5x}{3} = \frac{5(6)}{3}.$$

63. Let x be the number of months. Plan A costs $150x + 800$ and Plan B costs $200x + 200$. Plan B costs more in the long run. The number of months for which cost is the same is,
$$150x + 800 = 200x + 200$$
$$600 = 50x$$
$$\overline{}$$
$$x = 12 \text{ months.}$$

65. Setting the formulas equal to each other, one finds

$$.08aD = \frac{D(a+1)}{24}$$
$$.08a = \frac{(a+1)}{24}$$
$$1.92a = a+1$$
$$a = \frac{1}{.92}$$
$$a \approx 1.09$$

The dosage is the same if the child's age is 1 year.

67. Solve for a and b.
$$-3a + b = 9$$
$$2a + b = -1.$$

Multiply second equation by -1 and add to the first equation.

$$-3a + b = 9$$
$$-2a - b = 1$$
$$\overline{}$$
$$-5a = 10$$
$$a = -2$$

Substituting $a = -2$ into $2a + b = -1$, one finds $b = 3$. An equation of the line is $y = -2x + 3$.

69. Solve for a and b.
$$-2a + b = 3$$
$$4a + b = -7$$

Multiply first equation by -1 and add to the second equation.

$$2a - b = -3$$
$$4a + b = -7$$
$$\overline{}$$
$$6a = -10$$
$$a = -\frac{5}{3}$$

Substituting $a = -\frac{5}{3}$ into $-2a + b = 3$, one gets $b = -\frac{1}{3}$. An equation of the line is $y = -\frac{5}{3}x - \frac{1}{3}$.

For Thought

1. True

2. False, since $(1, 1, 0)$ does not satisfy $-x - y + z = 4$.

3. True, adding the first two equations gives $0 = 6$ which is false.

4. False, since $(2, 3, -1)$ does not satisfy $x - y - z = 8$.

5. True

6. True, adding the first two equations gives an identity. Also, multiplying the first by -2 and then adding to the third equation gives an identity.

$$\begin{array}{cc} x - y + z = 1 & -2x + 2y - 2z = -2 \\ -x + y - z = -1 & 2x - 2y + 2z = 2 \end{array}$$
$$\overline{} \qquad \overline{}$$
$$\quad 0 = 0 \qquad\qquad\qquad 0 = 0$$

7. True, there are an infinite number of solutions as seen in number 6.

8. True

9. True, if $x = 1$ then $(x + 2, x, x - 1) = (3, 1, 0)$.

10. False, the value is $0.05x + 0.10y + 0.25z$ dollars.

5.2 Exercises

1. Points on the plane are $(5, 0, 0)$, $(0, 5, 0)$, and $(0, 0, 5)$.

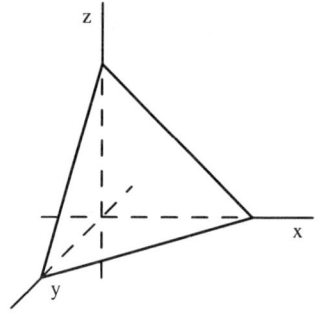

3. Points on the plane are $(3,0,0)$, $(0,3,0)$, and $(0,0,-3)$.

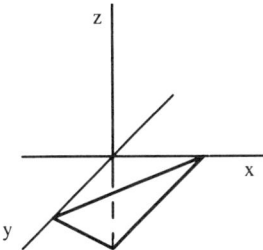

5. Add first and second equations and add last two equations.

$$x + y + z = 6 \qquad x + y + z = 6$$
$$2x - 2y - z = -5 \qquad 3x + y - z = 2$$
$$\overline{} \qquad \overline{}$$
$$3x - y = 1 \qquad 4x + 2y = 8$$

Divide $4x + 2y = 8$ by 2 and add to $3x - y = 1$.

$$2x + y = 4$$
$$3x - y = 1$$
$$\overline{}$$
$$5x = 5$$

So $x = 1$. From $3x - y = 1$, $3 - y = 1$ and $y = 2$. From $x + y + z = 6$, $1 + 2 + z = 6$ and $z = 3$. Solution set is $\{(1,2,3)\}$.

7. Multiply first equation by 2 and add to the second; Then multiply first equation by -3 and add to the third equation.

$$6x + 4y + 2z = 2 \qquad -9x - 6y - 3z = -3$$
$$x + y - 2z = -4 \qquad 2x - 3y + 3z = 1$$
$$\overline{} \qquad \overline{}$$
$$7x + 5y = -2 \qquad -7x - 9y = 2$$

Add $7x + 5y = -2$ to $-7x - 9y = -2$.

$$7x + 5y = -2$$
$$-7x - 9y = -2$$
$$\overline{}$$
$$-4y = -4$$

So $y = 1$. From $7x + 5y = -2$, $7x + 5 = -2$ and $x = -1$. From $3x + 2y + z = 1$, $-3 + 2 + z = 1$ and $z = 2$. Solution set is $\{(-1, 1, 2)\}$.

9. Add first equation to third equation. Multiply second equation by 2 and add to the first.

$$2x + y - 2z = -15 \qquad 8x - 4y + 2z = 30$$
$$x + 3y + 2z = -5 \qquad 2x + y - 2z = -15$$
$$\overline{} \qquad \overline{}$$
$$3x + 4y = -20 \qquad 10x - 3y = 15$$

Multiply $3x + 4y = -20$ by 3 and $10x - 3y = 15$ by 4 then add the equations.

$$9x + 12y = -60$$
$$40x - 12y = 60$$
$$\overline{}$$
$$49x = 0$$

So $x = 0$. From $3x + 4y = -20$, $4y = -20$ and $y = -5$. From $2x + y - 2z = -15$, $-5 - 2z = -15$ and $z = 5$. Solution set is $\{(0, -5, 5)\}$.

11. Multiply first equation by -2 and add to the second equation.

$$-2x + 4y - 6z = -10$$
$$2x - 4y + 6z = 3$$
$$\overline{}$$
$$0 = -7$$

This is false, so there is no solution.

13. Adding the two equations, we have $4x - 4z = -20$ or $z = x + 5$. From $x + 2y - 3z = -17$,

$$x + 2y - 3(x + 5) = -17$$
$$2y - 2x - 15 = -17$$
$$2y = 2x - 2$$
$$y = x - 1$$

The solution set is $\{(x, x - 1, x + 5) | x \text{ is any real number}\}$.

15. Adding the first two equations, we have $0 = 0$, which is an identity. Multiply second equation by 2 and add to the third equation.

$$-2x - 4y + 6z = -10$$
$$2x + 4y - 6z = 10$$
$$\overline{}$$
$$0 = 0$$

The system is dependent and solution set is $\{(x, y, z) | x + 2y - 3z = 5\}$.

17. Adding the two equations, $3x = 6$ or $x = 2$.
Substitute into the two equations.

$$2 + y - z = 2 \qquad 4 - y + z = 4$$
$$y - z = 0 \qquad -y + z = 0$$

In any case $y = z$. Solution set is $\{(2, y, y) | y \text{ is any real number}\}$.

19. If the second equation is multiplied by -1 and added to the first equation, the result is the third equation.

$$x + y = 5$$
$$-y + z = -2$$
$$\overline{}$$
$$x + z = 3$$

There are an infinite number of solutions. Since $y = 5 - x$ and $z = 3 - x$, the solution set is $\{(x, 5 - x, 3 - x) | x \text{ is any real number}\}$.

21. Multiply first equation by 2 and add to the second equation.

$$2x - 2y + 2z = 14$$
$$2y - 3z = -13$$
$$\overline{}$$
$$2x - z = 1$$

Multiply $2x - z = 1$ by -2 and add to the third equation.

$$-4x + 2z = -2$$
$$3x - 2z = -3$$
$$\overline{}$$
$$-x = -5$$

So $x = 5$. From $2x - z = 1$, $10 - z = 1$ and $z = 9$. From $2y - 3z = -13$, $2y - 27 = -13$ and $y = 7$. Solution set is $\{(5, 7, 9)\}$.

23. Multiply first equation by -2 and add to third equation. Then multiply first equation by -4 and add to the second.

$$-2x - 2y - 4z = -15 \qquad -4x - 4y - 8z = -30$$
$$5x + 2y + 5z = 21 \qquad 3x + 4y + z = 12$$
$$\overline{} \qquad \overline{}$$
$$3x + z = 6 \qquad -x - 7z = -18$$

Multiply $-x - 7z = -18$ by 3 and add to $3x + z = 6$.

$$-3x - 21z = -54$$
$$3x + z = 6$$
$$\overline{}$$
$$-20z = -48$$

So $z = 48/20 = 2.4$. From $3x + z = 6$, $3x + 2.4 = 6$ and $x = 1.2$. From $x + y + 2z = 7.5$, $1.2 + y + 4.8 = 7.5$ and $y = 1.5$. Solution set is $\{(1.2, 1.5, 2.4)\}$.

25. Multiply first equation above by -5 and add to 100 times the second one.

$$-5x - 5y - 5z = -45,000$$
$$5x + 6y + 9z = 71,000$$
$$\overline{}$$
$$y + 4z = 26,000$$

Substitute $z = 3y$ into $y + 4z = 26,000$.

$$y + 12y = 26,000$$
$$13y = 26,000$$
$$y = 2,000$$

From $z = 3y$, $z = 6,000$.
From $x + y + z = 9,000$, we have $x + 2000 + 6000 = 9000$ and $x = 1000$.
Solution set is $\{(1000, 2000, 6000)\}$.

27. Substitute $z = 4y - 5$ into $y = 3z + 2$

$$y = 3(4y - 5) + 2$$
$$y = 12y - 13$$
$$-11y = -13$$
$$y = 13/11$$

From $z = 4y - 5$, $z = 52/11 - 5 = -3/11$.
From $x = 2y - 1$, $x = 26/11 - 1 = 15/11$.
Solution set is $\{(15/11, 13/11, -3/11)\}$.

29. Substitute $(-1, -2), (2, 1), (-2, 1)$ into $y = ax^2 + bx + c$. So

$$a - b + c = -2$$
$$4a + 2b + c = 1$$
$$4a - 2b + c = 1$$

Multiply first one by -1 and add to the second and third equations.

$$-a + b - c = 2 \qquad -a + b - c = 2$$
$$4a + 2b + c = 1 \qquad 4a - 2b + c = 1$$
$$\overline{} \qquad \overline{}$$
$$3a + 3b = 3 \qquad 3a - b = 3$$

Multiply $3a + 3b = 3$ by -1 and add to $3a - b = 3$.

$$-3a - 3b = -3$$
$$3a - b = 3$$
$$\overline{}$$
$$-4b = 0$$

So $b = 0$. From $3a - b = 3$, $3a = 3$ and $a = 1$. From $a - b + c = -2$, $1 + c = -2$ and $c = -3$. Since the solution is $(a, b, c) = (1, 0, -3)$, the parabola is $y = x^2 - 3$.

31. Substitute $(0,0)$, $(1,3)$, $(2,2)$ into $y = ax^2 + bx + c$. So

$$c = 0$$
$$a + b + c = 3$$
$$4a + 2b + c = 2$$

Multiply second one by -2 and add to third equation.

$$-2a - 2b - 2c = -6$$
$$4a + 2b + c = 2$$
$$\overline{}$$
$$2a - c = -4$$

Substituting $c = 0$ into $2a - c = -4$, we get $2a = -4$ and $a = -2$. From $a + b + c = 3$, $-2 + b = 3$ and $b = 5$. Solution is $(a, b, c) = (-2, 5, 0)$, the parabola is $y = -2x^2 + 5x$.

33. Substitute $(0,4)$, $(-2,0)$, $(-3,1)$ into $y = ax^2 + bx + c$. So

$$c = 4$$
$$4a - 2b + c = 0$$
$$9a - 3b + c = 1$$

Multiply second and third equations by 3 and -2, respectively, then add the equations.

$$12a - 6b + 3c = 0$$
$$-18a + 6b - 2c = -2$$
$$\overline{}$$
$$-6a + c = -2$$

Substituting $c = 4$ into $-6a + c = -2$, $-6a + 4 = -2$ and $a = 1$. From $4a - 2b + c = 0$, $4 - 2b + 4 = 0$ and $b = 4$. Since $(a, b, c) = (1, 4, 4)$, the parabola is $y = x^2 + 4x + 4$.

35. $x + y + z = 1$

37. $x + \dfrac{1}{2}y - \dfrac{1}{2}z = 1$

39. Substitute $(12, 0.18)$, $(22, 0.23)$, $(30, 0.14)$ into $E = as^2 + bs + c$. This gives the following system

$$144a + 12b + c = 0.18$$
$$484a + 22b + c = 0.23$$
$$900a + 30b + c = 0.14$$

Subtract the first equation from the second equation. Also subtract the second from the third. Then

$$340a + 10b = 0.05$$
$$416a + 8b = -0.09$$

Multiply $416a + 8b = -0.09$ by 5 and add to -4 times $340a + 10b = 0.05$.

$$-1360a - 40b = -0.20$$
$$2080a + 40b = -0.45$$
$$\overline{}$$
$$720a = -0.65$$
$$a = -0.65/720$$
$$a \approx -0.0009$$

Since $416a + 8b = -0.09$,

$$8b = -0.09 - 416a$$
$$b = \dfrac{-0.09 - 416(-0.65/720)}{8}$$
$$b \approx 0.035694$$
$$b \approx 0.0357$$

Since $144a + 12b + c = 0.18$,

$$c \approx 0.18 - 144(-0.65/720) - 12(0.035694)$$
$$c \approx -0.1183$$

So $E = -0.0009s^2 + 0.0357s - 0.1183$. The speed that maximizes efficiency is

$$-\dfrac{b}{2a} \approx -\dfrac{0.0357}{2(-0.0009)} \approx 19.8 \text{ mph}.$$

41. Let x, y, and z be the amounts invested in stocks, bonds, and a mutual fund. Then

$$x + y + z = 25{,}000$$
$$0.08x + 0.10y + 0.06z = 1{,}860$$
$$2y = z$$

Multiply the first equation by -8 and add to 100 times the second.

$$-8x - 8y - 8z = -200,000$$
$$8x + 10y + 6z = 186,000$$
$$\overline{}$$
$$2y - 2z = -14,000$$

Substitute $z = 2y$ into $2y - 2z = -14,000$.
$$2y - 4y = -14,000$$
$$-2y = -14,000$$
$$y = 7,000$$

Since $z = 2y$, $z = 14,000$.
Since $x + y + z = 25,000$, $x = 4,000$.
Marita invested \$4,000 in stocks,
\$7,000 in bonds, and
\$14,000 in a mutual fund.

43. Let x, y, and z be the prices last year of a hamburger, fries, and a coke, respectively. So
$$x + y + z = 1.90$$
$$1.1x + 1.2y + 1.5z = 2.37$$
$$1.5z = 1.1x + 0.09$$

Multiply first equation by -12 and add to 100 times the second equation.
$$-12x - 12y - 12z = -22.8$$
$$11x + 12y + 15z = 23.7$$
$$\overline{}$$
$$-x + 3z = 0.9$$

Substitute $x = 3z - 0.9$ into $1.5z = 1.1x + 0.09$.
$$1.5z = 1.1(3z - 0.9) + 0.09$$
$$1.5z = 3.3z - 0.99 + 0.99$$
$$0.90 = 1.8z$$
$$0.50 = z$$

Since $x = 3z - 0.9$, $x = 1.50 - 0.90 = 0.60$.
From $x + y + z = 1.90$, we have $y = 0.80$.
The prices last year of a hamburger, fries and a coke are \$0.60, \$0.80, and \$0.50, respectively.

45. Let L_f and L_r be the weights on the left front tire and left rear tire, respectively. Let R_f and R_r be the weights on the right front tire and right rear tire, respectively. Since $1200(.51) = 612$ and $1200(.48) = 576$,
$$L_f + L_r = 612$$
$$R_r + L_r = 576$$
$$L_f, L_r, R_f, R_r \geq 280.$$

Three weight distributions are
$(L_f, L_r, R_f, R_r) = (332, 280, 292, 296)$,
$(L_f, L_r, R_f, R_r) = (322, 290, 302, 286)$, and
$(L_f, L_r, R_f, R_r) = (324, 288, 300, 288)$.

47. Let x, y, and z be the number of pennies, nickels, and dimes, respectively. So
$$x + y + z = 232$$
$$y + z = x$$
$$0.01x + 0.05y + 0.10z = 10.36$$

Multiply first equation by -1 and add to 10 times the third equation. Also combine first two equations.
$$-x - y - z = -232$$
$$0.1x + 0.5y + z = 103.6$$
$$\overline{}$$
$$-0.9x - 0.5y = -128.4$$

$$-x - y - z = -232$$
$$-x + y + z = 0$$
$$\overline{}$$
$$-2x = -232$$

Then $x = 116$. Substituting into $-0.9x - 0.5y = -128.4$,
$$-0.9(116) - 0.5y = -128.4$$
$$-104.4 - 0.5y = -128.4$$
$$24 = 0.5y$$
$$48 = y$$

From $x + y + z = 232$, $116 + 48 + z = 232$ and $z = 68$. Emma used 116 pennies, 48 nickels, and 68 dimes.

49. Let x, y, and z be the prices of a carton of milk, a cup of coffee, and a doughnut, respectively. So
$$3x + 4y + 7z = 5.45$$
$$4x + 2y + 8z = 5.30$$
$$2x + 5y + 6z = 5.15$$

Multiply third equation by -2 and add to second equation. Also, multiply first equation by -4 and add to 3 times the second inequation.

$$-4x - 10y - 12z = -10.30$$
$$4x + 2y + 8z = 5.30$$
$$\overline{}$$
$$-8y - 4z = -5$$

$$-12x - 16y - 28z = -21.80$$
$$12x + 6y + 24z = 15.90$$
$$\overline{}$$
$$-10y - 4z = -5.90$$

Multiply $-8y - 4z = -5$ by -1 and add to $-10y - 4z = -5.90$.

$$8y + 4z = 5$$
$$-10y - 4z = -5.90$$
$$\overline{}$$
$$-2y = -0.90$$

So $y = 0.45$. Substituting into $-8y - 4z = -5$, $-3.60 - 4z = -5$ and $z = 0.35$. Since $3x + 4y + 7z = 5.45$, we have $3x + 1.80 + 2.45 = 5.45$ and $x = 0.40$.

Alphonse's bill was $5(0.40) + 2(0.45) + 9(0.35) = \$6.05$. His change is $\$3.95$.

51. A system of equations is
$$4x = 6y$$
$$2(x + y) = 15(8)$$
$$6(x + y + 15 + 10) = 10z.$$

Rewriting the first two equations, one obtains
$$2x - 3y = 0$$
$$x + y = 60$$

Solving this smaller system, one finds $x = 36$ lb., $y = 24$ lb. Substituting into the third equation, one finds
$$z = \frac{6(x + y + 25)}{10} = \frac{6(60 + 25)}{10} = 51 \text{ lb.}$$

For Thought

1. True

2. False, when a line is tangent to a circle it intersects the circle at only one point.

3. True

4. False, they intersect at $(\pm 1, 0)$.

5. False, the intersections are at $(1, 1)$ and $(-1, -1)$.

6. False **7.** True

8. True **9.** True

10. False, two such numbers are $\dfrac{1}{2}\left(7 \pm \sqrt{45}\right)$.

5.3 Exercises

1. Substituting $y = x^2$ into $5x - y = 6$,
$$5x - x^2 = 6$$
$$x^2 - 5x = -6$$
$$x^2 - 5x + 6 = 0$$
$$(x - 3)(x - 2) = 0$$

If $x = 3, 2$ in $y = x^2$, then $y = 9, 4$. The solution set is $\{(2, 4), (3, 9)\}$.

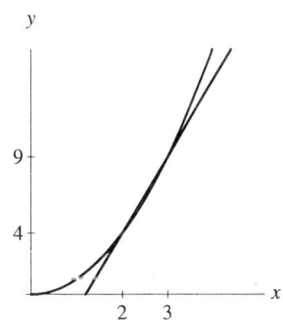

3. Substituting $y = \dfrac{x+1}{2}$ into $y = |x| - 1$,
$$\dfrac{x+1}{2} = |x| - 1$$
$$\dfrac{x}{2} + \dfrac{3}{2} = |x|$$
$$\dfrac{x}{2} + \dfrac{3}{2} = x \text{ or } \dfrac{x}{2} + \dfrac{3}{2} = -x$$
$$\dfrac{3}{2} = \dfrac{x}{2} \text{ or } \dfrac{3x}{2} = -\dfrac{3}{2}$$
$$x = 3 \text{ or } x = -1$$

Using $x = 3, -1$ in $y = \dfrac{x+1}{2}$, one finds $y = 2, 0$.
Solution set is $\{(3,2), (-1,0)\}$.

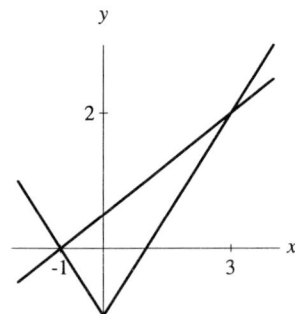

5. Equating $y = \sqrt{x}$ to $y = 2x$,
$$\sqrt{x} = 2x$$
$$x = 4x^2$$
$$x - 4x^2 = 0$$
$$x(1 - 4x) = 0$$

Using $x = 0, 1/4$ in $y = 2x$, $y = 0, 1/2$.
Solution set is $\{(0,0), (1/4, 1/2)\}$.

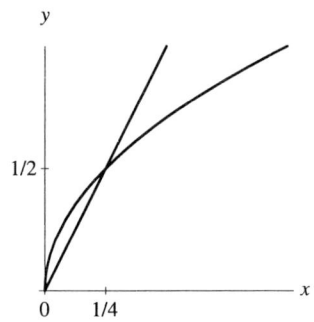

7. Equating $y = x^3$ to $y = 4x$,
$$x^3 = 4x$$
$$x^3 - 4x = 0$$
$$x(x^2 - 4) = 0$$
$$x = 0 \pm 2$$

Using $x = 0, 2, -2$ in $y = 4x$, we get $y = 0, 8, -8$.
Solution set is $\{(0,0), (2,8), (-2,-8)\}$.

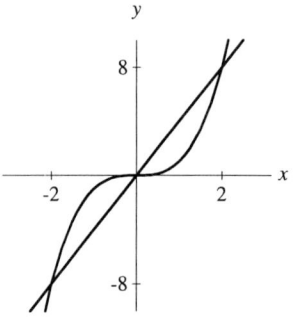

9. Substituting $x^2 = 2y$ into $x^2 + (y-2)^2 = 4$,
$$2y + (y-2)^2 = 4$$
$$2y + (y^2 - 4y + 4) = 4$$
$$y^2 - 2y = 0$$
$$y(y - 2) = 0$$

Using $y = 0, 2$ in $x^2 = 2y$, we have $x^2 = 0, 4$.
Solution set is $\{(0,0), (2,2), (-2,2)\}$.

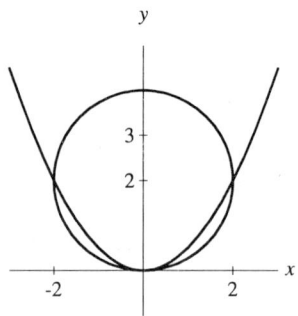

11. Equate $y = \log_2(x)$ to $y = \log_{1/3}(x)$ and use the base-changing formula.
$$\log_2(x) = \log_{1/3}(x)$$
$$\log_2(x) = \frac{\log_2(x)}{\log_2(1/3)}$$
$$\log_2(x) \cdot \log_2(1/3) = \log_2(x)$$
$$\log_2(x)\left[\log_2(1/3) - 1\right] = 0$$
$$\log_2(x) = 0$$
$$x = 1$$

Using $x = 1$ in $y = \log_2(x)$, $y = 0$.
Solution set is $\{(1, 0)\}$.

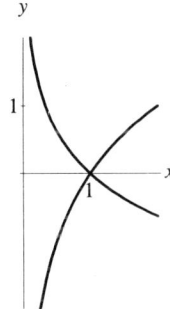

13. Substituting $y = x^2$ into $y = x^4 - x^2$,
$$x^2 = x^4 - x^2$$
$$2x^2 - x^4 = 0$$
$$x^2(2 - x^2) = 0$$
$$x = 0, \sqrt{2}, -\sqrt{2}$$

Using $x = 0, \sqrt{2}, -\sqrt{2}$ in $y = x^2$, we obtain $y = 0, 2, 2$.
Solution set is $\{(0,0), (\sqrt{2}, 2), (-\sqrt{2}, 2)\}$.

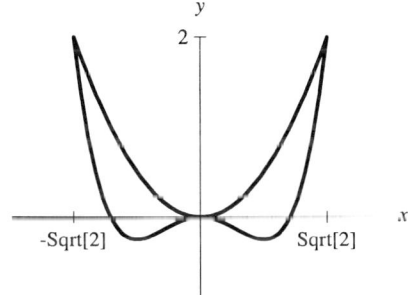

15. Substituting $y = -x - 4$ into $xy = 1$,
$$x(-x - 4) = 1$$
$$-x^2 - 4x = 1$$
$$x^2 + 4x = -1$$
$$x^2 + 4x + 4 = 3$$
$$(x + 2)^2 = 3$$
$$x = -2 \pm \sqrt{3}$$

Using $x = -2 + \sqrt{3}, -2 - \sqrt{3}$ in $y = -x - 4$, we find $y = -2 - \sqrt{3}, -2 + \sqrt{3}$.
Solution set is
$\{(-2 + \sqrt{3}, -2 - \sqrt{3}), (-2 - \sqrt{3}, -2 + \sqrt{3})\}$.

17. Substituting $x^2 = 2y^2 - 1$ into $2x^2 - y^2 = 1$,
$$2(2y^2 - 1) - y^2 = 1$$
$$3y^2 - 2 = 1$$
$$3y^2 = 3$$
$$y^2 = 1$$
$$y = \pm 1$$

Using $y = 1$ in $x^2 = 2y^2 - 1$, we get $x^2 = 1$ or $x = \pm 1$. Also, if $y = -1$ then $x = \pm 1$. Solution set is $\{(1,1), (1,-1), (-1,1), (-1,-1)\}$.

19. Multiply first equation by 2 and add to the second.
$$\frac{6}{x} - \frac{2}{y} = \frac{26}{10}$$
$$\frac{1}{x} + \frac{2}{y} = \frac{9}{10}$$
$$\overline{}$$
$$\frac{7}{x} = \frac{35}{10}$$

So $70 = 35x$ and $x = 2$.
Substituting into $\frac{3}{x} - \frac{1}{y} = \frac{13}{10}$,
$$\frac{3}{2} - \frac{1}{y} = \frac{13}{10}$$
$$-\frac{1}{y} = \frac{13}{10} - \frac{15}{10}$$
$$-\frac{1}{y} = -\frac{2}{10}$$
$$y = 5$$

Solution set is $\{(2, 5)\}$.

21. Substitute $y = 1 - x$ into $x^2 + xy - y^2 = -5$.
$$x^2 + x(1-x) - (1-x)^2 = -5$$
$$x^2 + x - x^2 - (1 - 2x + x^2) = -5$$
$$-x^2 + 3x - 1 = -5$$
$$x^2 - 3x + 1 = 5$$
$$x^2 - 3x - 4 = 0$$
$$(x-4)(x+1) = 0$$
$$x = 4, -1$$
Using $x = 4, -1$ in $y = 1 - x$, $y = -3, 2$.
Solution set is $\{(4, -3), (-1, 2)\}$.

23. Add the two given equations to obtain $xy = -2$.
Substitute $y = -2/x$ into $x^2 + 2xy - 2y^2 = -11$.
$$x^2 + 2x\left(-\frac{2}{x}\right) - 2 \cdot \frac{4}{x^2} = -11$$
$$x^2 - 4 - \frac{8}{x^2} = -11$$
$$x^4 + 7x^2 - 8 = 0$$
$$(x^2 + 8)(x^2 - 1) = 0$$
$$x = \pm 1$$
Using $x = 1, -1$ in $y = -2/x$, $y = -2, 2$.
Solution set is $\{(1, -2), (-1, 2)\}$.

25. Equate $y = 2^{x+1}$ to $y = 4^{-x}$.
$$2^{x+1} = (2^2)^{-x}$$
$$2^{x+1} = 2^{-2x}$$
$$x + 1 = -2x$$
$$3x = -1$$
Using $x = -1/3$ in $y = 2^{x+1}$, we get
$y = 2^{2/3}$. Solution set is $\left\{\left(-\frac{1}{3}, 2^{2/3}\right)\right\}$.

27. Equate $y = log_2(x)$ to $y = log_4(x+2)$ and use the base-changing formula.
$$log_2(x) = log_4(x+2)$$
$$log_2(x) = \frac{log_2(x+2)}{log_2(4)}$$
$$log_2(x) = \frac{log_2(x+2)}{2}$$
$$2 \cdot log_2(x) = log_2(x+2)$$
$$2 \cdot log_2(x) - log_2(x+2) = 0$$
$$log_2\left(\frac{x^2}{x+2}\right) = 0$$
$$\frac{x^2}{x+2} = 1$$
$$x^2 = x + 2$$
$$x^2 - x - 2 = 0$$
$$(x-2)(x+1) = 0$$
$$x = 2, -1$$
But $x = -1$ is an extraneous root since $log_2(-1)$ is undefined. Using $x = 2$ in $y = log_2(x)$, we have $y = 1$.
Solution set is $\{(2, 1)\}$.

29. Equate $y = log_2(x+2)$ to $y = 3 - log_2(x)$.
$$log_2(x+2) = 3 - log_2(x)$$
$$log_2(x+2) + log_2(x) = 3$$
$$log_2(x^2 + 2x) = 3$$
$$x^2 + 2x = 2^3$$
$$x^2 + 2x - 8 = 0$$
$$(x+4)(x-2) = 0$$
$$x = -4, 2$$
But $x = -4$ is an extraneous root since $log_2(-4)$ is undefined. Using $x = 2$ in $y = log_2(x+2)$, we get $y = log_2(4) = 2$.
Solution set is $\{(2, 2)\}$.

31. Equate $y = 3^x$ to $y = 2^x$.
$$3^x = 2^x$$
$$log(3^x) = log(2^x)$$
$$x \cdot log(3) = x \cdot log(2)$$
$$x[log(3) - log(2)] = 0$$
$$x = 0$$
Using $x = 0$ in $y = 3^x$,
$y = 3^0 = 1$. Solution set is $\{(0, 1)\}$.

33. From the graphs, the solution set is $\{(2,1),(0.3,-1.8)\}$.

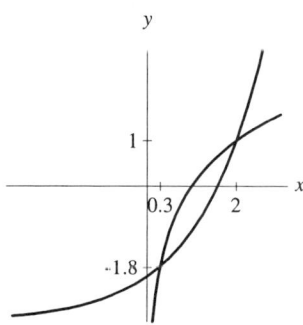

35. From the graphs, the solution set is $\{(1.9, 0.6),(0.1,-2.0)\}$.

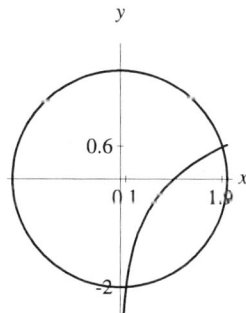

37. From the graphs, the solution set is $\{(2,2),(-.8,.6),(4,16)\}$.

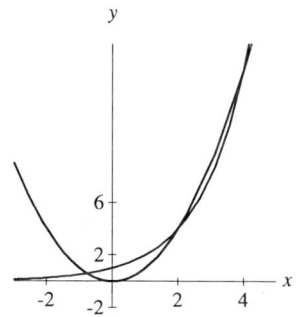

39. Let x and y be the lengths of the legs of the triangle. So
$$x^2 + y^2 = 15^2$$
$$\frac{1}{2}xy = 54$$
Substitute $y = \dfrac{108}{x}$ into $x^2 + y^2 = 225$ and use the quadratic formula.
$$x^2 + \frac{11664}{x^2} = 225$$
$$x^4 - 225x^2 + 11,664 = 0$$

$$x^2 = \frac{225 \pm \sqrt{225^2 - 4(11,664)}}{2}$$
$$x^2 = \frac{225 \pm 63}{2}$$
$$x^2 = 144, 81$$
$$x = 12, 9$$

Using $x = 12, 9$ in $y = \dfrac{108}{x}$, $y = 9, 12$.
The sides of the right triangle are 9m and 12m.

41. If x is the length of the hypotenuse then $\dfrac{x}{2}$ and $\dfrac{x\sqrt{3}}{2}$ are the lengths of the sides opposite the $30°$ and $60°$ angles. So
$$x + \frac{x}{2} + \frac{x\sqrt{3}}{2} = 12$$
$$2x + x + \sqrt{3}x = 24$$
$$(3 + \sqrt{3})x = 24$$
$$x = \frac{24}{3 + \sqrt{3}} \cdot \frac{3 - \sqrt{3}}{3 - \sqrt{3}}$$
$$x = 12 - 4\sqrt{3}$$

Substituting $x = 12 - 4\sqrt{3}$ in $\dfrac{x}{2}$ and $\dfrac{x\sqrt{3}}{2}$, then $6 - 2\sqrt{3}$ ft. and $6\sqrt{3} - 6$ft. are the lengths of the two sides and the hypotenuse is $12 - 4\sqrt{3}$ft.

43. The values of x and y must satisfy
$$6y = 6x$$
$$x(6 + y) = 7(4 + 12).$$
Since $x = y$ as seen from the first equation, upon substitution into the second equation one obtains
$$6x + x^2 = 112$$
$$x^2 + 6x - 112 = 0$$
$$(x + 14)(x - 8) = 0.$$
Then $x = 8$ in. and $y = 8$ oz. One must exclude the negative value $x = -14$.

45. Let x and y be the number of minutes it takes for pump A and pump B, respectively, to fill the vat. So
$$\frac{1}{x} + \frac{1}{y} = \frac{1}{8}$$
$$\frac{1}{x} - \frac{1}{y} = \frac{1}{12}$$

Adding the two equations,
$$\frac{2}{x} = \frac{3}{24} + \frac{2}{24}$$
$$\frac{2}{x} = \frac{5}{24}$$
$$5x = 48$$
$$x = 9.6$$

Substituting $x = \frac{48}{5}$ into $\frac{1}{x} + \frac{1}{y} = \frac{1}{8}$,
$$\frac{5}{48} + \frac{1}{y} = \frac{1}{8}$$
$$\frac{1}{y} = \frac{6}{48} - \frac{5}{48}$$
$$\frac{1}{y} = \frac{1}{48}$$
$$y = 48$$

Pump A can fill the vat by itself in $x = 9.6$ min. while Pump B will take $y = 48$ min.

47. Let x and y be two numbers.
$$x + y = 6$$
$$xy = 10$$

Substituting $y = 6 - x$ into $xy = 10$,
$$x(6 - x) = 10$$
$$6x - x^2 = 10$$
$$x^2 - 6x = -10$$
$$x^2 - 6x + 9 = -1$$
$$(x - 3)^2 = -1$$
$$x = 3 \pm i$$

If $x = 3 + i$ then $y = 6 - (3 + i) = 3 - i$.
If $x = 3 - i$ then $y = 6 - (3 - i) = 3 + i$.
The two numbers are $3 + i$ and $3 - i$.

49. Let l and w be the length and width.
$$20lw = 36,000$$
$$40l + 40w + 2lw = 7,200$$

Substitute $l = \frac{1800}{w}$ into
$$40l + 40w + 2lw = 7,200.$$

$$40 \cdot \frac{1800}{w} + 40w + 2 \cdot \frac{1800}{w} \cdot w = 7200$$
$$\frac{72000}{w} + 40w + 3600 = 7200$$
$$\frac{72000}{w} + 40w = 3600$$
$$40w^2 - 3600w + 72000 = 0$$
$$w^2 - 90w + 1800 = 0$$
$$(w - 60)(w - 30) = 0$$

Using $w = 30, 60$ in $l = \frac{1800}{w}$, $l = 60, 30$.
The length is 60 ft. and width is 30 ft.

51. From the graphs, the two models give the same population for $t = 0$, 9.66 years. The exponential population model is twice the linear model when $t \approx 29.5$ years.

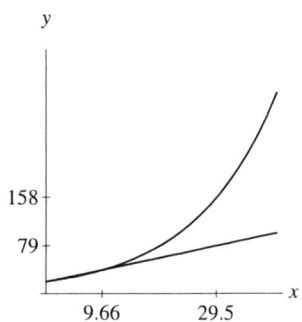

53. Let x be the time of the sunrise or the number of hours since midnight. Let S and B be Sally's and Bob's speed. The number of miles Sally and Bob drove are $(16 - x)S$ and $(21 - x)B$, respectively. Since they met at noon, the distance between Sally's house and Bob's house is $(12 - x)S + (12 - x)B$; or equivalently $(12 - x)(S + B)$. Then
$$(16 - x)S = (12 - x)(S + B)$$
$$(16 - x)S = (16 - x)(S + B) - 4(S + B)$$
$$0 = (16 - x)B - 4(S + B)$$
$$(16 - x)B = 4(S + B).$$
Likewise,
$$(21 - x)B = (21 - x)(S + B) - 9(S + B)$$
$$(21 - x)S = 9(S + B).$$
Combining, one obtains
$$\frac{(16 - x)B}{(21 - x)S} = \frac{4(S + B)}{9(S + B)}. \text{ Or, } \frac{(16 - x)9B}{(21 - x)4S} = 1.$$
Furthermore, since $(16 - x)S = (21 - x)B$

one finds $\frac{B}{S} = \frac{16-x}{21-x}$. Then

$$\frac{9(16-x)}{4(21-x)}\frac{B}{S} = 1$$

$$\frac{9}{4}\left(\frac{16-x}{21-x}\right)^2 = 1$$

$$\frac{16-x}{21-x} = \frac{2}{3} \quad \text{since } 16-x > 0$$

$$48 - 3x = 42 - 2x$$

$$x = 6.$$

The sunrise was at 6:00 a.m.

For Thought

1. True, $\frac{1}{x} + \frac{3}{x+1} = \frac{(x+1)+3x}{x(x+1)} = \frac{4x+1}{x(x+1)}$.

2. True, $x + \frac{3x}{x^2-1} = \frac{x(x^2-1)+3x}{x^2-1} = \frac{x^3+2x}{x^2-1}$.

3. False, by long division,
$$\frac{x^2}{x^2-9} = 1 + \frac{9}{x^2-9} = 1 + \frac{A}{x-3} + \frac{B}{x+3}.$$

4. True, $\frac{1}{2} + \frac{1}{2^3} = \frac{2^2+1}{2^3} = \frac{5}{8}$.

5. False, $\frac{3x-1}{x^3+x} = \frac{A}{x} + \frac{Bx+C}{x^2+1}$.

6. False, $\frac{1}{x^2-1} = \frac{1/2}{x-1} - \frac{1/2}{x+1}$.

7. True, by long division,
$$\begin{array}{r} x-1 \\ x^2+x-2 \overline{\smash{)}x^3+0x^2+0x+1} \\ \underline{x^3+x^2-2x} \\ -x^2+2x+1 \\ \underline{-x^2-x+2} \\ 3x-1 \end{array}$$

So $\frac{x^3+1}{x^2+x-2} = x - 1 + \frac{3x-1}{x^2+x-2}$.

8. False, $x^3 - 8 = (x-2)(x^2+2x+4)$.

9. True, $\frac{1}{x-1} + \frac{1}{x^2+x+1} = $
$$\frac{(x^2+x+1)+(x-1)}{x^3-1} = \frac{x^2+2x}{x^3-1}.$$

10. True, it is already in the form $\frac{Ax+B}{x^2+9}$.

5.4 Exercises

1. $\frac{3(x+1)+4(x-2)}{(x-2)(x+1)} = \frac{7x-5}{(x-2)(x+1)}$

3. $\frac{(x^2+2)-3(x-1)}{(x-1)(x^2+2)} = \frac{x^2-3x+5}{(x-1)(x^2+2)}$

5.
$$\frac{(2x+1)(x^2+3)+(x^3+2x+2)}{(x^2+3)^2} =$$
$$\frac{2x^3+x^2+6x+3)+(x^3+2x+2)}{(x^2+3)^2} =$$
$$\frac{3x^3+x^2+8x+5}{(x^2+3)^2} =$$

7.
$$\frac{(x-1)^2+(2x+3)(x-1)+(x^2+1)}{(x-1)^3} =$$
$$\frac{(x-2x+1)+(2x^2+x-3)+(x^2+1)}{(x-1)^3} =$$
$$\frac{4x^2-x-1}{(x-1)^3} =$$

9. Multiply equation by $(x-3)(x+3)$.
$$12 = A(x+3) + B(x-3)$$
$$12 = (A+B)x + (3A-3B)$$
$$A+B = 0 \text{ and } 3A - 3B = 12$$
Divide $3A - 3B = 12$ by 3 and add to $A + B = 0$.
$$A - B = 4$$
$$A + B = 0$$
$$\overline{}$$
$$2A = 4$$
Using $A = 2$ in $A + B = 0$, $B = -2$.
So $A = 2$ and $B = -2$.

11. $\frac{5x-1}{(x+1)(x-2)} = \frac{A}{x+1} + \frac{B}{x-2}$,
$$5x - 1 = A(x-2) + B(x+1)$$
$$5x - 1 = (A+B)x + (-2A+B)$$
$$A + B = 5 \text{ and } -2A + B = -1$$

Multiply $-2A + B = -1$ by -1 and add to $A + B = 5$.

$$2A - B = 1$$
$$A + B = 5$$
_____
$$3A = 6$$

Using $A = 2$ in $A + B = 5$, $B = 3$.

Answer is $\dfrac{2}{x+1} + \dfrac{3}{x-2}$

13. $\dfrac{2x+5}{(x+4)(x+2)} = \dfrac{A}{x+4} + \dfrac{B}{x+2}$,

$$2x + 5 = A(x+2) + B(x+4)$$
$$2x + 5 = (A+B)x + (2A + 4B)$$
$$A + B = 2 \text{ and } 2A + 4B = 5$$

Multiply $A + B = 2$ by -2 and add to $2A + 4B = 5$.

$$-2A - 2B = -4$$
$$2A + 4B = 5$$
_____
$$2B = 1$$

Using $B = 1/2$ in $A + B = 2$, $A = 3/2$.

Answer is $\dfrac{3/2}{x+4} + \dfrac{1/2}{x+2}$.

15. $\dfrac{2}{(x-3)(x+3)} = \dfrac{A}{x-3} + \dfrac{B}{x+3}$,

$$2 = A(x+3) + B(x-3)$$
$$2 = (A+B)x + (3A - 3B)$$
$$A + B = 0 \text{ and } 3A - 3B = 2$$

Multiply $A + B = 0$ by 3 and add to $3A - 3B = 2$.

$$3A + 3B = 0$$
$$3A - 3B = 2$$
_____
$$6A = 2$$

Using $A = 1/3$ in $A + B = 0$, $B = -1/3$.

Answer is $\dfrac{1/3}{x-3} + \dfrac{-1/3}{x+3}$.

17. $\dfrac{1}{x(x-1)} = \dfrac{A}{x} + \dfrac{B}{x-1}$,

$$1 = A(x-1) + Bx$$
$$1 = (A+B)x - A$$
$$A + B = 0 \text{ and } -A = 1$$

Using $A = -1$ in $A + B = 0$, $B = 1$.

Answer is $\dfrac{-1}{x} + \dfrac{1}{x-1}$.

19. Multiplying the equation by $(x+3)^2(x-2)$,

$$x^2 + x - 31 =$$
$$= A(x+3)(x-2) + B(x-2) + C(x+3)^2$$
$$= A(x^2 + x - 6) + B(x - 2) + C(x^2 + 6x + 9)$$
$$= (A+C)x^2 + (A+B+6C)x + (-6A - 2B + 9C)$$

Equate the coefficients and solve the system.

$$A + C = 1$$
$$A + B + 6C = 1$$
$$-6A - 2B + 9C = -31$$

Multiply $A + B + 6C = 1$ by 2 and add to $-6A - 2B + 9C = -31$.

$$2A + 2B + 12C = 2$$
$$-6A - 2B + 9C = -31$$
_____
$$-4A + 21C = -29$$

Multiply $A + C = 1$ by 4 and add to $-4A + 21C = -29$.

$$4A + 4C = 4$$
$$-4A + 21C = -29$$
_____
$$25C = -25$$

Using $C = -1$ in $A + C = 1$, $A = 2$.
From $A + B + 6C = 1$, $2 + B - 6 = 1$ and $B = 5$.
So $A = 2, B = 5$ and $C = -1$.

21. $\dfrac{-2x - 7}{(x+2)^2} = \dfrac{A}{x+2} + \dfrac{B}{(x+2)^2}$,

$$-2x - 7 = A(x+2) + B$$
$$-2x - 7 = Ax + (2A + B)$$
$$A = -2 \text{ and } 2A + B = -7$$

Using $A = -2$ in $2A + B = -7$, $B = -3$.

Answer is $\dfrac{-2}{x+2} + \dfrac{-3}{(x+2)^2}$.

5.4 PARTIAL FRACTIONS

23. Note that $x^3 + x^2 + x + 1 = x^2(x+1) + (x+1) = (x^2+1)(x+1)$. So

$$\frac{6x^2 - x + 1}{(x^2+1)(x+1)} = \frac{A}{x+1} + \frac{Bx+C}{x^2+1}$$

$$6x^2 - x + 1 = A(x^2+1) + (Bx+C)(x+1)$$

$$6x^2 - x + 1 = (A+B)x^2 + (B+C)x + (A+C)$$

Equating the coefficients,

$$A + B = 6$$
$$B + C = -1$$
$$A + C = 1$$

Multiply $A + B = 6$ by -1 and add to $B + C = -1$.

$$-A - B = -6$$
$$B + C = -1$$
$$\overline{-A + C = -7}$$

Adding $-A + C = -7$ and $A + C = 1$, $2C = -6$.

Using $C = -3$ in $B + C = -1$ and $A + C = 1$, $B = 2$ and $A = 4$. Answer is $\dfrac{4}{x+1} + \dfrac{2x-3}{x^2+1}$.

25.

$$\frac{3x^3 - x^2 + 19x - 9}{(x^2+9)^2} = \frac{Ax+B}{x^2+9} + \frac{Cx+D}{(x^2+9)^2},$$

$$3x^3 - x^2 + 19x - 9 = (Ax+B)(x^2+9) + (Cx+D)$$

$$= Ax^3 + Bx^2 + (9A+C)x + (9B+D)$$

So $A = 3$ and $B = -1$. From $9A + C = 19$ and $9B + D = -9$, $C = -8$ and $D = 0$.

Answer is $\dfrac{3x-1}{x^2+9} + \dfrac{-8x}{(x^2+9)^2}$.

27.

$$\frac{3x^2 + 17x + 14}{(x-2)(x^2+2x+4)} = \frac{A}{x-2} + \frac{Bx+C}{x^2+2x+4},$$

$$3x^2 + 17x + 14 = A(x^2+2x+4) + (Bx+C)(x-2)$$

$$= (A+B)x^2 + (2A-2B+C)x + (4A-2C)$$

Equating the coefficients,

$$A + B = 3$$
$$2A - 2B + C = 17$$
$$4A - 2C = 14$$

Multiply $A + B = 3$ by 2 and add to $2A - 2B + C = 17$.

$$2A + 2B = 6$$
$$2A - 2B + C = 17$$
$$\overline{4A + C = 23}$$

Multiplying $4A - 2C = 14$ by -1 and adding to $4A + C = 23$, $3C = 9$. So $C = 3$ and from $4A - 2C = 14$, $A = 5$. Using these on $2A - 2B + C = 17$, $B = -2$.

Answer is $\dfrac{5}{x-2} + \dfrac{-2x+3}{x^2+2x+4}$.

29. Divide $2x^3 + x^2 + 3x - 2$ by $x^2 - 1$ by long division.

$$\begin{array}{r} 2x + 1 \\ x^2 - 1 \overline{\smash{)}2x^3 + x^2 + 3x - 2} \\ \underline{2x^3 + 0x^2 - 2x } \\ x^2 + 5x - 2 \\ \underline{x^2 + 0x - 1} \\ 5x - 1 \end{array}$$

So $\dfrac{2x^3 + x^2 + 3x - 2}{x^2 - 1} = 2x + 1 + \dfrac{5x-1}{x^2-1}$.

Decompose $\dfrac{5x-1}{x^2-1} = \dfrac{A}{x-1} + \dfrac{B}{x+1}$.

$$5x - 1 = A(x+1) + B(x-1)$$
$$5x - 1 = (A+B)x + (A-B)$$

So $A + B = 5$ and $A - B = 1$.

Adding $A + B = 5$ and $A - B = -1$, $2A = 4$. Using $A = 2$ in $A + B = 5$, $B = 3$.

Answer is $2x + 1 + \dfrac{2}{x-1} + \dfrac{3}{x+1}$.

31. Since $\dfrac{3x^3 - 2x^2 + x - 2}{(x^2+x+1)^2} =$

$$\frac{Ax+B}{x^2+x+1} + \frac{Cx+D}{(x^2+x+1)^2},$$

$$3x^3 - 2x^2 + x - 2 = (Ax+B)(x^2+x+1) + (Cx+D)$$

$$= Ax^3 + (A+B)x^2 + (A+B+C)x + (B+D)$$

Equating the coefficients, $A = 3$.
Since $A + B = -2$, $B = -5$.
From $A + B + C = 1$ and $B + D = -2$, we have $C = 3$ and $D = 3$.

Answer is $\dfrac{3x-5}{x^2+x+1} + \dfrac{3x+3}{(x^2+x+1)^2}$.

33. Since $\dfrac{3x^3 + 4x^2 - 12x + 16}{(x-2)(x+2)(x^2+4)} =$
$\dfrac{A}{x-2} + \dfrac{B}{x+2} + \dfrac{Cx+D}{x^2+4}$, then
$3x^3 + 4x^2 - 12x + 16 =$
$= A(x+2)(x^2+4) + B(x-2)(x^2+4) +$
$\quad (Cx+D)(x^2-4)$
$= (A+B+C)x^3 + (2A-2B+D)x^2 +$
$\quad (4A+4B-4C)x + (8A-8B-4D)$

Equating the coefficients,
$$A + B + C = 3$$
$$2A - 2B + D = 4$$
$$4A + 4B - 4C = -12$$
$$8A - 8B - 4D = 16$$

Multiply first equation by -4 and add to the third. Multiply second equation by -4 and add to the fourth. Also multiply first equation by 2 and add to the second. So
$$-4A - 4B - 4C = -12$$
$$4A + 4B - 4C = -12$$
$$\overline{\qquad\qquad\qquad\qquad}$$
$$-8C = -24$$

$$-8A + 8B - 4D = -16$$
$$8A - 8B - 4D = 16$$
$$\overline{\qquad\qquad\qquad\qquad}$$
$$-8D = 0$$

$$2A + 2B + 2C = 6$$
$$2A - 2B + D = 4$$
$$\overline{\qquad\qquad\qquad\qquad}$$
$$4A + 2C + D = 10$$

So $C = 3$ and $D = 0$. From $4A + 2C + D = 10$, $A = 1$ and from $2A - 2B + D = 4$, $B = -1$.
Answer is $\dfrac{1}{x-2} + \dfrac{-1}{x+2} + \dfrac{3x}{x^2+4}$.

35. $\dfrac{5x^3 + x^2 + x - 3}{x^3(x-1)} = \dfrac{A}{x} + \dfrac{B}{x^2} + \dfrac{C}{x^3} + \dfrac{D}{x-1}$,
$5x^3 + x^2 + x - 3 =$
$= Ax^2(x-1) + Bx(x-1) + C(x-1) + Dx^3$
$= (A+D)x^3 + (-A+B)x^2 + (-B+C)x - C$

Equating the coefficients, $C = 3$.
From $-B + C = 1$, $B = 2$.
From $-A + B = 1$, $A = 1$.
From $A + D = 5$, $D = 4$.
Answer is $\dfrac{1}{x} + \dfrac{2}{x^2} + \dfrac{3}{x^3} + \dfrac{4}{x-1}$.

37. $\dfrac{6x^2 - 28x + 33}{(x-2)^2(x-3)} = \dfrac{A}{x-2} + \dfrac{B}{(x-2)^2} + \dfrac{C}{x-3}$,
$6x^2 - 28x + 33 =$
$= A(x-2)(x-3) + B(x-3) + C(x-2)^2$
$= (A+C)x^2 + (-5A + B - 4C)x +$
$\quad (6A - 3B + 4C)$.

Equating the coefficients,
$$A + C = 6$$
$$-5A + B - 4C = -28$$
$$6A - 3B + 4C = 33$$

Multiply second equation by 3 and add to the third.
$$-15A + 3B - 12C = -84$$
$$6A - 3B + 4C = 33$$
$$\overline{\qquad\qquad\qquad\qquad}$$
$$-9A - 8C = -51$$

Multiply $A + C = 6$ by 8 and add to $-9A - 8C = -51$
$$-9A - 8C = -51$$
$$8A + 8C = 48$$
$$\overline{\qquad\qquad\qquad\qquad}$$
$$-A = -3$$

Using $A = 3$ in $A + C = 6$, $C = 3$.
From $6A - 3B + 4C = 33$, $B = -1$.
Answer is $\dfrac{3}{x-2} + \dfrac{-1}{(x-2)^2} + \dfrac{3}{x-3}$.

39. By synthetic division,

-5	1	4	-11	-30
		-5	5	30
	1	-1	-6	0

$x^3 + 4x^2 - 11x - 30 = (x+5)(x^2 - x - 6)$
$\qquad\qquad\qquad\qquad = (x+5)(x+2)(x-3)$

Decomposing, $\dfrac{9x^2 + 21x - 24}{(x+5)(x+2)(x-3)} =$

5.5 INEQUALITIES AND SYSTEMS OF INEQUALITIES IN TWO VARIABLES

$$= \frac{A}{x+5} + \frac{B}{x+2} + \frac{C}{x-3}.$$

So $9x^2 + 21x - 24 =$
$$= A(x+2)(x-3) + B(x+5)(x-3) +$$
$$C(x+5)(x+2)$$

Substituting $x = -2, 3, -5$,

$-30 = -15B$ $\quad$ $120 = 40C$ $\quad$ $96 = -24A$
$2 = B$ $\quad\quad\quad$ $3 = C$ $\quad\quad\quad$ $4 = A$

Answer is $\dfrac{4}{x+5} + \dfrac{2}{x+2} + \dfrac{3}{x-3}.$

41. Note that $x^3 - 3x^2 + 3x - 1 = (x-1)^3$. So

$$\frac{x^2 - 2}{(x-1)^3} = \frac{A}{x-1} + \frac{B}{(x-1)^2} + \frac{C}{(x-1)^3}.$$

$x^2 - 2 = A(x-1)^2 + B(x-1) + C$
$\quad\quad\quad = Ax^2 + (-2A+B)x + (A-B+C)$

So $A = 1$. Since $-2A + B = 0$, $B = 2$.
Since $A - B + C = -2$, $1 - 2 + C = -2$
and $C = -1$.

Answer is $\dfrac{1}{x-1} + \dfrac{2}{(x-1)^2} + \dfrac{-1}{(x-1)^3}.$

43. $\dfrac{x}{(ax+b)^2} = \dfrac{A}{ax+b} + \dfrac{B}{(ax+b)^2},$

$x = A(ax+b) + B$
$\quad = aAx + (bA + B)$

So $aA = 1$ and $A = 1/a$. Since $bA + B = 0$,
$\dfrac{b}{a} + B = 0$ and $B = -b/a$.

Answer is $\dfrac{1/a}{ax+b} + \dfrac{-b/a}{(ax+b)^2}.$

45. $\dfrac{x+c}{x(ax+b)} = \dfrac{A}{x} + \dfrac{B}{ax+b},$

$x + c = A(ax+b) + Bx$
$\quad\quad = (aA + B)x + bA$

So $bA = c$ and $A = \dfrac{c}{b}$. Since $aA + B = 1$,

$\dfrac{ac}{b} + B = 1$ and $B = 1 - \dfrac{ac}{b}$.

Answer is $\dfrac{c/b}{x} + \dfrac{1 - ac/b}{ax+b}.$

47. $\dfrac{1}{x^2(ax+b)} = \dfrac{A}{x} + \dfrac{B}{x^2} + \dfrac{C}{ax+b},$

$1 = Ax(ax+b) + B(ax+b) + Cx^2$
$1 = (aA+C)x^2 + (bA+aB)x + bB.$

So $bB = 1$ and $B = \dfrac{1}{b}$. Since $bA + aB = 0$,

$bA + \dfrac{a}{b} = 0$ and $A = -\dfrac{a}{b^2}.$

Since $aA + C = 0$, $-\dfrac{a^2}{b^2} + C = 0$ and $C = \dfrac{a^2}{b^2}.$

Answer is $\dfrac{-a/b^2}{x} + \dfrac{1/b}{x^2} + \dfrac{a^2/b^2}{ax+b}.$

For Thought

1. False $\quad$ 2. False $\quad$ 3. True

4. False, because $x^2 + y^2 > 5$ is the region outside of a circle of radius $\sqrt{5}$.

5. True, $(-2, 1)$ satisfies both equations in system (a).

6. True

7. False, $(-2, 0)$ does not satisfy $y < x + 2$.

8. False, $(-1, 2)$ lies on the line $y - 3x = 5$.

9. True

10. True

5.5 Exercises

1. c $\quad$ 3. d

5. $y < 2x$

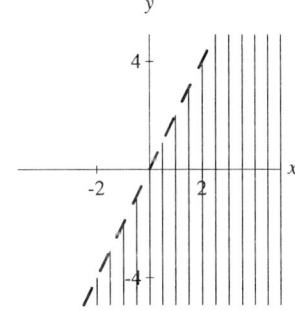

7. $x + y > 3$

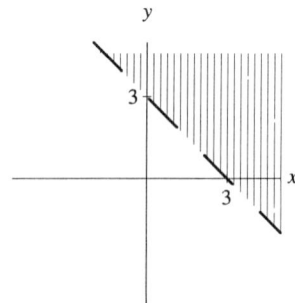

9. $2x - y \leq 4$

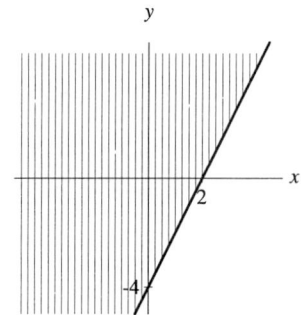

11. $y < -3x - 4$

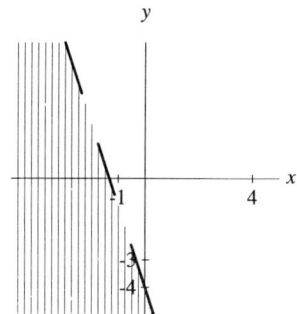

13. $x - 3 \geq 0$

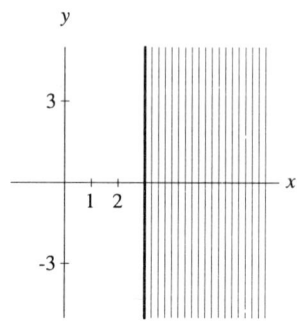

15. $20x - 30y \leq 6000$

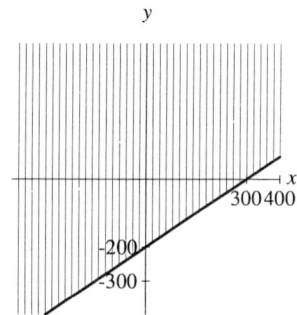

17. $y < 3$

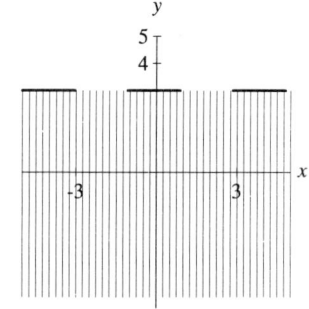

19. $y > -x^2$

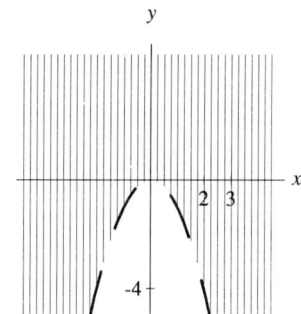

21. $x^2 + y^2 \geq 1$

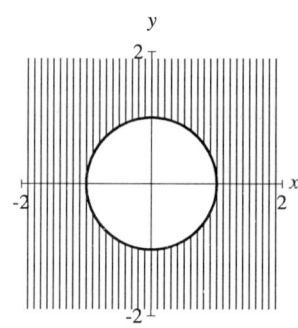

5.5 INEQUALITIES AND SYSTEMS OF INEQUALITIES IN TWO VARIABLES

23. $x > |y|$

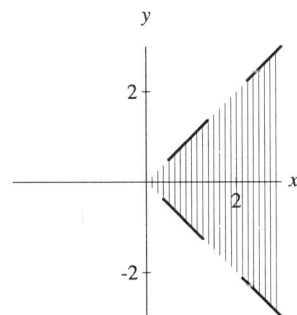

25. $x \geq y^2$

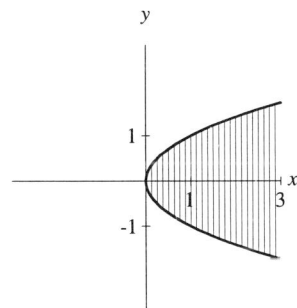

27. $y \geq x^3$

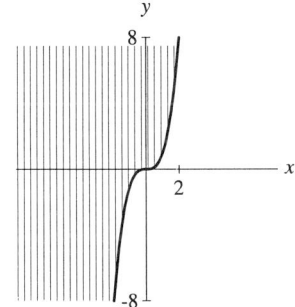

29. $y > 2^x$

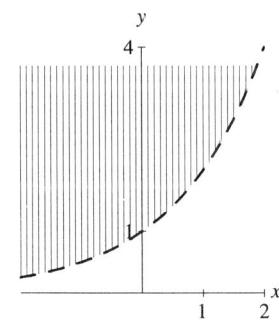

31. $y > x - 4, y < -x - 2$

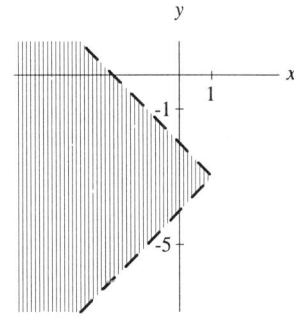

33. $3x - 4y \leq 12, x + y \geq -3$

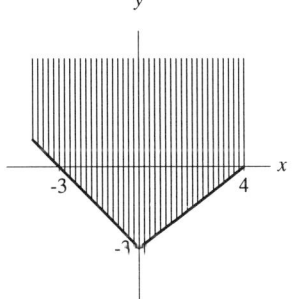

35. $3x - y < 4, y < 3x + 5$

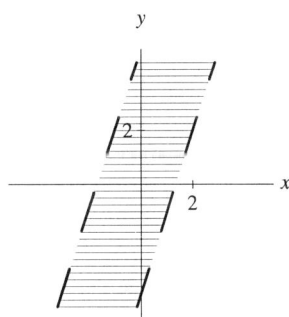

37. No solution to the system $y + x < 0, y > 3 - x$

39. $x + y < 5, y \geq 2$

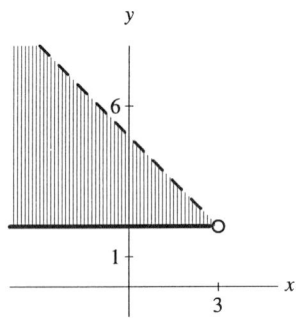

41. $y < x - 3, x \leq 4$

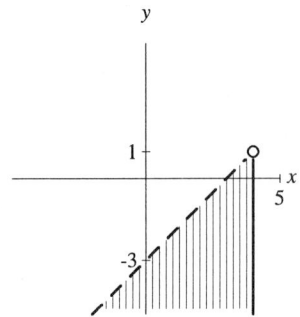

43. $y > x^2 - 3, y < x + 1$

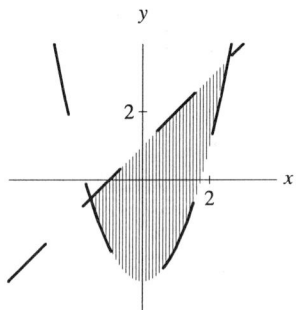

45. $x^2 + y^2 \geq 4, x^2 + y^2 \leq 16$

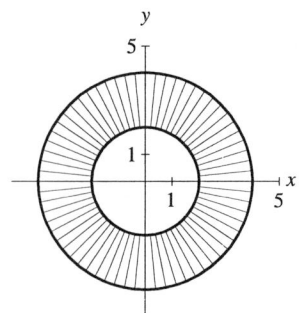

47. $(x-3)^2 + y^2 < 25, (x+3)^2 + y^2 < 25$

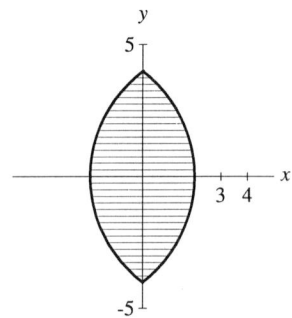

49. $x^2 + y^2 > 4, |x| \leq 4$

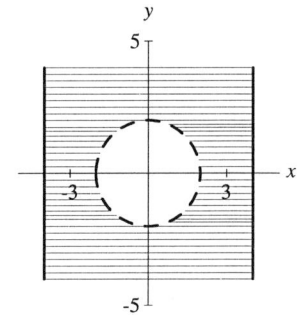

51. $y > |2x| - 4, y \leq \sqrt{4 - x^2}$

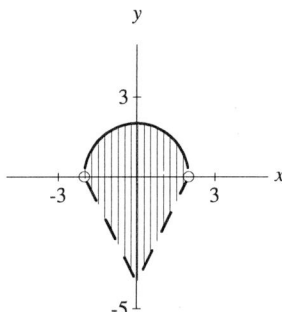

53. $|x - 1| < 2, |y - 1| < 4$

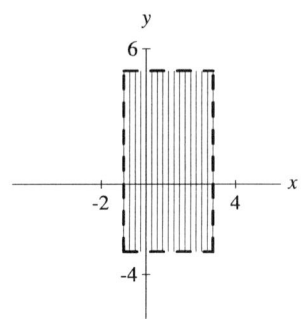

55. $x \geq 0, y \geq 0, x + y \leq 4$

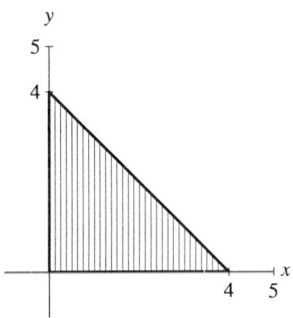

57. $x \geq 0, y \geq 0, x + y \geq 4, y \geq -2x + 6$

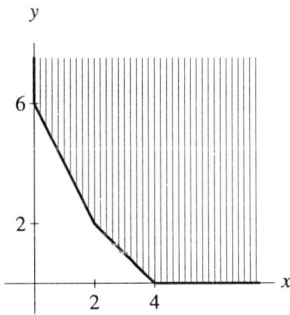

59. $x^2 + y^2 \geq 9, x^2 + y^2 \leq 25, y \geq |x|$

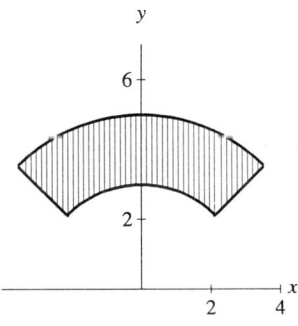

61. $y > (x-1)^3, y > 1, x + y > -2$

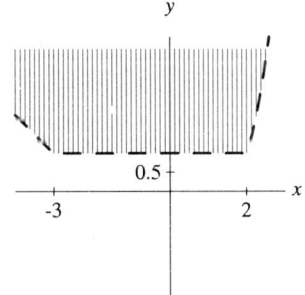

63. $y > 2^x, y < 6 - x^2, x + y > 0$

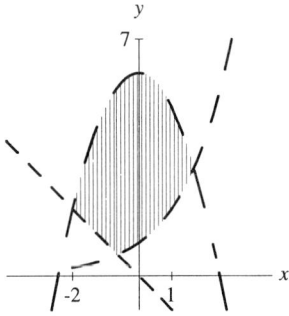

65. $x \geq 0, y \geq 0, y \leq -\dfrac{2}{3}x + 5, y \leq -3x + 12$

67. $x \geq 0, y \geq 0, y \geq -\dfrac{1}{2}x + 3, y \geq -\dfrac{3}{2}x + 5$

69. System is $\begin{aligned} |x| &< 2 \\ |y| &< 2 \end{aligned}$

71. Since a circle of radius 9 with center at the origin is given by $x^2 + y^2 = 81$, the system is

$$x^2 + y^2 < 81$$
$$x > 0$$
$$y > 0$$

73. $(-1.17, 1.84)$ is a solution.

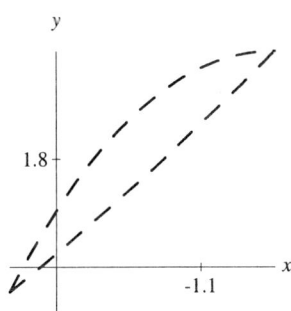

75. $(150, 22.4)$ is a solution.

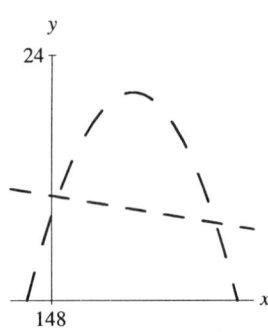

77. Let w and h be the width and height, respectively. Then $50 + 2w + 2h \leq 130$. The system is

$$w + h \leq 40$$
$$w, h \geq 0$$

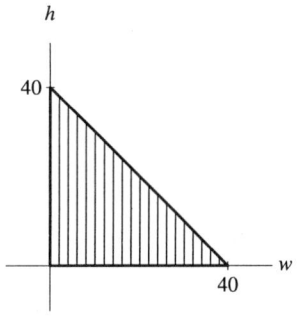

79. Let x and y be the number of mid-size and full-size cars, respectively. Divide $10,000x + 15,000y \leq 1,500,000$ by 1000. The system is

$$x + 1.5y \leq 150$$
$$x \geq 0$$
$$y \geq 0$$

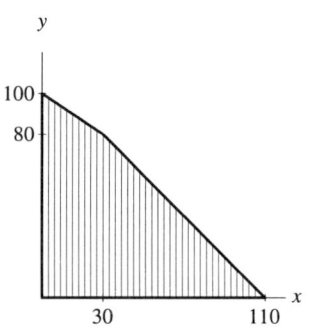

81. Let x and y be the number of $50 tickets and $100 tickets, respectively. Simplifying $y \leq 0.2(x + y)$, $0.8y \leq 0.2x$, so $y \leq \frac{1}{4}x$. The system is

$$y \leq \frac{1}{4}x$$
$$x + y \leq 500$$
$$x \geq 0, \quad y \geq 0$$

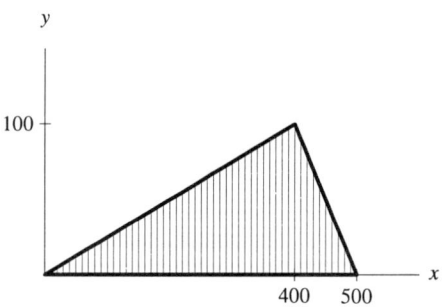

For Thought

1. False, $x \geq 0$ include points on the x-axis and the first & fourth quadrants.

2. False, $y \geq 2$ include points on or above the line $y = 2$.

3. False

4. False, since x-intercept is $(6,0)$ and y-intercept is $(0,4)$.

5. True

6. True

7. False

8. True, since $R(1,3) = 30(1) + 15(3) = 75$.

9. False, since $C(0,5) = 7(0) + 9(5) + 3 = 48$.

10. True

5.6 Exercises

1. Vertices are $(0,0), (0,4), (4,0)$

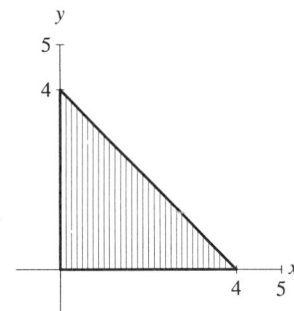

3. Vertices are $(0,0), (1,3), (1,0), (0,3)$

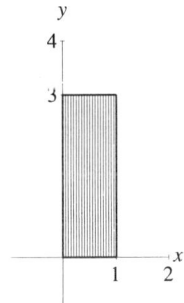

5. Vertices are $(0,0), (2,2), (0,4), (3,0)$

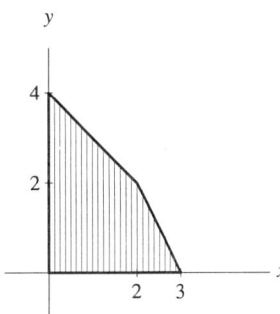

7. Vertices are $(3,0), (1,2), (0,4)$

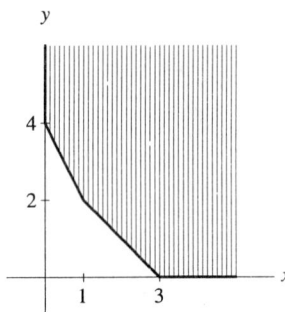

9. Vertices are $(1,3), (4,0), (0,6)$

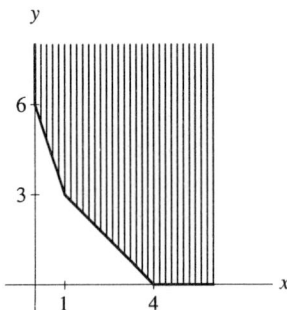

11. Vertices are $(1,5), (6,0), (0,8)$

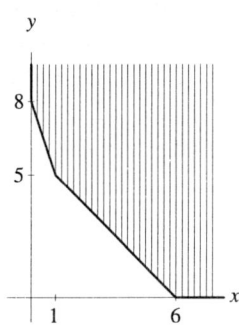

13. The value of $T(x,y) = 2x + 3y$ on the vertices are $T(0,0) = 0$, $T(0,4) = 12$, $T(3,3) = 15$, and $T(5,0) = 10$. The maximum value is 15.

15. The value of $H(x,y) = 2x + 2y$ on the vertices are $T(0,6) = 12$, $T(2,2) = 8$, and $T(5,0) = 10$. The minimum value is 8.

17. The values of $P(x,y) = 5x + 9y$ on the vertices are $P(0,0) = 0$, $P(6,0) = 30$, $P(0,3) = 27$. Maximum value is 30.

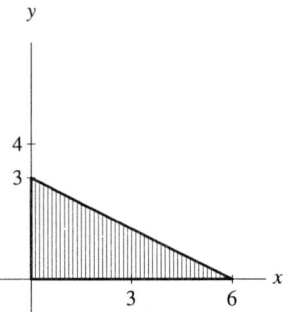

19. The values of $C(x,y) = 10x + 20y$ on the vertices are $C(0,8) = 160$, $C(5,3) = 110$, and $C(10,0) = 100$. Minimum value is 100.

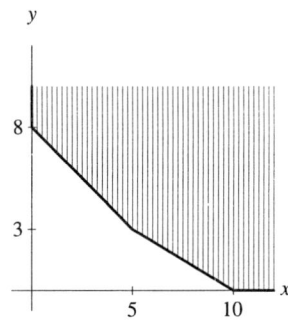

21. Let x and y be the number of bird houses and mail boxes, respectively.

$$\text{Max } 12x+20y$$
$$3x+4y \le 48$$
$$x+2y \le 20$$
$$x,y \ge 0$$

The values of $R(x,y) = 12x+20y$ on the vertices are $R(0,0) = 0$, $R(0,10) = 200$, $R(8,6) = 216$, and $R(16,0) = 192$. To maximize revenue, they must sell 8 bird houses and 6 mailboxes.

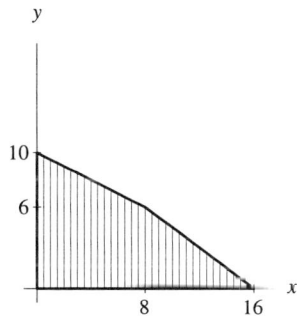

23. Let x and y be the number of bird houses and mail boxes, respectively. The values of $R(x,y) = 18x + 20y$ on the vertices are $R(0,0) = 0$, $R(0,10) = 200$, $R(8,6) = 264$, and $R(16,0) = 288$. To maximize revenue, they must sell 16 bird houses and 0 mailboxes.

25. Let x and y be the number of small and large truck loads, respectively.

$$\text{Min } 70x+60y$$
$$12x + 20y \ge 120$$
$$x+y \ge 8$$
$$x,y \ge 0$$

The values of $C(x,y) = 70x + 60y$ on the vertices are $C(0,8) = 480$, $C(10,0) = 700$, and $C(5,3) = 530$. To minimize costs, they must make 8 large truck loads and 0 small truck loads.

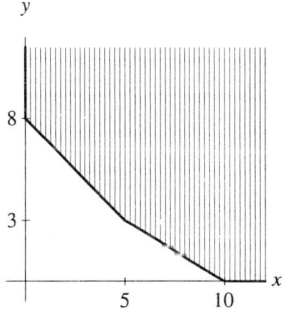

27. Let x and y be the number of small and large truck loads, respectively. The values of $C(x,y) = 70x + 75y$ on the vertices are $C(0,8) = 600$, $C(10,0) = 700$, and $C(5,3) = 575$. To minimize costs, they must make 5 small truck loads and 3 large truck loads.

Review Exercises

1. Solution is $(3,5)$.

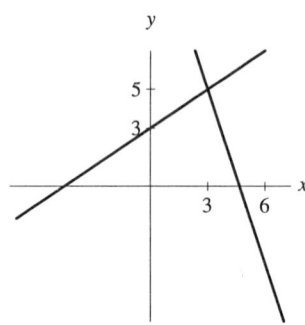

3. Solution is $(-1,3)$.

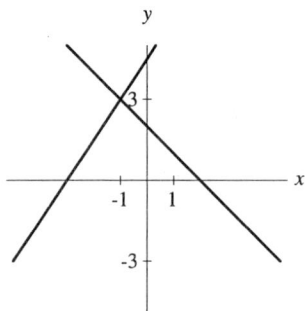

5. Substitute $y = x$ into $3x - 5y = 19$.
$$3x - 5x = 19$$
$$-2x = 19$$
$$x = -19/2$$

Independent and the solution set is $\{(-19/2, -19/2)\}$.

7. Multiply $4x - 3y = 6$ by 2 and $3x + 2y = 9$ by 3 add the equations.
$$8x - 6y = 12$$
$$9x + 6y = 27$$
$$\overline{}$$
$$17x = 39$$

Using $x = 39/17$ in $4x - 3y = 6$,
$$\frac{156}{17} - 3y = 6$$
$$\frac{54}{17} = 3y$$
$$\frac{18}{17} = y$$

Independent and the solution set is $\{(39/17, 18/17)\}$.

9. Substitute $y = -3x + 1$ into $6x + 2y = 2$.
$$6x + 2(-3x + 1) = 2$$
$$6x - 6x + 2 = 2$$
$$2 = 2$$

Dependent and the solution set is $\{(x,y) : y = -3x + 1\}$.

11. Multiply $3x - 4y = 12$ by 2 and add to the second equation.
$$6x - 8y = 24$$
$$-6x + 8y = 9$$
$$\overline{}$$
$$0 = 33$$

Inconsistent and there is no solution.

13. Add first and second equations. Multiply $2x + y + z = 1$ by -3 and add to the third equation.

$$x + y - z = 8 \qquad -6x - 3y - 3z = -3$$
$$2x + y + z = 1 \qquad x + 2y + 3z = -5$$
$$\overline{} \qquad \overline{}$$
$$3x + 2y = 9 \qquad -5x - y = -8$$

Multiply $-5x - y = -8$ by 2 and add to $3x + 2y = 9$.
$$-10x - 2y = -16$$
$$3x + 2y = 9$$
$$\overline{}$$
$$-7x = -7$$

Using $x = 1$ in $3x + 2y = 9$, $3 + 2y = 9$ or $y = 3$. From $x + y - z = 8$, $1 + 3 - z = 8$ or $z = -4$. Solution set is $\{(1, 3, -4)\}$.

15. Multiply first equation by -2 and add to the second equation. Multiply first equation by -2 and add to the third one.

$$-2x - 2y - 2z = -2 \qquad -2x - 2y - 2z = -2$$
$$2x - y + 2z = 2 \qquad 2x + 2y + 2z = 2$$
$$\overline{} \qquad \overline{}$$
$$-3y = 0 \qquad 0 = 0$$
$$y = 0$$

Using $y = 0$ in $x + y + z = 1$, $x + z = 1$ and $z = 1 - x$. Solution set is $\{(x, 0, 1 - x) : x \text{ is any real number}\}$.

REVIEW EXERCISES

17. Multiply first equation by -1 and add to the third equation.

$$-x - y - z = -1$$
$$x + y + z = 4$$
$$\overline{}$$
$$0 = 3$$

There is no solution since $0 = 3$ is false.

19. Substitute $x = y^2$ into $x^2 + y^2 = 4$ and use the quadratic formula.

$$y^4 + y^2 = 4$$
$$y^4 + y^2 - 4 = 0$$
$$y^2 = \frac{-1 + \sqrt{17}}{2}$$
$$y = \pm\sqrt{\frac{-1 + \sqrt{17}}{2}}$$

So $x = y^2 = \dfrac{-1 + \sqrt{17}}{2}$.

Solution set is $\left(\dfrac{-1 + \sqrt{17}}{2}, \pm\sqrt{\dfrac{-1 + \sqrt{17}}{2}}\right)$

21. Equate $y = x^2$ to $y = |x|$.

$$x^2 = \sqrt{x^2}$$
$$x^4 = x^2$$
$$x^2(x^2 - 1) = 0$$
$$x = 0, \pm 1$$

Using $x = 0, 1, -1$ in $y = x^2$, $y = 0, 1, 1$, respectively. Solution set is $\{(0,0),(1,1),(-1,1)\}$.

23.
$$\frac{7x - 7}{(x - 3)(x + 4)} = \frac{A}{x - 3} + \frac{B}{x + 4},$$

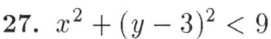
$$7x - 7 = (A + B)x + (4A - 3B)$$

Equating the coefficients,

$$A + B = 7$$
$$4A - 3B = -7$$

The solution of this system is $A = 2$, $B = 5$.

Answer is $\dfrac{2}{x - 3} + \dfrac{5}{x + 4}$.

25.
$$x^3 - 3x^2 + 4x - 12 = x^2(x - 3) + 4(x - 3)$$
$$= (x^2 + 4)(x - 3)$$

So $\dfrac{7x^2 - 7x + 23}{(x - 3)(x^2 + 4)} = \dfrac{A}{x - 3} + \dfrac{Bx + C}{x^2 + 4}$

$$7x^2 - 7x + 23 = A(x^2 + 4) + (Bx + C)(x - 3)$$
$$= (A+B)x^2 + (-3B + C)x + (4A - 3C)$$

Equating the coefficients,

$$A + B = 7$$
$$-3B + C = -7$$
$$4A - 3C = 23$$

The solution of this system is $A = 5$, $B = 2$, and $C = -1$. Answer is $\dfrac{5}{x - 3} + \dfrac{2x - 1}{x^2 + 4}$

27. $x^2 + (y - 3)^2 < 9$

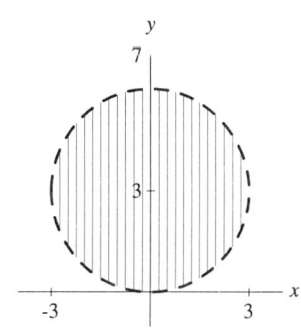

29. $x \leq (y - 1)^2$

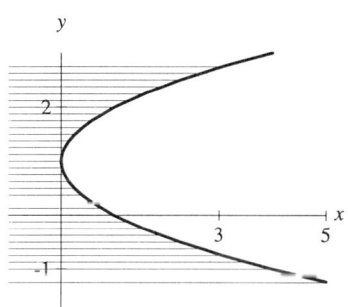

31. $2x - 3y \geq 6$, $x \leq 2$

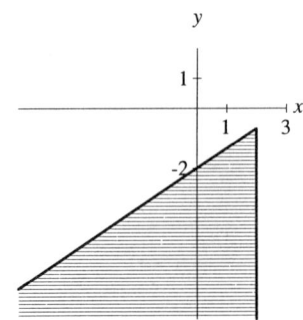

33. $y \geq 2x^2 - 6$, $x^2 + y^2 \leq 9$

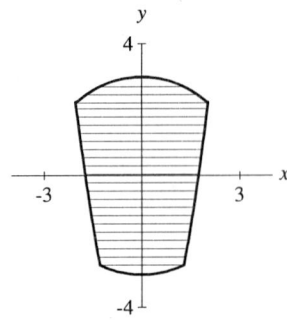

35. $x \geq 0$, $y \geq 1$, $x + 2y \leq 10$, $3x + 4y \leq 24$

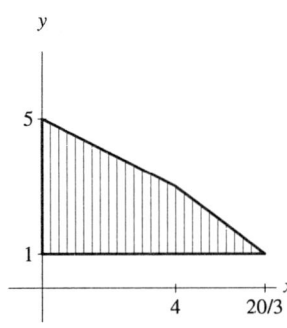

37. $x \geq 0$, $y \geq 0$, $x + 6y \geq 60$, $x + y \geq 35$

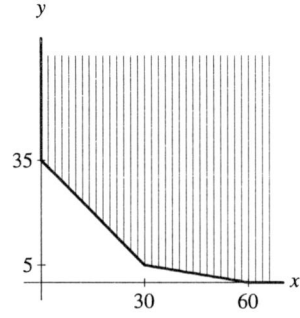

39. Substitute $(-2, 3)$ and $(4, -1)$ into $y = mx + b$.

$$-2m + b = 3$$
$$4m + b = -1$$

Multiply first equation by -1 and add to the second equation.

$$2m - b = -3$$
$$4m + b = -1$$
$$\overline{}$$
$$6m = -4$$

Using $m = -2/3$ in $-2m + b = 3$, $4/3 + b = 3$ and $b = 5/3$. Equation of line is $y = -\dfrac{2}{3}x + \dfrac{5}{3}$.

41. Substitute $(1, 4)$, $(3, 20)$, and $(-2, 25)$ into $y = ax^2 + bx + c$.

$$a + b + c = 4$$
$$9a + 3b + c = 20$$
$$4a - 2b + c = 25$$

The solution of the above system is $a = 3$, $b = -4$, $c = 5$. Parabola is given by $y = 3x^2 - 4x + 5$.

43. Let x and y be the number of tacos and burritos.

$$x + 2y = 181$$
$$2x + 3y = 300$$

Solving the above system, $x = 57$ tacos and $y = 62$ burritos.

45. Let x, y and z be the selling price of a daisy, carnation, and a rose, respectively. Then

$$5x + 3y + 2z = 3.05$$
$$3x + y + 4z = 2.75$$
$$4x + 2y + z = 2.10$$

Solving the above system, $x = 0.30$, $y = 0.25$, and $z = 0.40$. Esther's economy special sells for $x + y + z = \$0.95$.

47. The value of $C(x,y) = 0.42x + 0.84y$ on the vertices are $C(0,35) = 29.4$, $C(30,5) = 16.8$, and $C(60,0) = 25.2$. Minimum value is 16.8 .

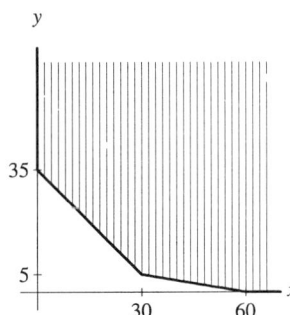

49. Let x and y be the number of barrels of oil from a pipeline and barges, respectively.

$$\text{Min } 20x + 18y$$
$$x + y \geq 12,000,000$$
$$x \leq 12,000,000$$
$$x \geq 6,000,000$$
$$y \leq 8,000,000$$
$$x, y \geq 0$$

The values of $C(x, y) = 20x + 18y$ on the vertices are
$C(12\text{million}, 0) = 240\text{million}$,
$C(12\text{million}, 8\text{million}) = 384\text{million}$,
$C(6\text{million}, 8\text{million}) = 264\text{million}$, and
$C(6\text{million}, 6\text{million}) = 228\text{million}$.
To minimize cost, purchase 6 million barrels from each source.

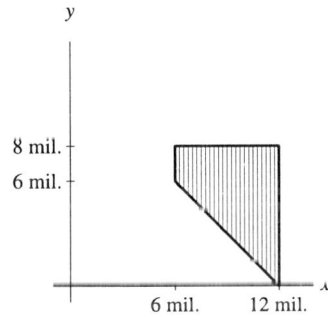

Chapter 5 Test

1. Solution set is $(-3, 4)$.

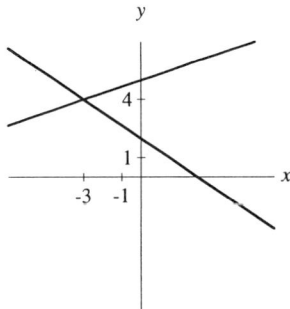

2. Substitute $y = 4 - 2x$ into $3x - 4y = 9$.
$$3x - 4(4 - 2x) = 9$$
$$3x - 16 + 8x = 9$$
$$11x = 25$$
$$x = 25/11$$
Using $x = 25/11$ in $y = 4 - 2x$,
we have $y = 4 - 50/11 = -6/11$.
Solution set is $\{(25/11, -6/11)\}$.

3. Multiply $10x - 3y = 22$ by 2 and $7x + 2y = 40$ by 3. Adding the equations,
$$20x - 6y = 44$$
$$21x + 6y = 120$$
$$\overline{}$$
$$41x = 164$$
$$x = 4$$
Using $x = 4$ in $10x - 3y = 22$, $40 - 3y = 22$ and $y = 6$. Solution set is $\{(4, 6)\}$.

4. Substitute $x = 6 - y$ into $3x + 3y = 4$.
$$3(6 - y) + 3y = 4$$
$$18 - 3y + 3y = 4$$
$$18 = 4$$
The system is inconsistent.

5. Substitute $y = \dfrac{1}{2}x + 3$ into $x - 2y = -6$.
$$x - 2\left(\dfrac{1}{2}x + 3\right) = -6$$
$$x - x - 6 = -6$$
$$-6 = -6$$
The system is dependent.

6. Substitute $y = 2x - 1$ into $y = 3x + 20$.
$$2x - 1 = 3x + 20$$
$$-21 = x$$
Using $x = -21$ in $y = 2x - 1$, $y = -43$.
The system is independent.

7. Substitute $y = -x + 2$ into $y = -x + 5$.
$$-x + 2 = -x + 5$$
$$2 = 5$$
The system is inconsistent.

8. Add the two equations.
$$2x - y + z = 4$$
$$-x + 2y - z = 6$$
$$\overline{}$$
$$x + y = 10$$
Using $y = 10 - x$ in $2x - y + z = 4$,
$2x - (10 - x) + z = 4$ and $z = 14 - 3x$.
Solution set is
$\{(x, 10 - x, 14 - 3x) : x \text{ is any real number}\}$.

9. Add the first two equations. Also, multiply second equation by 3 and add to the third.

$$x - 2y - z = 2 \qquad 6x + 9y + 3z = -3$$
$$2x + 3y + z = -1 \qquad 3x - y - 3z = -4$$
$$\overline{} \qquad \overline{}$$
$$3x + y = 1 \qquad 9x + 8y = -7$$

Multiply $3x + y = 1$ by -3 and add to $9x + 8y = -7$.
$$-9x - 3y = -3$$
$$9x + 8y = -7$$
$$\overline{}$$
$$5y = -10$$
Using $y = -2$ in $3x + y = 1$, $3x - 2 = 1$ or $x = 1$. From $x - 2y - z = 2$, $1 + 4 - z = 2$ or $z = 3$. Solution set is $\{(1, -2, 3)\}$.

10. Add second and third equations.
$$x + y - z = 4$$
$$-x - y + z = 2$$
$$\overline{}$$
$$0 = 6$$
Inconsistent and no solution.

11. Multiply $x^2 + y^2 = 16$ by -1 and add to $x^2 - 4y^2 = 16$.
$$-x^2 - y^2 = -16$$
$$x^2 - 4y^2 = 16$$
$$\overline{}$$
$$-5y^2 = 0$$
$$y = 0$$
Using $y = 0$ in $x^2 + y^2 = 16$, $x^2 = 16$ and $x = \pm 4$. Solution set is $\{(4, 0), (-4, 0)\}$.

12. Substitute $y = x^2 - 5x$ into $x + y = -2$.
$$x + (x^2 - 5x) = -2$$
$$x^2 - 4x = -2$$
$$x^2 - 4x + 4 = -2 + 4$$
$$(x - 2)^2 = 2$$
$$x = 2 \pm \sqrt{2}$$
Using $x = 2 + \sqrt{2}$ and $x = 2 - \sqrt{2}$ in $y = -2 - x$, we have $y = -4 - \sqrt{2}$ and $y = -4 + \sqrt{2}$, respectively. Solution set is
$\{(2 + \sqrt{2}, -4 - \sqrt{2}), (2 - \sqrt{2}, -4 + \sqrt{2})\}$.

13.
$$\frac{2x + 10}{(x - 4)(x + 2)} = \frac{A}{x - 4} + \frac{B}{x + 2}$$
$$2x + 10 = A(x + 2) + B(x - 4)$$
$$2x + 10 = (A + B)x + (2A - 4B)$$
Equating the coefficients,
$$A + B = 2$$
$$2A - 4B = 10$$
Solution of above system is $A = 3, B = -1$.
Answer is $\dfrac{3}{x - 4} + \dfrac{-1}{x + 2}$.

14.
$$\frac{4x^2 + x - 2}{x^2(x - 1)} = \frac{A}{x} + \frac{B}{x^2} + \frac{C}{x - 1},$$
$$4x^2 + x - 2 = Ax(x - 1) + B(x - 1) + Cx^2$$
$$= (A + C)x^2 + (-A + B)x - B$$
Equating the coefficients,
$$A + C = 4$$
$$-A + B = 1$$
$$-B = -2$$

Solution of above system is $B = 2, A = 1,$ and $C = 3$. Answer is $\dfrac{1}{x} + \dfrac{2}{x^2} + \dfrac{3}{x-1}$.

15. $2x - y < 8$

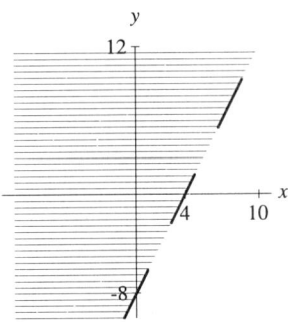

16. $x + y \leq 5, x - y < 0$

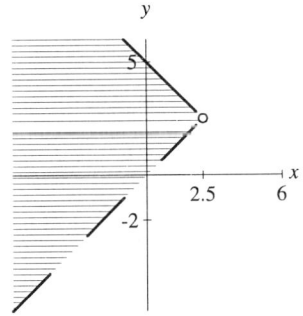

17. $x^2 + y^2 \leq 9, y \leq 1 - x^2$

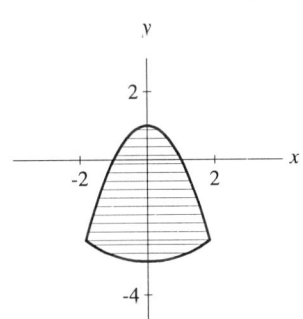

18. Let x and y be the number of male and female students, respectively. Then
$$\frac{1}{3}x + \frac{1}{4}y = 15$$
$$x + y = 52$$
Solution of above system is $x = 24$ males and $y = 28$ females.

19. Let x and y be the number of TV commercials and newspaper ads, respectively. The linear program is
$$\text{Max } 14{,}000x + 6{,}000y$$
$$9{,}000x + 3{,}000y \leq 99{,}000$$
$$x + y \leq 23$$
$$x, y \geq 0$$
Values of $N(x,y) = 14{,}000x + 6{,}000y$ on the vertices are $N(0,0) = 0$, $N(0,23) = 138{,}000$, $N(5,18) = 178{,}000$, and $N(11,0) = 154{,}000$. To obtain maximum audience exposure, hospital must have 5 TV commercials and 18 newspaper ads.

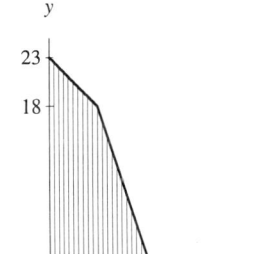

Tying It All Together

1. Multiply equation by $24(x+5)$.
$$24(x-2) = 11(x+5)$$
$$24x - 48 = 11x + 55$$
$$13x = 103$$
Solution set is $\{103/13\}$.

2. Multiply equation by $24x(x+5)$.
$$24(x+5) + 24x(x-2) = 11x(x+5)$$
$$24x + 120 + 24x^2 - 48x = 11x^2 + 55x$$
$$13x^2 - 79x + 120 = 0$$
$$(13x - 40)(x - 3) = 0$$
Solution set is $\{40/13, 3\}$.

3.
$$5 - 3x - 6 - 2x + 4 = 7$$
$$3 - 5x = 7$$
$$-5x = 4$$
Solution set is $\{-4/5\}$.

4. Solve an equivalent statement without absolute values.
$$3 - 2x = 5 \text{ or } 3 - 2x = -5$$
$$-2x = 2 \text{ or } -2x = -8$$
$$x = -1 \text{ or } x = 4$$
Solution set is $\{4, -1\}$.

5. Square both sides of the equation.
$$3 - 2x = 25$$
$$-2x = 22$$
Solution set is $\{-11\}$.

6. Isolate x^2 in one side and take square roots.
$$3x^2 = 4$$
$$x^2 = \frac{4}{3}$$
$$x = \pm \frac{2}{\sqrt{3}}$$
$$x = \pm \frac{2}{\sqrt{3}} \cdot \frac{\sqrt{3}}{\sqrt{3}}$$
Solution set is $\left\{\pm \frac{2\sqrt{3}}{3}\right\}$.

7. Multiply equation by x^2.
$$(x - 2)^2 = x^2$$
$$x^2 - 4x + 4 = x^2$$
$$-4x = -4$$
Solution set is $\{1\}$.

8. Since $2^{x-1} = 9$, $x - 1 = \log_2(9)$ by definition of a logarithm.
Solution set is $\{1 + \log_2(9)\}$.

9. Simplify left-hand side as a single logarithm.
$$\log((x+1)(x+4)) = 1$$
$$x^2 + 5x + 4 = 10^1$$
$$x^2 + 5x - 6 = 0$$
$$(x+6)(x-1) = 0$$
$$x = -6, 1$$

But $\log(x+1)$ is undefined when $x = -6$.
Solution set is $\{1\}$.

10. Raise both sides of equation to the power $-3/2$.
$$x = \pm \left(\frac{1}{4}\right)^{-3/2}$$
$$x = \pm (4)^{3/2}$$
$$x = \pm (4^{1/2})^3$$
$$x = \pm (2)^3$$
Solution set is $\{\pm 8\}$.

11. Use the quadratic formula.
$$x^2 - 3x - 6 = 0$$
$$x = \frac{3 \pm \sqrt{33}}{2}$$
Solution set is $\left\{\frac{3 \pm \sqrt{33}}{2}\right\}$.

12.
$$(x - 3)^2 = \frac{1}{2}$$
$$x - 3 = \pm \frac{\sqrt{2}}{2}$$
$$x = \frac{6}{2} \pm \frac{\sqrt{2}}{2}$$
Solution set is $\left\{\frac{6 \pm \sqrt{2}}{2}\right\}$.

13. Since $3 - 2x > 0$, $3 > 2x$ and $x < 3/2$. Solution set is $(-\infty, 3/2)$ and its graph is <===) ------ >

14. Solution set is $(-\infty, 3/2) \cup (3/2, \infty)$ and graph is <===)(===>

15. The sign graph of $(x-3)(x+3) \geq 0$ is

```
- - - - - - - - - 0 + + + +
- - - - - 0 + + + + + + + +
<-------|---------|------->
       -3         3
```

So the solution set is $(-\infty, -3] \cup [3, \infty)$ and the graph is <==] ------[==>

16. Note that $x^2 + 2x - 8 \leq 27$ is equivalent to $x^2 + 2x - 35 \leq 0$. The sign graph of $(x+7)(x-5) \leq 0$ is

```
- - - - - - - - 0 + + + +
- - - - 0 + + + + + + + +
<─────────┼─────────┼─────────>
         -7         5
```

So the solution set is $[-7, 5]$ and the graph is $\longleftarrow ---[===]---\longrightarrow$ with -7 and 5 marked.

17. $3 - 2x > y$

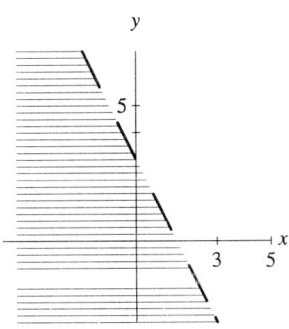

18. $|3 - 2x| > y$

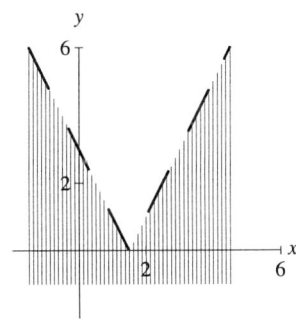

19. $x^2 \geq 9$

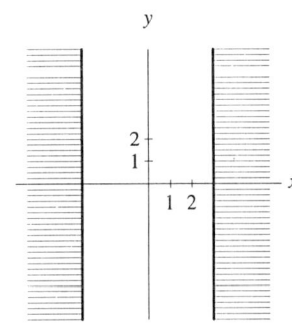

20. $(x-2)(x+4) \leq y$

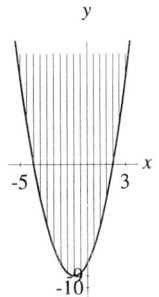

For Thought

1. False, augmented matrix is a 2×3 matrix.

2. False, required matrix is $\begin{bmatrix} 1 & -1 & | & 4 \\ 3 & 1 & | & 5 \end{bmatrix}$.

3. True 4. True

5. True, row operation done is $R_1 + R_2 \to R_2$.

6. True, row operation done is $-R_1 + R_2 \to R_2$.

7. False, since it corresponds to $\begin{matrix} x = 2 \\ y = 7 \end{matrix}$

8. True, $0 \cdot x + 0 \cdot y = 7$ has no solution.

9. False, system is dependent. 10. True

6.1 Exercises

1. 1×3 3. 1×1 5. 3×2

7. $\begin{bmatrix} 1 & -2 & | & 4 \\ 3 & 2 & | & -5 \end{bmatrix}$

9. $\begin{bmatrix} 1 & -1 & -1 & | & 4 \\ 1 & 3 & -1 & | & 1 \\ 0 & 2 & -5 & | & -6 \end{bmatrix}$

11. $\begin{matrix} 3x + 4y = -2 \\ 3x - 5y = 0 \end{matrix}$

13.
$$5x = 6$$
$$-4x + 2z = -1$$
$$4x + 4y = 7$$

15. Multiply R_1 by $\dfrac{1}{2}$.

17. Multiply R_2 by $-\dfrac{1}{6}$.

19. On $\begin{bmatrix} 1 & 1 & | & 5 \\ -2 & 1 & | & -1 \end{bmatrix}$ use $2R_1 + R_2 \to R_2$ to get

$\begin{bmatrix} 1 & 1 & | & 5 \\ 0 & 3 & | & 9 \end{bmatrix}$, use $\dfrac{1}{3}R_2 \to R_2$ to get

$\begin{bmatrix} 1 & 1 & | & 5 \\ 0 & 1 & | & 3 \end{bmatrix}$, use $-R_2 + R_1 \to R_1$ to get

$\begin{bmatrix} 1 & 0 & | & 2 \\ 0 & 1 & | & 3 \end{bmatrix}$, solution is $(2, 3)$.

21. Rewrite system as $\begin{matrix} 2x + y = 4 \\ x - y = 8 \end{matrix}$

On $\begin{bmatrix} 2 & 1 & | & 4 \\ 1 & -1 & | & 8 \end{bmatrix}$ use $R_1 - 2R_2 \to R_2$ to get

$\begin{bmatrix} 2 & 1 & | & 4 \\ 0 & 3 & | & -12 \end{bmatrix}$, use $\dfrac{1}{3}R_2 \to R_2$ to get

$\begin{bmatrix} 2 & 1 & | & 4 \\ 0 & 1 & | & -4 \end{bmatrix}$, use $-R_2 + R_1 \to R_1$ to get

$\begin{bmatrix} 2 & 0 & | & 8 \\ 0 & 1 & | & -4 \end{bmatrix}$, use $\dfrac{1}{2}R_1 \to R_1$ to get

$\begin{bmatrix} 1 & 0 & | & 4 \\ 0 & 1 & | & -4 \end{bmatrix}$, solution is $(4, -4)$.

23. On $\begin{bmatrix} 2 & -1 & | & 3 \\ 3 & 2 & | & 15 \end{bmatrix}$ use $-R_1 + R_2 \to R_1$ to get

$\begin{bmatrix} 1 & 3 & | & 12 \\ 3 & 2 & | & 15 \end{bmatrix}$, use $-3R_1 + R_2 \to R_2$ to get

$\begin{bmatrix} 1 & 3 & | & 12 \\ 0 & 7 & | & 21 \end{bmatrix}$, use $\dfrac{1}{7}R_2 \to R_2$ to get

$\begin{bmatrix} 1 & 3 & | & 12 \\ 0 & 1 & | & 3 \end{bmatrix}$, use $-3R_2 + R_1 \to R_1$ to get

$\begin{bmatrix} 1 & 0 & | & 3 \\ 0 & 1 & | & 3 \end{bmatrix}$, solution is $(3, 3)$.

25. On $\begin{bmatrix} 0.4 & -0.2 & | & 0 \\ 1 & 1.5 & | & 2 \end{bmatrix}$ use $5R_1 \to R_1$ and $2R_2 \to R_2$ to get

$\begin{bmatrix} 2 & -1 & | & 0 \\ 2 & 3 & | & 4 \end{bmatrix}$, use $R_1 - R_2 \to R_2$ to get

$\begin{bmatrix} 2 & -1 & | & 0 \\ 0 & -4 & | & -4 \end{bmatrix}$, use $-\dfrac{1}{4}R_2 \to R_2$ to get

$\begin{bmatrix} 2 & -1 & | & 0 \\ 0 & 1 & | & 1 \end{bmatrix}$, use $R_1 + R_2 \to R_1$ to get

$\begin{bmatrix} 2 & 0 & | & 1 \\ 0 & 1 & | & 1 \end{bmatrix}$, use $\dfrac{1}{2}R_1 \to R_1$ to get

$\begin{bmatrix} 1 & 0 & | & 0.5 \\ 0 & 1 & | & 1 \end{bmatrix}$, solution is $(0.5, 1)$.

6.1 SOLVING LINEAR SYSTEMS USING MATRICES

27. On $\begin{bmatrix} 3 & -5 & | & 7 \\ -3 & 5 & | & 4 \end{bmatrix}$ use $R_1 + R_2 \to R_2$ to get

$\begin{bmatrix} 3 & -5 & | & 7 \\ 0 & 0 & | & 11 \end{bmatrix}$, inconsistent and no solution.

29. On $\begin{bmatrix} 0.5 & 1.5 & | & 2 \\ 3 & 9 & | & 12 \end{bmatrix}$ use $2R_1 \to R_1$ to get

$\begin{bmatrix} 1 & 3 & | & 4 \\ 3 & 9 & | & 12 \end{bmatrix}$, use $-3R_1 + R_2 \to R_2$ to get

$\begin{bmatrix} 1 & 3 & | & 4 \\ 0 & 0 & | & 0 \end{bmatrix}$, solution set is $\{(u,v) : u+3v = 4\}$.

31.

On $\begin{bmatrix} 1 & 1 & 1 & | & 6 \\ 1 & -1 & -1 & | & 0 \\ 0 & 2 & -1 & | & 3 \end{bmatrix}$ use $R_1 - R_2 \to R_2$ to get

$\begin{bmatrix} 1 & 1 & 1 & | & 6 \\ 0 & 2 & 2 & | & 6 \\ 0 & 2 & -1 & | & 3 \end{bmatrix}$, use $\frac{1}{2}R_2 \to R_2$ to get

$\begin{bmatrix} 1 & 1 & 1 & | & 6 \\ 0 & 1 & 1 & | & 3 \\ 0 & 2 & -1 & | & 3 \end{bmatrix}$, use $-R_2 + R_1 \to R_1$ to get

$\begin{bmatrix} 1 & 0 & 0 & | & 3 \\ 0 & 1 & 1 & | & 3 \\ 0 & 2 & -1 & | & 3 \end{bmatrix}$, use $-2R_2 + R_3 \to R_3$ to get

$\begin{bmatrix} 1 & 0 & 0 & | & 3 \\ 0 & 1 & 1 & | & 3 \\ 0 & 0 & -3 & | & -3 \end{bmatrix}$, use $-\frac{1}{3}R_3 \to R_3$ to get

$\begin{bmatrix} 1 & 0 & 0 & | & 3 \\ 0 & 1 & 1 & | & 3 \\ 0 & 0 & 1 & | & 1 \end{bmatrix}$, use $-R_3 + R_2 \to R_2$ to get

$\begin{bmatrix} 1 & 0 & 0 & | & 3 \\ 0 & 1 & 0 & | & 2 \\ 0 & 0 & 1 & | & 1 \end{bmatrix}$, solution is $(3, 2, 1)$.

33.

Rewrite system as $\begin{aligned} 2x + y - z &= 2 \\ x + 2y - z &= 2 \\ x - y + 2z &= 2. \end{aligned}$

On $\begin{bmatrix} 2 & 1 & -1 & | & 2 \\ 1 & 2 & -1 & | & 2 \\ 1 & -1 & 2 & | & 2 \end{bmatrix}$, use $R_1 - R_2 \to R_1$

and $R_2 - R_3 \to R_3$ to get

$\begin{bmatrix} 1 & -1 & 0 & | & 0 \\ 1 & 2 & -1 & | & 2 \\ 0 & 3 & -3 & | & 0 \end{bmatrix}$, use $R_1 - R_2 \to R_2$

and $\frac{1}{3}R_3 \to R_3$ to get

$\begin{bmatrix} 1 & -1 & 0 & | & 0 \\ 0 & -3 & 1 & | & -2 \\ 0 & 1 & -1 & | & 0 \end{bmatrix}$, use $R_1 + R_3 \to R_1$

and $R_3 + R_2 \to R_2$ to get

$\begin{bmatrix} 1 & 0 & -1 & | & 0 \\ 0 & -2 & 0 & | & -2 \\ 0 & 1 & -1 & | & 0 \end{bmatrix}$, use $-\frac{1}{2}R_2 \to R_2$ to get

$\begin{bmatrix} 1 & 0 & -1 & | & 0 \\ 0 & 1 & 0 & | & 1 \\ 0 & 1 & -1 & | & 0 \end{bmatrix}$, use $-R_3 + R_2 \to R_3$ to get

$\begin{bmatrix} 1 & 0 & -1 & | & 0 \\ 0 & 1 & 0 & | & 1 \\ 0 & 0 & 1 & | & 1 \end{bmatrix}$, use $R_1 + R_3 \to R_1$ to get

$\begin{bmatrix} 1 & 0 & 0 & | & 1 \\ 0 & 1 & 0 & | & 1 \\ 0 & 0 & 1 & | & 1 \end{bmatrix}$, solution is $(1, 1, 1)$.

35.

Interchange rows of $\begin{bmatrix} 2 & -2 & 1 & | & -2 \\ 1 & 1 & -3 & | & 3 \\ 1 & -3 & 1 & | & -5 \end{bmatrix}$ to get

$\begin{bmatrix} 1 & 1 & -3 & | & 3 \\ 1 & -3 & 1 & | & -5 \\ 2 & -2 & 1 & | & -2 \end{bmatrix}$, use $R_1 - R_2 \to R_2$

and $R_1 + R_2 - R_3 \to R_3$ to get

$\begin{bmatrix} 1 & 1 & -3 & | & 3 \\ 0 & 4 & -4 & | & 8 \\ 0 & 0 & -3 & | & 0 \end{bmatrix}$, use $\frac{1}{4}R_2 \to R_2$

and $-\frac{1}{3}R_3 \to R_3$ to get

$\begin{bmatrix} 1 & 1 & -3 & | & 3 \\ 0 & 1 & -1 & | & 2 \\ 0 & 0 & 1 & | & 0 \end{bmatrix}$, use $R_2 + R_3 \to R_2$ to get

$\begin{bmatrix} 1 & 1 & -3 & | & 3 \\ 0 & 1 & 0 & | & 2 \\ 0 & 0 & 1 & | & 0 \end{bmatrix}$, use $R_1 - R_2 \to R_1$ to get

$\begin{bmatrix} 1 & 0 & -3 & | & 1 \\ 0 & 1 & 0 & | & 2 \\ 0 & 0 & 1 & | & 0 \end{bmatrix}$, use $3R_3 + R_1 \to R_1$ to get

$\begin{bmatrix} 1 & 0 & 0 & | & 1 \\ 0 & 1 & 0 & | & 2 \\ 0 & 0 & 1 & | & 0 \end{bmatrix}$, solution is $(a,b,c) = (1,2,0)$.

37.

Rewrite system as
$$x - 3y + z = 0$$
$$x - y - 3z = 4$$
$$x + y + 2z = -1.$$

On $\begin{bmatrix} 1 & -3 & 1 & | & 0 \\ 1 & -1 & -3 & | & 4 \\ 1 & 1 & 2 & | & -1 \end{bmatrix}$, use $R_1 - R_2 \to R_2$

and $R_1 - R_3 \to R_3$ to get

$\begin{bmatrix} 1 & -3 & 1 & | & 0 \\ 0 & -2 & 4 & | & -4 \\ 0 & -4 & -1 & | & 1 \end{bmatrix}$, use $-\frac{1}{2}R_2 \to R_2$

and $-R_3 \to R_3$ to get

$\begin{bmatrix} 1 & -3 & 1 & | & 0 \\ 0 & 1 & -2 & | & 2 \\ 0 & 4 & 1 & | & -1 \end{bmatrix}$, use $R_1 - R_3 \to R_1$ to get

$\begin{bmatrix} 1 & -7 & 0 & | & 1 \\ 0 & 1 & -2 & | & 2 \\ 0 & 4 & 1 & | & -1 \end{bmatrix}$, use $-4R_2 + R_3 \to R_3$

to get

$\begin{bmatrix} 1 & -7 & 0 & | & 1 \\ 0 & 1 & -2 & | & 2 \\ 0 & 0 & 9 & | & -9 \end{bmatrix}$, use $\frac{1}{9}R_3 \to R_3$ to get

$\begin{bmatrix} 1 & -7 & 0 & | & 1 \\ 0 & 1 & -2 & | & 2 \\ 0 & 0 & 1 & | & -1 \end{bmatrix}$, use $2R_3 + R_2 \to R_2$

to get

$\begin{bmatrix} 1 & -7 & 0 & | & 1 \\ 0 & 1 & 0 & | & 0 \\ 0 & 0 & 1 & | & -1 \end{bmatrix}$, use $R_1 + 7R_2 \to R_1$ to get

$\begin{bmatrix} 1 & 0 & 0 & | & 1 \\ 0 & 1 & 0 & | & 0 \\ 0 & 0 & 1 & | & -1 \end{bmatrix}$, solution is $(1, 0, -1)$.

39.

On $\begin{bmatrix} 1 & -2 & 3 & | & 1 \\ 2 & -4 & 6 & | & 2 \\ -3 & 6 & -9 & | & -3 \end{bmatrix}$, use $\frac{1}{2}R_2 \to R_2$

and $-\frac{1}{3}R_3 \to R_3$ to get

$\begin{bmatrix} 1 & -2 & 3 & | & 1 \\ 1 & -2 & 3 & | & 1 \\ 1 & -2 & 3 & | & 1 \end{bmatrix}$, use $R_2 - R_1 \to R_2$

and $R_3 - R_1 \to R_3$ to get

$\begin{bmatrix} 1 & -2 & 3 & | & 1 \\ 0 & 0 & 0 & | & 0 \\ 0 & 0 & 0 & | & 0 \end{bmatrix}$. Dependent system and

solution set is $\{(x,y,z) : x - 2y + 3z = 1\}$.

41.

On $\begin{bmatrix} 1 & -1 & 1 & | & 2 \\ 2 & 1 & -1 & | & 1 \\ 2 & -2 & 2 & | & 5 \end{bmatrix}$, use $R_1 - \frac{1}{2}R_3 \to R_3$

to get $\begin{bmatrix} 1 & -1 & 1 & | & 2 \\ 2 & 1 & -1 & | & 1 \\ 0 & 0 & 0 & | & -1/2 \end{bmatrix}$, inconsistent

and no solution.

43.

On $\begin{bmatrix} 1 & 1 & -1 & | & 3 \\ 3 & 1 & 1 & | & 7 \\ 1 & -1 & 3 & | & 1 \end{bmatrix}$, use $R_1 - R_3 \to R_3$

and $3R_1 - R_2 \to R_2$ to get

$\begin{bmatrix} 1 & 1 & -1 & | & 3 \\ 0 & 2 & -4 & | & 2 \\ 0 & 2 & -4 & | & 2 \end{bmatrix}$, use $R_2 - R_3 \to R_3$ to get

$\begin{bmatrix} 1 & 1 & -1 & | & 3 \\ 0 & 2 & -4 & | & 2 \\ 0 & 0 & 0 & | & 0 \end{bmatrix}$, use $\frac{1}{2}R_2 \to R_2$ to get

$\begin{bmatrix} 1 & 1 & -1 & | & 3 \\ 0 & 1 & -2 & | & 1 \\ 0 & 0 & 0 & | & 0 \end{bmatrix}$, use $-R_2 + R_1 \to R_1$ to get

$\begin{bmatrix} 1 & 0 & 1 & | & 2 \\ 0 & 1 & -2 & | & 1 \\ 0 & 0 & 0 & | & 0 \end{bmatrix}$. Substitute $z = 2 - x$

into $y = 1 + 2z = 1 + 2(2 - x) = 5 - 2x$. Solution set is $\{(x, 5 - 2x, 2 - x) : x \text{ is any real number}\}$.

45.

On $\begin{bmatrix} 2 & -1 & 3 & | & 1 \\ 1 & 1 & -1 & | & 4 \end{bmatrix}$, use $-2R_2 + R_1 \to R_2$

to get

$\begin{bmatrix} 2 & -1 & 3 & | & 1 \\ 0 & -3 & 5 & | & -7 \end{bmatrix}$, use $-3R_1 + R_2 \to R_1$

to get

$\begin{bmatrix} -6 & 0 & -4 & | & -10 \\ 0 & -3 & 5 & | & -7 \end{bmatrix}$, use $-\frac{1}{6}R_1 \to R_1$ and $-\frac{1}{3}R_2 \to R_2$ to get

$\begin{bmatrix} 1 & 0 & 2/3 & | & 5/3 \\ 0 & 1 & -5/3 & | & 7/3 \end{bmatrix}$. Note $x = \frac{5}{3} - \frac{2z}{3}$ and $y = \frac{7}{3} + \frac{5z}{3}$. Solution set is

$\left\{ \left(\frac{5-2z}{3}, \frac{5z+7}{3}, z \right) : z \text{ is any real number} \right\}$.

47.

On $\begin{bmatrix} 1 & -1 & 1 & -1 & | & 2 \\ -1 & 2 & -1 & -1 & | & -1 \\ 2 & -1 & -1 & 1 & | & 4 \\ 1 & 3 & -2 & -3 & | & 6 \end{bmatrix}$,

use $2R_2 + R_3 \to R_3$ and $R_2 + R_4 \to R_4$ to get

$\begin{bmatrix} 1 & -1 & 1 & -1 & | & 2 \\ 0 & 1 & 0 & -2 & | & 1 \\ 0 & 3 & -3 & -1 & | & 2 \\ 0 & 5 & -3 & -4 & | & 5 \end{bmatrix}$, use $-3R_2 + R_3 \to R_3$

and $-5R_2 + R_4 \to R_4$ to get

$\begin{bmatrix} 1 & -1 & 1 & -1 & | & 2 \\ 0 & 1 & 0 & -2 & | & 1 \\ 0 & 0 & -3 & 5 & | & -1 \\ 0 & 0 & -3 & 6 & | & 0 \end{bmatrix}$, use $R_1 + R_2 \to R_1$

and $-\frac{1}{2}R_4 \to R_4$ to get

$\begin{bmatrix} 1 & 0 & 1 & -3 & | & 3 \\ 0 & 1 & 0 & -2 & | & 1 \\ 0 & 0 & -3 & 5 & | & -1 \\ 0 & 0 & 1 & -2 & | & 0 \end{bmatrix}$, use $R_3 \to R_4$

and $R_4 \to R_3$ to get

$\begin{bmatrix} 1 & 0 & 1 & -3 & | & 3 \\ 0 & 1 & 0 & -2 & | & 1 \\ 0 & 0 & 1 & -2 & | & 0 \\ 0 & 0 & -3 & 5 & | & -1 \end{bmatrix}$, use $-3R_3 - R_4 \to R_4$

to get

$\begin{bmatrix} 1 & 0 & 1 & -3 & | & 3 \\ 0 & 1 & 0 & -2 & | & 1 \\ 0 & 0 & 1 & -2 & | & 0 \\ 0 & 0 & 0 & 1 & | & 1 \end{bmatrix}$, use $R_1 - R_3 \to R_1$ to get

$\begin{bmatrix} 1 & 0 & 0 & -1 & | & 3 \\ 0 & 1 & 0 & -2 & | & 1 \\ 0 & 0 & 1 & -2 & | & 0 \\ 0 & 0 & 0 & 1 & | & 1 \end{bmatrix}$, solution is

$(x, y, z, w) = (4, 3, 2, 1)$

49. Let x and y be the number of hours at Burgers and the Soap Opera, respectively.

Augmented matrix is $A = \begin{bmatrix} 5 & 5.5 & | & 311 \\ 1 & 1 & | & 60 \end{bmatrix}$.

On A use $-5R_2 + R_1 \to R_2$ to get

$\begin{bmatrix} 5 & 5.5 & | & 311 \\ 0 & 0.5 & | & 11 \end{bmatrix}$, use $\frac{1}{5}R_1 \to R_1$

and $2R_2 \to R_2$ to get

$\begin{bmatrix} 1 & 1.1 & | & 62.2 \\ 0 & 1 & | & 22 \end{bmatrix}$, use $-1.1R_2 + R_1 \to R_1$

to get

$\begin{bmatrix} 1 & 0 & | & 38 \\ 0 & 1 & | & 32 \end{bmatrix}$. Mike worked $x = 38$ hours

at Burgers and $y = 22$ hours at Soap Opera.

51. Let x, y, and z be the amounts invested in a mutual fund, in treasury bills, and in bonds, respectively. Augmented matrix is

$A = \begin{bmatrix} 1 & 1 & 1 & | & 40,000 \\ 0.08 & 0.09 & 0.12 & | & 3,660 \\ 1 & -1 & -1 & | & 0 \end{bmatrix}$.

On A use $100R_2 \to R_2$ to get

$\begin{bmatrix} 1 & 1 & 1 & | & 40,000 \\ 8 & 9 & 12 & | & 366,000 \\ 1 & -1 & -1 & | & 0 \end{bmatrix}$, use $R_2 - 8R_1 \to R_2$

to get

$\begin{bmatrix} 1 & 1 & 1 & | & 40,000 \\ 0 & 1 & 4 & | & 46,000 \\ 1 & -1 & -1 & | & 0 \end{bmatrix}$, use $R_1 - R_3 \to R_1$

and $R_1 - R_4 \to R_4$ to get

$\begin{bmatrix} 1 & 0 & -3 & | & -6,000 \\ 0 & 1 & 4 & | & 46,000 \\ 0 & 2 & 2 & | & 40,000 \end{bmatrix}$, use $\frac{1}{2}R_3 \to R_3$ to get

$\begin{bmatrix} 1 & 0 & -3 & | & -6,000 \\ 0 & 1 & 4 & | & 46,000 \\ 0 & 1 & 1 & | & 20,000 \end{bmatrix}$, use $R_2 - R_3 \to R_3$

to get

$\begin{bmatrix} 1 & 0 & -3 & | & -6,000 \\ 0 & 1 & 4 & | & 46,000 \\ 0 & 0 & 3 & | & 26,000 \end{bmatrix}$, use $R_2 - R_3 \to R_2$

and $\dfrac{1}{3}R_3 \to R_3$ to get

$$\begin{bmatrix} 1 & 0 & -3 & | & -6,000 \\ 0 & 1 & 1 & | & 20,000 \\ 0 & 0 & 1 & | & 8,666.67 \end{bmatrix}, \text{ use } 3R_3 + R_1 \to R_1$$

and $-R_3 + R_2 \to R_2$ to get

$$\begin{bmatrix} 1 & 0 & 0 & | & 20,000 \\ 0 & 1 & 0 & | & 11,333.33 \\ 0 & 0 & 1 & | & 8,666.67 \end{bmatrix}. \text{ Investments were}$$

$x = \$20,000$ in a mutual fund, $y = \$11,333.33$ in treasury bills and $z = \$8,666.67$ in bonds.

53.

Augmented matrix is

$$A = \begin{bmatrix} -1 & -1 & 1 & | & 4 \\ 1 & 1 & 1 & | & 2 \\ 8 & 2 & 1 & | & 7 \end{bmatrix}.$$

On A use $R_1 \to R_2$ and $R_2 \to R_1$ to get

$$\begin{bmatrix} 1 & 1 & 1 & | & 2 \\ -1 & -1 & 1 & | & 4 \\ 8 & 2 & 1 & | & 7 \end{bmatrix}, \text{ use } 8R_1 - R_3 \to R_3$$

and $\dfrac{8R_2 + R_3}{-3} \to R_2$ to get

$$\begin{bmatrix} 1 & 1 & 1 & | & 2 \\ 0 & 2 & -3 & | & -13 \\ 0 & 6 & 7 & | & 9 \end{bmatrix}, \text{ use } -3R_2 + R_3 \to R_3$$

to get

$$\begin{bmatrix} 1 & 1 & 1 & | & 2 \\ 0 & 2 & -3 & | & -13 \\ 0 & 0 & 16 & | & 48 \end{bmatrix}, \text{ use } \dfrac{1}{2}R_2 \to R_2$$

and $\dfrac{1}{16}R_3 \to R_3$ to get

$$\begin{bmatrix} 1 & 1 & 1 & | & 2 \\ 0 & 1 & -3/2 & | & -13/2 \\ 0 & 0 & 1 & | & 3 \end{bmatrix}, \text{ use } \dfrac{3}{2}R_3 + R_2 \to R_2$$

to get

$$\begin{bmatrix} 1 & 1 & 1 & | & 2 \\ 0 & 1 & 0 & | & -2 \\ 0 & 0 & 1 & | & 3 \end{bmatrix}, \text{ use } -R_3 + R_1 \to R_1 \text{ to get}$$

$$\begin{bmatrix} 1 & 1 & 0 & | & -1 \\ 0 & 1 & 0 & | & -2 \\ 0 & 0 & 1 & | & 3 \end{bmatrix}, \text{ use } R_1 - R_2 \to R_1 \text{ to get}$$

$$\begin{bmatrix} 1 & 0 & 0 & | & 1 \\ 0 & 1 & 0 & | & -2 \\ 0 & 0 & 1 & | & 3 \end{bmatrix}. \text{ Answer is}$$

$a = 1, b = -2,$ and $c = 3$

55. Since the number of cars entering M.L. King Dr. and Washington St. is 750 and $x + y$ is the number of cars leaving the intersection of M.L. King Dr. and Washington St. then $x + y = 750$.

On the intersection of M.L. King Dr. and JFK Blvd., the number of cars entering this intersection is $450 + x$ and the number of cars leaving is $700 + z$. So $450 + x = 700 + z$.

Simplifying, one gets $y = 750 - x$ and $z = x - 250$; and since y and z are non-negative, $250 \le x \le 750$. The values of $x, y,$ and z that realizes this traffic flow must satisfy

$$y = 750 - x$$
$$z = x - 250$$
$$250 \le x \le 750$$

When $z = 50$, $50 = x - 250$ or $x = 300$ and $y = 750 - 300 = 450$.

For Thought

1. True

2. False, since the orders of matrices A and C are different.

3. False, $A + B = \begin{bmatrix} 2 \\ 6 \end{bmatrix}$.

4. True, $C + D = \begin{bmatrix} 1-3 & 1+5 \\ 3+1 & 3-2 \end{bmatrix} = \begin{bmatrix} -2 & 6 \\ 4 & 1 \end{bmatrix}$.

5. True, $A - B = \begin{bmatrix} 1-1 \\ 3-3 \end{bmatrix} = \begin{bmatrix} 0 \\ 0 \end{bmatrix}$.

6. False, $3B = 3\begin{bmatrix} 1 \\ 3 \end{bmatrix} = \begin{bmatrix} 3 \\ 9 \end{bmatrix}$.

7. False, $-A = -\begin{bmatrix} 1 \\ 3 \end{bmatrix} = \begin{bmatrix} -1 \\ -3 \end{bmatrix}$.

6.2 OPERATIONS WITH MATRICES

8. False, matrices of different orders cannot be added. 9. False, matrices of different orders cannot be subtracted.

10. False, $C + 2D = \begin{bmatrix} 1 & 1 \\ 3 & 3 \end{bmatrix} + \begin{bmatrix} -6 & 10 \\ 2 & -4 \end{bmatrix} =$
$= \begin{bmatrix} -5 & 11 \\ 5 & -7 \end{bmatrix}$.

6.2 Exercises

1. $x = 2$, $y = 5$

3. Since $2x = 6$ and $4y = 10$, $x = 3$ and $y = 5/2$. Also $3z = z + y$ and so $z = y/2 = (5/2) \div 2 = 5/4$. Then $x = 3$, $y = 5/2$, and $z = 5/4$.

5. $\begin{bmatrix} 5 \\ 6 \end{bmatrix}$

7. $\begin{bmatrix} 0.4 & 0.15 \\ 0.7 & 1.1 \end{bmatrix}$

9. $\begin{bmatrix} 1/2 & 3/2 \\ 3 & -12 \end{bmatrix}$

11. $\begin{bmatrix} 10 & 0 \\ -2 & 16 \end{bmatrix}$

13. undefined

15.
$2\begin{bmatrix} -\sqrt{2} \\ \sqrt{5} \\ 3\sqrt{3} \end{bmatrix} - \begin{bmatrix} -2\sqrt{2} \\ 2\sqrt{5} \\ -\sqrt{3} \end{bmatrix} = \begin{bmatrix} -2\sqrt{2} + 2\sqrt{2} \\ 2\sqrt{5} - 2\sqrt{5} \\ 6\sqrt{3} + \sqrt{3} \end{bmatrix}$
$= \begin{bmatrix} 0 \\ 0 \\ 7\sqrt{3} \end{bmatrix}$

17.
$\begin{bmatrix} 2a \\ 2b \end{bmatrix} + \begin{bmatrix} 6a \\ 12b \end{bmatrix} + \begin{bmatrix} 5a \\ -15b \end{bmatrix} = \begin{bmatrix} 13a \\ -b \end{bmatrix}$

19.
$\begin{bmatrix} 0.4x - 0.6x & 0.4y - 0.9y \\ 0.8x - 1.5x & 3.2y + 0.3y \end{bmatrix} =$
$\begin{bmatrix} -x & -0.5y \\ -0.7x & 3.5y \end{bmatrix} =$

21. $\begin{bmatrix} -1 & 13 \\ -9 & 3 \\ 6 & -2 \end{bmatrix}$

23. $\begin{bmatrix} 3x & 2y & -z \\ -6x & 3y & 7z \\ 0 & -7y & -7z \end{bmatrix}$

25. $\begin{bmatrix} -5 & -1 \\ 10 & 4 \end{bmatrix}$

27. undefined

29.
$\begin{bmatrix} -5 & -1 \\ 10 & 4 \end{bmatrix} + \begin{bmatrix} -3 & -4 \\ 2 & -5 \end{bmatrix} =$
$\begin{bmatrix} -8 & -5 \\ 12 & -1 \end{bmatrix}$

31.
$\begin{bmatrix} -4 & 1 \\ 3 & 0 \end{bmatrix} + \begin{bmatrix} -4 & -6 \\ 9 & -1 \end{bmatrix} =$
$\begin{bmatrix} -8 & -5 \\ 12 & -1 \end{bmatrix}$

33. $\begin{bmatrix} -5 \\ 7 \end{bmatrix}$

35. $\begin{bmatrix} -5 \\ 7 \end{bmatrix}$

37. $\begin{bmatrix} -8 & 2 \\ 6 & 0 \end{bmatrix} - \begin{bmatrix} -1 & -2 \\ 7 & 4 \end{bmatrix} = \begin{bmatrix} -7 & 4 \\ -1 & -4 \end{bmatrix}$

39. $\begin{bmatrix} -8 \\ 10 \end{bmatrix} - \begin{bmatrix} -3 \\ 6 \end{bmatrix} = \begin{bmatrix} -5 \\ 4 \end{bmatrix}$

41. Equate corresponding entries.
$$x + y = 5$$
$$x - y = 1$$
Adding the two equations, $2x = 6$. Substituting $x = 3$ into $x + y = 5$, $3 + y = 5$ and $y = 2$. Solution is $(3, 2)$.

43. Equate corresponding entries.

$$2x + 3y = 7$$
$$x - 4y = -13$$

Multiply second equation by -2 and add to the first one.

$$2x + 3y = 7$$
$$-2x + 8y = 26$$
$$\overline{}$$
$$11y = 33$$
$$y = 3$$

Substitute $y = 3$ into $x - 4y = -13$ to get $x - 12 = -13$ and $x = -1$. Solution is $(-1, 3)$.

45. Equating corresponding entries,

$$x + y + z = 8$$
$$x - y - z = -7$$
$$x - y + z = 2$$

Adding the first and second equations, $2x = 1$ and $x = 0.5$. Multiply second equation by -1 and add to the third.

$$-x + y + z = 7$$
$$x - y + z = 2$$
$$\overline{}$$
$$2z = 9$$
$$z = 4.5$$

Substitute $x = 0.5$ and $z = 4.5$ into $x + y + z = 8$ to get $y + 5 = 8$ and $y = 3$. Solution is $(x, y, z) = (0.5, 3, 4.5)$.

47. The matrices for January, February and March are, respectively,

$$J = \begin{bmatrix} 120 \\ 30 \\ 4 \end{bmatrix}, F = \begin{bmatrix} 130 \\ 70 \\ 50 \end{bmatrix}, \text{ and } M = \begin{bmatrix} 140 \\ 60 \\ 45 \end{bmatrix}.$$

The sum $J + F + M = \begin{bmatrix} \$390 \\ \$160 \\ \$135 \end{bmatrix}$ represents the total expenses on food, clothing and utilities for the three months.

49. The supply matrix for the first week is

$$S = \begin{bmatrix} 40 & 80 \\ 30 & 90 \\ 80 & 200 \end{bmatrix}.$$ Next week's supply matrix is

$$S + 0.5S = \begin{bmatrix} 40 & 80 \\ 30 & 90 \\ 80 & 200 \end{bmatrix} + \begin{bmatrix} 20 & 40 \\ 15 & 45 \\ 40 & 100 \end{bmatrix} = \begin{bmatrix} 60 & 120 \\ 45 & 135 \\ 120 & 300 \end{bmatrix}.$$

51. yes, yes

53. yes, yes

55. yes, yes

57. $\begin{bmatrix} 0 & 0 \\ 0 & 0 \end{bmatrix}$

For Thought

1. True   **2.** True   **3.** False, they cannot be multiplied since the number of columns of A is not the same as the number of rows of C.

4. False, the order of CA is 2×1.   **5.** True

6. True, $BC = [7 \cdot 2 + 9 \cdot 4 \quad 7 \cdot 3 + 9 \cdot 5] =$
$= [14 + 36 \quad 21 + 45] = [50 \quad 66]$.

7. True, $AB = \begin{bmatrix} 1 \\ 6 \end{bmatrix} [7 \quad 9] = \begin{bmatrix} 1 \cdot 7 & 1 \cdot 9 \\ 6 \cdot 7 & 6 \cdot 9 \end{bmatrix} =$
$= \begin{bmatrix} 7 & 9 \\ 42 & 54 \end{bmatrix}.$

8. True, $\begin{bmatrix} 2 & 3 \\ 4 & 5 \end{bmatrix} \begin{bmatrix} 2 & -1 \\ 0 & 3 \end{bmatrix} =$
$\begin{bmatrix} 4 + 0 & -2 + 9 \\ 8 + 0 & -4 + 15 \end{bmatrix} = \begin{bmatrix} 4 & 7 \\ 8 & 11 \end{bmatrix}.$

9. True, $BA = [7 \quad 9] \begin{bmatrix} 1 \\ 6 \end{bmatrix} = [7 + 54] = [61]$.

10. False, since $EC = \begin{bmatrix} 2 & -1 \\ 0 & 3 \end{bmatrix} \begin{bmatrix} 2 & 3 \\ 4 & 5 \end{bmatrix} =$
$= \begin{bmatrix} 4 - 4 & 6 - 5 \\ 0 + 12 & 0 + 15 \end{bmatrix} = \begin{bmatrix} 0 & 1 \\ 12 & 15 \end{bmatrix}$ and from Number 8 one has $EC \neq CE$.

6.3 Exercises

1. 3×5 **3.** 1×1

5. 5×5 **7.** 3×3 **9.** undefined

11.
$$AB = \begin{bmatrix} 2\cdot 2 & 2\cdot 3 & 2\cdot 4 \\ -3\cdot 2 & -3\cdot 3 & -3\cdot 4 \\ 1\cdot 2 & 1\cdot 3 & 1\cdot 4 \end{bmatrix} =$$
$$\begin{bmatrix} 4 & 6 & 8 \\ -6 & -9 & -12 \\ 2 & 3 & 4 \end{bmatrix}$$

13. $\begin{bmatrix} 2+0+0 & 2+3+0 & 2+3+4 \end{bmatrix} =$
$\begin{bmatrix} 2 & 5 & 9 \end{bmatrix}$

15. undefined

17.
$$EC = \begin{bmatrix} 2+4+1 & 3+5+0 \\ 0+4+1 & 0+5+0 \\ 0+0+1 & 0+0+0 \end{bmatrix} =$$
$$\begin{bmatrix} 7 & 8 \\ 5 & 5 \\ 1 & 0 \end{bmatrix}$$

19.
$$DC = \begin{bmatrix} 2-4+1 & 6-5+0 \\ 0+12+2 & 0+15+0 \end{bmatrix} =$$
$$\begin{bmatrix} 1 & 1 \\ 14 & 15 \end{bmatrix}$$

21. undefined

23.
$$EA = \begin{bmatrix} 2-3+1 \\ 0-3+1 \\ 0+0+1 \end{bmatrix} = \begin{bmatrix} 0 \\ -2 \\ 1 \end{bmatrix}$$

25. undefined

27.
$$\begin{bmatrix} 2\cdot 1+0\cdot 0 & 2\cdot 1+0\cdot 1 \\ 3\cdot 1+1\cdot 0 & 3\cdot 1+1\cdot 1 \end{bmatrix} = \begin{bmatrix} 2 & 2 \\ 3 & 4 \end{bmatrix}$$

29.
$$\begin{bmatrix} 7\cdot 3+4\cdot(-5) & 7\cdot(-4)+4\cdot 7 \\ 5\cdot 3+3\cdot(-5) & 5\cdot(-4)+3\cdot 7 \end{bmatrix} =$$
$$\begin{bmatrix} 1 & 0 \\ 0 & 1 \end{bmatrix}$$

31.
$$\begin{bmatrix} -0.5\cdot 1+4\cdot 0 & 0.5\cdot 0+4\cdot 1 \\ 9\cdot 1+0.7\cdot 0 & 9\cdot 0+0.7\cdot 1 \end{bmatrix} =$$
$$\begin{bmatrix} -0.5 & 4 \\ 9 & 0.7 \end{bmatrix}$$

33. $\begin{bmatrix} -2a+6a & -6b+3b \end{bmatrix} = \begin{bmatrix} 4a & -3b \end{bmatrix}$

35.
$$\begin{bmatrix} -2a+0\cdot 1 & 5a+0\cdot 4 & 3a+0\cdot 6 \\ 0\cdot(-2)+b & 0\cdot 5+4b & 0\cdot 3+6b \end{bmatrix} =$$
$$\begin{bmatrix} -2a & 5a & 3a \\ b & 4b & 6b \end{bmatrix}$$

37. $\begin{bmatrix} 1\cdot 1+2\cdot 0+3\cdot 1 & 1\cdot 0+2\cdot 1+3\cdot 0 & 1\cdot 1+2\cdot 1+3\cdot 1 \end{bmatrix} =$
$\begin{bmatrix} 4 & 2 & 6 \end{bmatrix} =$

39. $\begin{bmatrix} -1\cdot(-5)+0\cdot 1+3\cdot 4 \end{bmatrix} = \begin{bmatrix} 17 \end{bmatrix}$

41.
$$\begin{bmatrix} x^2 & xy \\ xy & y^2 \end{bmatrix}$$

43.
$$\begin{bmatrix} -1\cdot \sqrt{2}+2\cdot 0+3\sqrt{2} \\ 3\cdot \sqrt{2}+4\cdot 0+4\sqrt{2} \end{bmatrix} = \begin{bmatrix} 2\sqrt{2} \\ 7\sqrt{2} \end{bmatrix}$$

45.
$$\begin{bmatrix} (1/2)\cdot(-8)+(1/3)\cdot(-5) & 6+(1/3)\cdot 15 \\ (1/4)\cdot(-8)+(1/5)\cdot(-5) & 3+(1/5)\cdot 15 \end{bmatrix}$$
$$= \begin{bmatrix} -4-(5/3) & 6+5 \\ -2-1 & 3+3 \end{bmatrix} = \begin{bmatrix} -17/3 & 11 \\ -3 & 6 \end{bmatrix}$$

47. undefined

49.
$$\begin{bmatrix} 9+0-7 & 8+0-8 & 10+0-4 \\ 0+3+0 & 0+5+0 & 0+2+0 \\ 9+3+7 & 8+5+8 & 10+2+4 \end{bmatrix} =$$
$$\begin{bmatrix} 2 & 0 & 6 \\ 3 & 5 & 2 \\ 19 & 21 & 16 \end{bmatrix}$$

51.
$$\begin{bmatrix} 1-0.6-0.6 & 1.5+1.2-1 \\ 0.8+0.4+1.8 & 1.2-0.8+3 \\ 0.4+0.6-2.4 & 0.6-1.2-4 \end{bmatrix} =$$
$$\begin{bmatrix} -0.2 & 1.7 \\ 3 & 3.4 \\ -1.4 & -4.6 \end{bmatrix}$$

53. System of equations is
$$2x - 3y = 0$$
$$x + 2y = 7$$

Multiply second equation by -2 and add to the first one.
$$2x - 3y = 0$$
$$-2x - 4y = -14$$
$$\overline{}$$
$$-7y = -14$$

Substituting $y = 2$ into $x + 2y = 7$, $x + 4 = 7$ and $x = 3$. Solution is $(3, 2)$.

55. System of equations is
$$2x + 3y = 5$$
$$4x + 6y = 9$$

Multiply first equation by -2 and add to the second one.
$$-4x - 6y = -10$$
$$4x + 6y = 9$$
$$\overline{}$$
$$0 = -1$$

No solution since $0 = -1$ is a contradiction.

57. System of equations is given by
$$x + y + z = 4$$
$$y + z = 5$$
$$z = 6$$

Substitute $z = 6$ into $y + z = 5$ to get $y + 6 = 5$ and $y = -1$. From $x + y + z = 4$, we have $x - 1 + 6 = 4$ and $x = -1$. Solution is $(x, y, z) = (-1, -1, 6)$.

59.
$$\begin{bmatrix} 2 & 3 \\ 4 & -1 \end{bmatrix} \begin{bmatrix} x \\ y \end{bmatrix} = \begin{bmatrix} 9 \\ 6 \end{bmatrix}$$

61.
$$\begin{bmatrix} 1 & 2 & -1 \\ 3 & -1 & 3 \\ 2 & 1 & -4 \end{bmatrix} \begin{bmatrix} x \\ y \\ z \end{bmatrix} = \begin{bmatrix} 3 \\ 1 \\ 0 \end{bmatrix}$$

63.
$$A = \begin{bmatrix} \$24,000 & \$40,000 \\ \$38,000 & \$70,000 \end{bmatrix}, Q = \begin{bmatrix} 4 \\ 7 \end{bmatrix}, \text{ and}$$

matrix product $AQ = \begin{bmatrix} \$376,000 \\ \$642,000 \end{bmatrix}$

represents the costs for labor and material for building 4 economy houses and 7 deluxe models.

65. False

67. True

69. True

For Thought

1. True, $AB = \begin{bmatrix} 10 - 9 & -6 + 6 \\ 15 - 15 & -9 + 10 \end{bmatrix} =$
$$= \begin{bmatrix} 1 & 0 \\ 0 & 1 \end{bmatrix} = \begin{bmatrix} 10 - 9 & 15 - 15 \\ -6 + 6 & -9 + 10 \end{bmatrix} = BA.$$

2. True, $AB = BA = I$ by number 1.

3. True, by number 2 the inverse of B is A.

4. False, AC is undefined, although AC is defined. **5.** True

6. False, a non-square matrix has no inverse.

7. False, the coefficient matrix is $\begin{bmatrix} 2 & 3 \\ 3 & 1 \end{bmatrix}$.

8. True, since $A^{-1} = B$ then $A^{-1}D =$
$$= \begin{bmatrix} 5 & -3 \\ -3 & 2 \end{bmatrix} \begin{bmatrix} 11 \\ 19 \end{bmatrix} = \begin{bmatrix} -2 \\ 5 \end{bmatrix}.$$

9. False, $(-2, 5)$ does not satisfy $3x + y = 19$.

10. False, the solution is $\begin{bmatrix} 2 & 3 \\ 3 & 1 \end{bmatrix}^{-1} \begin{bmatrix} 3 \\ -7 \end{bmatrix}$.

6.4 Exercises

1.
$$I\begin{bmatrix} -3 & 5 \\ 12 & 6 \end{bmatrix} = \begin{bmatrix} -3 & 5 \\ 12 & 6 \end{bmatrix}$$

3.
$$\begin{bmatrix} -8+9 & -6+6 \\ 12-12 & 9-8 \end{bmatrix} = \begin{bmatrix} 1 & 0 \\ 0 & 1 \end{bmatrix}$$

5.
$$\begin{bmatrix} 5-4 & -4+4 \\ 5-5 & -4+5 \end{bmatrix} = \begin{bmatrix} 1 & 0 \\ 0 & 1 \end{bmatrix}$$

7.
$$\begin{bmatrix} 3 & 5 & 1 \\ 4 & 5 & 7 \\ 4 & 9 & 2 \end{bmatrix} I = \begin{bmatrix} 3 & 5 & 1 \\ 4 & 5 & 7 \\ 4 & 9 & 2 \end{bmatrix}$$

9.
$$\begin{bmatrix} 0+0+1 & 1+0-1 & -3+0-3 \\ 0+0+0 & 1+0+0 & -3+3+0 \\ 0+0+0 & 0+0+0 & 0+1+0 \end{bmatrix} =$$
$$\begin{bmatrix} 1 & 0 & 0 \\ 0 & 1 & 0 \\ 0 & 0 & 1 \end{bmatrix}$$

11.
$$\begin{bmatrix} .5+.5+0 & -.5+.5+0 & .5-.5+0 \\ 0+.5-.5 & 0+.5+.5 & 0-.5+.5 \\ .5+0-.5 & -.5+0+.5 & .5+0+.5 \end{bmatrix} =$$
$$\begin{bmatrix} 1 & 0 & 0 \\ 0 & 1 & 0 \\ 0 & 0 & 1 \end{bmatrix}$$

13. Yes, since $\begin{bmatrix} 3 & 1 \\ 11 & 4 \end{bmatrix}\begin{bmatrix} 4 & -1 \\ -11 & 3 \end{bmatrix} =$
$\begin{bmatrix} 12-11 & -3+3 \\ 44-44 & -11+12 \end{bmatrix} = \begin{bmatrix} 1 & 0 \\ 0 & 1 \end{bmatrix}$ and
similarly $\begin{bmatrix} 4 & -1 \\ -11 & 3 \end{bmatrix}\begin{bmatrix} 3 & 1 \\ 11 & 4 \end{bmatrix} = I$.

15. No, since $\begin{bmatrix} 1/2 & -1 \\ 3 & -12 \end{bmatrix}\begin{bmatrix} 4 & 2 \\ 1 & 1 \end{bmatrix} =$
$\begin{bmatrix} 2-1 & 1-1 \\ 12-12 & 6-12 \end{bmatrix} = \begin{bmatrix} 1 & 0 \\ 0 & -6 \end{bmatrix} \neq I$.

17. No, since only square matrices may have an inverse.

19. On $\begin{bmatrix} 1 & 4 & | & 1 & 0 \\ 0 & 2 & | & 0 & 1 \end{bmatrix}$, use $R_1 - 2R_2 \to R_1$ to get
$\begin{bmatrix} 1 & 0 & | & 1 & -2 \\ 0 & 2 & | & 0 & 1 \end{bmatrix}$, use $\frac{1}{2}R_2 \to R_2$ to get
$\begin{bmatrix} 1 & 0 & | & 1 & -2 \\ 0 & 1 & | & 0 & 1/2 \end{bmatrix}$. So $A^{-1} = \begin{bmatrix} 1 & -2 \\ 0 & 1/2 \end{bmatrix}$.

21. On $\begin{bmatrix} 1 & 6 & | & 1 & 0 \\ 1 & 9 & | & 0 & 1 \end{bmatrix}$, use $R_2 - R_1 \to R_2$ to get
$\begin{bmatrix} 1 & 6 & | & 1 & 0 \\ 0 & 3 & | & -1 & 1 \end{bmatrix}$, use $-2R_2 + R_1 \to R_1$ to get
$\begin{bmatrix} 1 & 0 & | & 3 & 0 \\ 0 & 3 & | & -1 & 1 \end{bmatrix}$, use $\frac{1}{3}R_2 \to R_2$ to get
$\begin{bmatrix} 1 & 0 & | & 3 & -2 \\ 0 & 1 & | & -1/3 & 1/3 \end{bmatrix}$.
$A^{-1} = \begin{bmatrix} 3 & -2 \\ -1/3 & 1/3 \end{bmatrix}$.

23. On $\begin{bmatrix} -2 & -3 & | & 1 & 0 \\ 3 & 4 & | & 0 & 1 \end{bmatrix}$, use $R_2 + R_1 \to R_1$ to get
$\begin{bmatrix} 1 & 1 & | & 1 & 1 \\ 3 & 4 & | & 0 & 1 \end{bmatrix}$, use $-3R_1 + R_2 \to R_2$ to get
$\begin{bmatrix} 1 & 1 & | & 1 & 1 \\ 0 & 1 & | & -3 & -2 \end{bmatrix}$, use $R_1 - R_2 \to R_1$ to get
$\begin{bmatrix} 1 & 0 & | & 4 & 3 \\ 0 & 1 & | & -3 & -2 \end{bmatrix}$. So $A^{-1} = \begin{bmatrix} 4 & 3 \\ -3 & -2 \end{bmatrix}$.

25. On $\begin{bmatrix} 1 & -5 & | & 1 & 0 \\ -1 & 3 & | & 0 & 1 \end{bmatrix}$, use $R_1 + R_2 \to R_2$ to get
$\begin{bmatrix} 1 & -5 & | & 1 & 0 \\ 0 & -2 & | & 1 & 1 \end{bmatrix}$, use $-\frac{1}{2}R_2 \to R_2$ to get
$\begin{bmatrix} 1 & -5 & | & 1 & 0 \\ 0 & 1 & | & -1/2 & -1/2 \end{bmatrix}$, use $5R_2 + R_1 \to R_1$
to get $\begin{bmatrix} 1 & 0 & | & -3/2 & -5/2 \\ 0 & 1 & | & -1/2 & -1/2 \end{bmatrix}$.
So $A^{-1} = \begin{bmatrix} -3/2 & -5/2 \\ -1/2 & -1/2 \end{bmatrix}$.

27. On $\begin{bmatrix} -1 & 5 & | & 1 & 0 \\ 2 & -10 & | & 0 & 1 \end{bmatrix}$, use $R_1 + R_2 \to R_1$ and $\frac{1}{2}R_2 \to R_2$ to get

$\begin{bmatrix} 1 & -5 & | & 1 & 1 \\ 1 & -5 & | & 0 & 1/2 \end{bmatrix}$, use $R_1 - R_2 \to R_1$ to get

$\begin{bmatrix} 0 & 0 & | & 1 & 1/2 \\ 1 & -5 & | & 0 & 1/2 \end{bmatrix}$. So A has no inverse.

29. On $\begin{bmatrix} 1 & 1 & 0 & | & 1 & 0 & 0 \\ 0 & -1 & -1 & | & 0 & 1 & 0 \\ 1 & 0 & -1 & | & 0 & 0 & 1 \end{bmatrix}$, use

$R_2 + R_1 \to R_1$ and $R_1 - R_3 \to R_3$ to get

$\begin{bmatrix} 1 & 0 & -1 & | & 1 & 1 & 0 \\ 0 & -1 & -1 & | & 0 & 1 & 0 \\ 0 & 1 & 1 & | & 1 & 0 & 1 \end{bmatrix}$, use $R_2 + R_3 \to R_2$

to get

$\begin{bmatrix} 1 & 0 & -1 & | & 1 & 1 & 0 \\ 0 & 0 & 0 & | & 1 & 1 & 1 \\ 0 & 1 & 1 & | & 1 & 0 & 1 \end{bmatrix}$. So A has no inverse.

31. On $\begin{bmatrix} 1 & 1 & 1 & | & 1 & 0 & 0 \\ 1 & -1 & -1 & | & 0 & 1 & 0 \\ 1 & -1 & 1 & | & 0 & 0 & 1 \end{bmatrix}$, use

$R_1 + R_2 \to R_2$ and $R_2 - R_3 \to R_3$ to get

$\begin{bmatrix} 1 & 1 & 1 & | & 1 & 0 & 0 \\ 2 & 0 & 0 & | & 1 & 1 & 0 \\ 0 & 0 & -2 & | & 0 & 1 & -1 \end{bmatrix}$, use

$-2R_1 + R_2 \to R_2$ and $-\frac{1}{2}R_3 \to R_3$ to get

$\begin{bmatrix} 1 & 1 & 1 & | & 1 & 0 & 0 \\ 0 & -2 & -2 & | & -1 & 1 & 0 \\ 0 & 0 & 1 & | & 0 & -1/2 & 1/2 \end{bmatrix}$, use

$-\frac{1}{2}R_2 \to R_2$ to get

$\begin{bmatrix} 1 & 1 & 1 & | & 1 & 0 & 0 \\ 0 & 1 & 1 & | & 1/2 & -1/2 & 0 \\ 0 & 0 & 1 & | & 0 & -1/2 & 1/2 \end{bmatrix}$, use

$R_1 - R_2 \to R_1$ to get

$\begin{bmatrix} 1 & 0 & 0 & | & 1/2 & 1/2 & 0 \\ 0 & 1 & 1 & | & 1/2 & -1/2 & 0 \\ 0 & 0 & 1 & | & 0 & -1/2 & 1/2 \end{bmatrix}$, use

$R_2 - R_3 \to R_2$ to get

$\begin{bmatrix} 1 & 0 & 0 & | & 1/2 & 1/2 & 0 \\ 0 & 1 & 0 & | & 1/2 & 0 & -1/2 \\ 0 & 0 & 1 & | & 0 & -1/2 & 1/2 \end{bmatrix}$.

So $A^{-1} = \begin{bmatrix} 1/2 & 1/2 & 0 \\ 1/2 & 0 & -1/2 \\ 0 & -1/2 & 1/2 \end{bmatrix}$.

33. On $\begin{bmatrix} 0 & 2 & 0 & | & 1 & 0 & 0 \\ 3 & 3 & 2 & | & 0 & 1 & 0 \\ 2 & 5 & 1 & | & 0 & 0 & 1 \end{bmatrix}$, use

$R_2 - R_3 \to R_3$ to get

$\begin{bmatrix} 0 & 2 & 0 & | & 1 & 0 & 0 \\ 3 & 3 & 2 & | & 0 & 1 & 0 \\ 1 & -2 & 1 & | & 0 & 1 & -1 \end{bmatrix}$, use

$R_3 \to R_1$ and $R_1 \to R_3$ to get

$\begin{bmatrix} 1 & -2 & 1 & | & 0 & 1 & -1 \\ 3 & 3 & 2 & | & 0 & 1 & 0 \\ 0 & 2 & 0 & | & 1 & 0 & 0 \end{bmatrix}$, use

$R_3 + R_1 \to R_1$ and $R_2 - R_3 \to R_2$ to get

$\begin{bmatrix} 1 & 0 & 1 & | & 1 & 1 & -1 \\ 3 & 1 & 2 & | & -1 & 1 & 0 \\ 0 & 2 & 0 & | & 1 & 0 & 0 \end{bmatrix}$, use

$-3R_1 + R_2 \to R_2$ to get

$\begin{bmatrix} 1 & 0 & 1 & | & 1 & 1 & -1 \\ 0 & 1 & -1 & | & -4 & -2 & 3 \\ 0 & 2 & 0 & | & 1 & 0 & 0 \end{bmatrix}$, use

$-2R_2 + R_3 \to R_3$ to get

$\begin{bmatrix} 1 & 0 & 1 & | & 1 & 1 & -1 \\ 0 & 1 & -1 & | & -4 & -2 & 3 \\ 0 & 0 & 2 & | & 9 & 4 & -6 \end{bmatrix}$, use

$\frac{1}{2}R_3 \to R_3$ to get

$\begin{bmatrix} 1 & 0 & 1 & | & 1 & 1 & -1 \\ 0 & 1 & -1 & | & -4 & -2 & 3 \\ 0 & 0 & 1 & | & 9/2 & 2 & -3 \end{bmatrix}$, use

$R_2 + R_3 \to R_2$ to get

$\begin{bmatrix} 1 & 0 & 1 & | & 1 & 1 & -1 \\ 0 & 1 & 0 & | & 1/2 & 0 & 0 \\ 0 & 0 & 1 & | & 9/2 & 2 & -3 \end{bmatrix}$, use

$R_1 - R_3 \to R_1$ to get

6.4 INVERSE OF MATRICES

$$\begin{bmatrix} 1 & 0 & 0 & | & -7/2 & -1 & 2 \\ 0 & 1 & 0 & | & 1/2 & 0 & 0 \\ 0 & 0 & 1 & | & 9/2 & 2 & -3 \end{bmatrix}.$$

So $A^{-1} = \begin{bmatrix} -7/2 & -1 & 2 \\ 1/2 & 0 & 0 \\ 9/2 & 2 & -3 \end{bmatrix}$.

35. Since coefficient matrix is $A = \begin{bmatrix} 1 & 6 \\ 1 & 9 \end{bmatrix}$,

$A^{-1} \begin{bmatrix} -3 \\ -6 \end{bmatrix} = \begin{bmatrix} 3 & -2 \\ -1/3 & 1/3 \end{bmatrix} \begin{bmatrix} -3 \\ -6 \end{bmatrix} =$

$= \begin{bmatrix} 3 \\ -1 \end{bmatrix}$. Solution is $(x,y) = (3,-1)$.

37. Since coefficient matrix is $A = \begin{bmatrix} 1 & 6 \\ 1 & 9 \end{bmatrix}$,

$A^{-1} \begin{bmatrix} 4 \\ 5 \end{bmatrix} = \begin{bmatrix} 3 & -2 \\ -1/3 & 1/3 \end{bmatrix} \begin{bmatrix} 4 \\ 5 \end{bmatrix} =$

$= \begin{bmatrix} 2 \\ 1/3 \end{bmatrix}$. Solution is $(x,y) = (2, 1/3)$.

39. Since coefficient matrix is $A = \begin{bmatrix} -2 & -3 \\ 3 & 4 \end{bmatrix}$,

$A^{-1} \begin{bmatrix} 1 \\ -1 \end{bmatrix} = \begin{bmatrix} 4 & 3 \\ -3 & -2 \end{bmatrix} \begin{bmatrix} 1 \\ -1 \end{bmatrix} =$

$= \begin{bmatrix} 1 \\ -1 \end{bmatrix}$. Solution is $(x,y) = (1,-1)$.

41. Since coefficient matrix is $A = \begin{bmatrix} 1 & -5 \\ -1 & 3 \end{bmatrix}$,

$A^{-1} \begin{bmatrix} -5 \\ 1 \end{bmatrix} = \begin{bmatrix} -3/2 & -5/2 \\ -1/2 & -1/2 \end{bmatrix} \begin{bmatrix} -5 \\ 1 \end{bmatrix} =$

$= \begin{bmatrix} 5 \\ 2 \end{bmatrix}$. Solution is $(x,y) = (5,2)$.

43. Since coefficient matrix is $A = \begin{bmatrix} 1 & 1 & 1 \\ 1 & -1 & -1 \\ 1 & -1 & 1 \end{bmatrix}$,

$A^{-1} \begin{bmatrix} 3 \\ -1 \\ 5 \end{bmatrix} =$

$= \begin{bmatrix} 1/2 & 1/2 & 0 \\ 1/2 & 0 & -1/2 \\ 0 & -1/2 & 1/2 \end{bmatrix} \begin{bmatrix} 3 \\ -1 \\ 5 \end{bmatrix} = \begin{bmatrix} 1 \\ -1 \\ 3 \end{bmatrix}$.

Solution is $(x,y,z) = (1,-1,3)$.

45. Since coefficient matrix is $A = \begin{bmatrix} 0 & 1 & 0 \\ 3 & 3 & 2 \\ 2 & 5 & 1 \end{bmatrix}$,

$A^{-1} \begin{bmatrix} 6 \\ 16 \\ 19 \end{bmatrix} =$

$= \begin{bmatrix} -7/2 & -1 & 2 \\ 1/2 & 0 & 0 \\ 9/2 & 2 & -3 \end{bmatrix} \begin{bmatrix} 6 \\ 16 \\ 19 \end{bmatrix} = \begin{bmatrix} 1 \\ 3 \\ 2 \end{bmatrix}$.

Solution is $(x,y,z) = (1,3,2)$.

47. Coefficient matrix of $\begin{array}{l} 0.3x + 0.1y = 3 \\ 2x + 4y = 7 \end{array}$

is $A = \begin{bmatrix} 0.3 & 0.1 \\ 2 & 4 \end{bmatrix}$. Note $A^{-1} \begin{bmatrix} 3 \\ 7 \end{bmatrix} =$

$= \begin{bmatrix} 4 & -0.1 \\ -2 & 0.3 \end{bmatrix} \begin{bmatrix} 3 \\ 7 \end{bmatrix} = \begin{bmatrix} 11.3 \\ -3.9 \end{bmatrix}$.

Solution is $(x,y) = (11.3, -3.9)$.

49. Use the Gauss-Jordan method. On augmented

matrix $\begin{bmatrix} 1 & -1 & 1 & | & 5 \\ 2 & -1 & 3 & | & 1 \\ 0 & 1 & 1 & | & -9 \end{bmatrix}$, use

$R_2 - R_1 \to R_1$ to get

$\begin{bmatrix} 1 & 0 & 2 & | & -4 \\ 2 & -1 & 3 & | & 1 \\ 0 & 1 & 1 & | & -9 \end{bmatrix}$, use

$2R_1 + R_2 \to R_2$ to get

$\begin{bmatrix} 1 & 0 & 2 & | & 4 \\ 0 & -1 & -1 & | & 9 \\ 0 & 1 & 1 & | & -9 \end{bmatrix}$, use

$R_2 + R_3 \to R_3$ and $-R_2 \to R_2$ to get

$\begin{bmatrix} 1 & 0 & 2 & | & -4 \\ 0 & 1 & 1 & | & -9 \\ 0 & 0 & 0 & | & 0 \end{bmatrix}$. Since

$y + z = -9$ and $x + 2z = -4$, the solution set is

$\{(-2z - 4, -z - 9, z) : z \text{ is any real number }\}$

51.

Note coefficient matrix is $A = \begin{bmatrix} 1 & 1 & 1 \\ 2 & 4 & 1 \\ 1 & 3 & 6 \end{bmatrix}$ and

$A^{-1} \begin{bmatrix} 1 \\ 2 \\ 3 \end{bmatrix} = \begin{bmatrix} 7/4 & -1/4 & -1/4 \\ -11/12 & 5/12 & 1/12 \\ 1/6 & -1/6 & 1/6 \end{bmatrix} \begin{bmatrix} 1 \\ 2 \\ 3 \end{bmatrix}$

$= \begin{bmatrix} 1/2 \\ 1/6 \\ 1/3 \end{bmatrix}$. Solution is

$(x, y, z) = (1/2, 1/6, 1/3)$.

53.

On $\begin{bmatrix} 1 & 0 & 1 & | & 1 & 0 & 0 \\ 0 & 2 & 2 & | & 0 & 1 & 0 \\ 2 & 1 & 0 & | & 0 & 0 & 1 \end{bmatrix}$, use

$2R_1 - R_3 \to R_3$ and $\frac{1}{2}R_2 \to R_2$ to get

$\begin{bmatrix} 1 & 0 & 1 & | & 1 & 0 & 0 \\ 0 & 1 & 1 & | & 0 & 1/2 & 0 \\ 0 & -1 & 2 & | & 2 & 0 & -1 \end{bmatrix}$, use

$R_2 + R_3 \to R_3$ to get

$\begin{bmatrix} 1 & 0 & 1 & | & 1 & 0 & 0 \\ 0 & 1 & 1 & | & 0 & 1/2 & 0 \\ 0 & 0 & 3 & | & 2 & 1/2 & -1 \end{bmatrix}$, use

$\frac{1}{3}R_3 \to R_3$ to get

$\begin{bmatrix} 1 & 0 & 1 & | & 1 & 0 & 0 \\ 0 & 1 & 1 & | & 0 & 1/2 & 0 \\ 0 & 0 & 1 & | & 2/3 & 1/6 & -1/3 \end{bmatrix}$, use

$R_2 - R_3 \to R_2$ and $-R_2 \to R_2$ to get

$\begin{bmatrix} 1 & 0 & 0 & | & 1/3 & -1/6 & 1/3 \\ 0 & 1 & 0 & | & -2/3 & 1/3 & 1/3 \\ 0 & 0 & 1 & | & 2/3 & 1/6 & -1/3 \end{bmatrix}$.

So $A^{-1} = \begin{bmatrix} 1/3 & -1/6 & 1/3 \\ -2/3 & 1/3 & 1/3 \\ 2/3 & 1/6 & -1/3 \end{bmatrix}$

55.

On $\begin{bmatrix} 0 & 4 & 2 & | & 1 & 0 & 0 \\ 0 & 3 & 2 & | & 0 & 1 & 0 \\ 1 & -1 & 1 & | & 0 & 0 & 1 \end{bmatrix}$, use

$R_1 \to R_3$ and $R_3 \to R_1$ to get

$\begin{bmatrix} 1 & -1 & 1 & | & 0 & 0 & 1 \\ 0 & 3 & 2 & | & 0 & 1 & 0 \\ 0 & 4 & 2 & | & 1 & 0 & 0 \end{bmatrix}$, use

$R_3 - R_2 \to R_3$ to get

$\begin{bmatrix} 1 & -1 & 1 & | & 0 & 0 & 1 \\ 0 & 1 & 0 & | & 1 & -1 & 0 \\ 0 & 4 & 2 & | & 1 & 0 & 0 \end{bmatrix}$, use

$R_2 - R_1 \to R_1$ to get

$\begin{bmatrix} 1 & 0 & 1 & | & 1 & -1 & 1 \\ 0 & 1 & 0 & | & 1 & -1 & 0 \\ 0 & 4 & 2 & | & 1 & 0 & 0 \end{bmatrix}$, use

$-4R_2 + R_3 \to R_3$ to get

$\begin{bmatrix} 1 & 0 & 1 & | & 1 & -1 & 1 \\ 0 & 1 & 0 & | & 1 & -1 & 0 \\ 0 & 0 & 2 & | & -3 & 4 & 0 \end{bmatrix}$, use

$\frac{1}{2}R_3 \to R_3$ to get

$\begin{bmatrix} 1 & 0 & 1 & | & 1 & -1 & 1 \\ 0 & 1 & 0 & | & 1 & -1 & 0 \\ 0 & 0 & 1 & | & -3/2 & 2 & 0 \end{bmatrix}$, use

$R_1 - R_3 \to R_1$ to get

$\begin{bmatrix} 1 & 0 & 0 & | & 5/2 & -3 & 1 \\ 0 & 1 & 0 & | & 1 & -1 & 0 \\ 0 & 0 & 1 & | & -3/2 & 2 & 0 \end{bmatrix}$.

So $A^{-1} = \begin{bmatrix} -5/2 & -3 & 1 \\ 1 & -1 & 0 \\ -3/2 & 2 & 0 \end{bmatrix}$

57.

On $\begin{bmatrix} 1 & 2 & 3 & 4 & | & 1 & 0 & 0 & 0 \\ 0 & 1 & 2 & 3 & | & 0 & 1 & 0 & 0 \\ 0 & 0 & 1 & 2 & | & 0 & 0 & 1 & 0 \\ 0 & 0 & 0 & 1 & | & 0 & 0 & 0 & 1 \end{bmatrix}$, use

$R_1 - 2R_2 \to R_1$ to get

$\begin{bmatrix} 1 & 0 & -1 & -2 & | & 1 & -2 & 0 & 0 \\ 0 & 1 & 2 & 3 & | & 0 & 1 & 0 & 0 \\ 0 & 0 & 1 & 2 & | & 0 & 0 & 1 & 0 \\ 0 & 0 & 0 & 1 & | & 0 & 0 & 0 & 1 \end{bmatrix}$, use

$-2R_4 + R_3 \to R_3$ to get

$\begin{bmatrix} 1 & 0 & -1 & -2 & | & 1 & -2 & 0 & 0 \\ 0 & 1 & 2 & 3 & | & 0 & 1 & 0 & 0 \\ 0 & 0 & 1 & 0 & | & 0 & 0 & 1 & -2 \\ 0 & 0 & 0 & 1 & | & 0 & 0 & 0 & 1 \end{bmatrix}$, use

$-2R_3 + R_2 \to R_2$ and $-2R_4 + R_3 \to R_3$ to get

$\begin{bmatrix} 1 & 0 & -1 & -2 & | & 1 & -2 & 0 & 0 \\ 0 & 1 & 0 & 3 & | & 0 & 1 & -2 & 4 \\ 0 & 0 & 1 & -2 & | & 0 & 0 & 1 & -4 \\ 0 & 0 & 0 & 1 & | & 0 & 0 & 0 & 1 \end{bmatrix}$, use

6.4 INVERSE OF MATRICES

$R_1 + R_3 \to R_1$ to get

$$\left[\begin{array}{cccc|cccc} 1 & 0 & 0 & -4 & 1 & -2 & 1 & -4 \\ 0 & 1 & 0 & 3 & 0 & 1 & -2 & 4 \\ 0 & 0 & 1 & -2 & 0 & 0 & 1 & -4 \\ 0 & 0 & 0 & 1 & 0 & 0 & 0 & 1 \end{array}\right], \text{ use}$$

$2R_4 + R_3 \to R_3$, $R_2 - 3R_4 \to R_2$ and $4R_4 + R_1 \to R_1$ to get

$$\left[\begin{array}{cccc|cccc} 1 & 0 & 0 & 0 & 1 & -2 & 1 & 0 \\ 0 & 1 & 0 & 0 & 0 & 1 & -2 & 1 \\ 0 & 0 & 1 & 0 & 0 & 0 & 1 & -2 \\ 0 & 0 & 0 & 1 & 0 & 0 & 0 & 1 \end{array}\right].$$

So $A^{-1} = \begin{bmatrix} 1 & -2 & 1 & 0 \\ 0 & 1 & -2 & 1 \\ 0 & 0 & 1 & -2 \\ 0 & 0 & 0 & 1 \end{bmatrix}$.

59.

$$A^{-1} = \begin{bmatrix} -55/6 & 5/2 & 5/3 \\ 35/12 & -5/4 & 5/6 \\ 40/3 & 0 & -10/3 \end{bmatrix} \text{ and }$$

$A^{-1} \begin{bmatrix} 27 \\ 9 \\ 16 \end{bmatrix} = \begin{bmatrix} -165 \\ 97.5 \\ 240 \end{bmatrix}$. Solution is

$(x, y, z) = (-165, 97.5, 240)$.

61.

$$A^{-1} = \begin{bmatrix} 68/133 & -36/133 & 8/19 \\ 10/19 & -12/19 & 6/19 \\ 127/133 & -122/133 & 6/19 \end{bmatrix} \text{ and }$$

$A^{-1} \begin{bmatrix} 16 \\ 24 \\ -8 \end{bmatrix} \approx \begin{bmatrix} -1.6842 \\ -9.2632 \\ -9.2632 \end{bmatrix}$. Solution is

$(x, y, z) \approx (-1.6842, -9.2632, -9.2632)$.

63.

Since $AA^{-1} = \begin{bmatrix} a & 7 \\ 3 & b \end{bmatrix} \begin{bmatrix} -b & 7 \\ 3 & -a \end{bmatrix} =$

$= \begin{bmatrix} 21 - ab & 0 \\ 0 & 21 - ab \end{bmatrix}$, $21 - ab = 1$ and

$ab = 20$. List of permissible pairs (a, b) are

$$\begin{array}{c|cccccc} a & 1 & 2 & 4 & 5 & 10 & 20 \\ b & 20 & 10 & 5 & 4 & 2 & 1 \end{array}.$$

Matrices are $\begin{bmatrix} 1 & 7 \\ 3 & 20 \end{bmatrix}, \begin{bmatrix} 2 & 7 \\ 3 & 10 \end{bmatrix},$

$\begin{bmatrix} 4 & 7 \\ 3 & 5 \end{bmatrix}, \begin{bmatrix} 5 & 7 \\ 3 & 4 \end{bmatrix},$

$\begin{bmatrix} 10 & 7 \\ 3 & 2 \end{bmatrix},$ and $\begin{bmatrix} 20 & 7 \\ 3 & 1 \end{bmatrix}.$

65. Let x and y be the costs of a dozen eggs and a magazine before taxes. So

$$0.08x + 0.05y = 0.15$$
$$x + y = 270 - 0.15.$$

If $A = \begin{bmatrix} .08 & .05 \\ 1 & 1 \end{bmatrix}$ then $A^{-1} \begin{bmatrix} .15 \\ 2.55 \end{bmatrix} =$

$= \begin{bmatrix} 100/3 & -5/3 \\ -100/3 & 8/3 \end{bmatrix} \begin{bmatrix} .15 \\ 2.55 \end{bmatrix} = \begin{bmatrix} .75 \\ 1.80 \end{bmatrix}.$

Eggs costs $0.75 a dozen and magazine costs $1.80.

67. Let x and y be the costs of one load of plywood and a load insulation, respectively. So

$$4x + 6y = 2500$$
$$3x + 5y = 1950.$$

If $A = \begin{bmatrix} 4 & 6 \\ 3 & 5 \end{bmatrix}$ then $A^{-1} \begin{bmatrix} 2500 \\ 1950 \end{bmatrix} =$

$= \begin{bmatrix} 5/2 & -3 \\ -3/2 & 2 \end{bmatrix} \begin{bmatrix} 2500 \\ 1950 \end{bmatrix} = \begin{bmatrix} 400 \\ 150 \end{bmatrix}.$

One load of plywood costs $400 and a load of insulation costs $150.

69.

One computes $A^{-1} = \begin{bmatrix} 2 & -1 \\ -5 & 3 \end{bmatrix}$. To decode the message, we find

$\begin{bmatrix} 2 & -1 \\ -5 & 3 \end{bmatrix} \begin{bmatrix} 36 \\ 65 \end{bmatrix} = \begin{bmatrix} 7 \\ 15 \end{bmatrix} = \begin{bmatrix} g \\ o \end{bmatrix},$

$\begin{bmatrix} 2 & -1 \\ -5 & 3 \end{bmatrix} \begin{bmatrix} 49 \\ 83 \end{bmatrix} = \begin{bmatrix} 15 \\ 4 \end{bmatrix} = \begin{bmatrix} o \\ d \end{bmatrix},$

$\begin{bmatrix} 2 & -1 \\ -5 & 3 \end{bmatrix} \begin{bmatrix} 12 \\ 24 \end{bmatrix} = \begin{bmatrix} 0 \\ 12 \end{bmatrix} = \begin{bmatrix} space \\ l \end{bmatrix},$

$\begin{bmatrix} 2 & -1 \\ -5 & 3 \end{bmatrix} \begin{bmatrix} 66 \\ 111 \end{bmatrix} = \begin{bmatrix} 21 \\ 3 \end{bmatrix} = \begin{bmatrix} u \\ c \end{bmatrix},$

$\begin{bmatrix} 2 & -1 \\ -5 & 3 \end{bmatrix} \begin{bmatrix} 33 \\ 55 \end{bmatrix} = \begin{bmatrix} 11 \\ 0 \end{bmatrix} = \begin{bmatrix} k \\ space \end{bmatrix}.$

The message is 'good luck.'

71. Let x and y be the amounts invested in the Asset Manager Fund and Magellan Fund, respectively. Since $.86(60,000) = 51,600$, then

$$x + y = 60,000$$
$$.76x + .90y = 51,600.$$

If $A = \begin{bmatrix} 1 & 1 \\ .76 & .90 \end{bmatrix}$, then

$$A^{-1} = \begin{bmatrix} 45/7 & -50/7 \\ -38/7 & 50/7 \end{bmatrix} \text{ and}$$

$A^{-1} \begin{bmatrix} 60,000 \\ 51,600 \end{bmatrix} = \begin{bmatrix} 17,143 \\ 42,857 \end{bmatrix}$. In the Asset Manager Fund, \$17,143 was invested and in the Magellan Fund \$42,857 was invested.

73. Let $x, y,$ and z be the prices of an animal totem, a trade-bead necklace, and a tribal mask, respectively. System is

$$24x + 33y + 12z = 202.23$$
$$19x + 40y + 22z = 209.38$$
$$30x + 9y + 19z = 167.66$$

The inverse of the coefficient matrix A is

$$A^{-1} = \frac{1}{11,007} \begin{bmatrix} 562 & -519 & 246 \\ 299 & 96 & -300 \\ 1029 & 774 & 333 \end{bmatrix}.$$

Since $A^{-1} \begin{bmatrix} 202.23 \\ 209.38 \\ 167.66 \end{bmatrix} \approx \begin{bmatrix} \$4.20 \\ \$2.75 \\ \$0.89 \end{bmatrix}$, the solution is $(x, y, z) \approx (\$4.20, \$2.75, \$0.89)$.

For Thought

1. False, $|A| = 12 - (-5) = 17$.

2. True, $|A| \neq 0$.

3. True, $|B| = 4 \cdot 5 - (-2)(-10) = 0$.

4. False, $|B| = 0$.

5. True, determinant of coefficient matrix A is nonzero.

6. True, in general $|LM| = |L||M|$ for any square matrices L and M of the same size.

7. False, system is not linear.   8. True

9. False, $\begin{vmatrix} 2 & 0.1 \\ 100 & 5 \end{vmatrix} = 2 \cdot 5 - 100(0.1) = 0.$

10. False, because a 2×2 matrix is not equal to the number 27.

6.5 Exercises

1. $\begin{vmatrix} 1 & 3 \\ 0 & 2 \end{vmatrix} = 1 \cdot 2 - 0 \cdot 3 = 2$

3. $\begin{vmatrix} 3 & 4 \\ 2 & 9 \end{vmatrix} = 3(9) - 2(4) = 19$

5. $\begin{vmatrix} -0.3 & -0.5 \\ -0.7 & 0.2 \end{vmatrix} = (-0.3)(0.2) - (-0.7)(-0.5)$
$= -0.41$

7. $\begin{vmatrix} 1/8 & -3/8 \\ 2 & -1/4 \end{vmatrix} = (1/8)(-1/4) - (2)(-3/8)$
$= -1/32 + 3/4 = 23/32$

9. $\begin{vmatrix} 0.02 & 0.4 \\ 1 & 20 \end{vmatrix} = (.02)(20) - (.4)(1) = 0$

11. $\begin{vmatrix} 3 & -5 \\ -9 & 15 \end{vmatrix} = (3)(15) - (-9)(-5) = 0$

13. Note $D = \begin{vmatrix} 2 & -1 \\ 1 & 3 \end{vmatrix} = 7, D_x = \begin{vmatrix} -11 & -1 \\ 12 & 3 \end{vmatrix} =$
$= -21$, and $D_y = \begin{vmatrix} 2 & -11 \\ 1 & 12 \end{vmatrix} = 35$. So

$$x = \frac{D_x}{D} = -\frac{21}{7} = -3 \text{ and } y = \frac{D_y}{D} = \frac{35}{7} = 5.$$

15. Rewrite system as

$$x - y = 6$$
$$x + y = 5$$

Note $D = \begin{vmatrix} 1 & -1 \\ 1 & 1 \end{vmatrix} = 2, D_x = \begin{vmatrix} 6 & -1 \\ 5 & 1 \end{vmatrix} =$

$= 11$, and $D_y = \begin{vmatrix} 1 & 6 \\ 1 & 5 \end{vmatrix} = -1$. So
$$x = \frac{D_x}{D} = \frac{11}{2} \text{ and } y = \frac{D_y}{D} = -\frac{1}{2}.$$

17.
Note $D = \begin{vmatrix} 1/2 & -1/3 \\ 1/4 & 1/2 \end{vmatrix} = 1/3$,

$D_x = \begin{vmatrix} 4 & -1/3 \\ 6 & 1/2 \end{vmatrix} = 4$, and

$D_y = \begin{vmatrix} 1/2 & 4 \\ 1/4 & 6 \end{vmatrix} = 2$. So

$$x = \frac{D_x}{D} = \frac{4}{1/3} = 12 \text{ and } y = \frac{D_y}{D} = \frac{2}{1/3} = 6.$$

19.
Note $D = \begin{vmatrix} 0.2 & 0.12 \\ 1 & 1 \end{vmatrix} = .08$,

$D_x = \begin{vmatrix} 148 & .12 \\ 900 & 1 \end{vmatrix} = 40$, and

$D_y = \begin{vmatrix} .2 & 148 \\ 1 & 900 \end{vmatrix} = 32$. So

$$x = \frac{D_x}{D} = \frac{40}{.08} = 500 \text{ and}$$
$$y = \frac{D_y}{D} = \frac{32}{.08} = 400.$$

21. Cramer's rule does not apply since
$$D = \begin{vmatrix} 3 & 1 \\ -6 & -2 \end{vmatrix} = 0.$$ Dividing the second equation by -2, one gets the first equation. Solution set is $\{(x,y) : 3x + y = 6\}$.

23. Adding the two equations, one gets $0 = 19$. Inconsistent and no solution.

25. Use Cramer's Rule on simplified system,
$$x - y = 3$$
$$3x - y = -9.$$

Note, $D = \begin{vmatrix} 1 & -1 \\ 3 & -1 \end{vmatrix} = 2$,

$D_x = \begin{vmatrix} 3 & -1 \\ -9 & -1 \end{vmatrix} = -12$, and

$D_y = \begin{vmatrix} 1 & 3 \\ 3 & -9 \end{vmatrix} = -18$. So
$$x = \frac{D_x}{D} = \frac{-12}{2} = -6 \text{ and}$$
$$y = \frac{D_y}{D} = \frac{-18}{2} = -9.$$

27. Note, $D = \begin{vmatrix} \sqrt{2} & \sqrt{3} \\ 3\sqrt{2} & -2\sqrt{3} \end{vmatrix} = -5\sqrt{6}$,

$D_x = \begin{vmatrix} 4 & \sqrt{3} \\ -3 & -2\sqrt{3} \end{vmatrix} = -5\sqrt{3}$, and

$D_y = \begin{vmatrix} \sqrt{2} & 4 \\ 3\sqrt{2} & -3 \end{vmatrix} = -15\sqrt{2}$.

So $x = \frac{D_x}{D} = \frac{-5\sqrt{3}}{-5\sqrt{6}} = \frac{1}{\sqrt{2}} =$
$= \frac{\sqrt{2}}{2}$ and $y = \frac{D_y}{D} = \frac{-15\sqrt{2}}{-5\sqrt{6}} = \frac{3}{\sqrt{3}} = \sqrt{3}$.

29. Multiply second equation by -1 and add to the first one.
$$x^2 + y^2 = 25$$
$$-x^2 + y = -5$$
$$\overline{}$$
$$y^2 + y = 20$$
$$y^2 + y - 20 = 0$$
$$(y+5)(y-4) = 0$$

If $y = -5$ then $x^2 = 0$ and $x = 0$.
If $y = 4$ then $x^2 = 9$ and $x = \pm 3$.
Solutions are $(0, -5), (\pm 3, 4)$.

31. Solving for x in the second equation, $x = 4 + 2y$. Substituting into first equation,
$$4 + 2y - 2y - y^2$$
$$4 = y^2$$
$$\pm 2 = y$$

Using $y = 2$ in $x = 4 + 2y$, $x = 8$.
Similarly, if $y = -2$ then $x = 0$.
Solutions are $(8, 2), (0, -2)$.

33.
Invertible, since $\begin{vmatrix} 4 & 0.5 \\ 2 & 3 \end{vmatrix} = 12 - 1 = 11 \neq 0$

35.
Not invertible, since $\begin{vmatrix} 3 & -4 \\ 9 & -12 \end{vmatrix} = -36 + 36 = 0$

37. Note, $D = \begin{vmatrix} 3.47 & 23.09 \\ 12.48 & 3.98 \end{vmatrix} = -274.3526$,

$D_x = \begin{vmatrix} 5978.95 & 23.09 \\ 2765.34 & 3.98 \end{vmatrix} = -40,055.4796$, and

$D_y = \begin{vmatrix} 3.47 & 5978.95 \\ 12.48 & 2765.34 \end{vmatrix} = -65,021.5662$.

Then $x = \dfrac{D_x}{D} = 146$ and $y = \dfrac{D_y}{D} = 237$.

39. Let x and y be the number of boys and girls, respectively. So

$$0.44x + 0.35y = 231$$
$$x + y = 615$$

Note, $D = \begin{vmatrix} .44 & .35 \\ 1 & 1 \end{vmatrix} = .09$,

$D_x = \begin{vmatrix} 231 & .35 \\ 615 & 1 \end{vmatrix} = 15.75$, and

$D_y = \begin{vmatrix} .44 & 231 \\ 1 & 615 \end{vmatrix} = 39.6$.

There were $x = \dfrac{D_x}{D} = \dfrac{15.75}{.09} = 175$ boys and $y = \dfrac{D_y}{D} = \dfrac{39.6}{.09} = 440$ girls.

41. Let x and y be the ten's digit and unit's digits in Sarah's age, respectively. So $10x + y$ was her age when she went to bed and $10y + x$ was her age when she awoke. Simplifying $10y + x = 72 + 10x + y$, $x - y = -8$. System is

$$x + y = 10$$
$$x - y = -8$$

Note, $D = \begin{vmatrix} 1 & 1 \\ 1 & -1 \end{vmatrix} = -2$,

$D_x = \begin{vmatrix} 10 & 1 \\ -8 & -1 \end{vmatrix} = -2$, and

$D_y = \begin{vmatrix} 1 & 10 \\ 1 & -8 \end{vmatrix} = -18$.

So $x = \dfrac{D_x}{D} = \dfrac{-2}{-2} = 1$ and $y = \dfrac{D_y}{D} = \dfrac{-18}{-2} = 9$. Sarah's age when she went to bed was 19 years old.

43. Let x be the number of degrees in each of the equal angles and let y be the measurement of the third angle. System is

$$2x + y = 180$$
$$x - y = 2$$

Note, $D = \begin{vmatrix} 2 & 1 \\ 1 & -1 \end{vmatrix} = -3$,

$D_x = \begin{vmatrix} 180 & 1 \\ 2 & -1 \end{vmatrix} = -182$, and

$D_y = \begin{vmatrix} 2 & 180 \\ 1 & 2 \end{vmatrix} = -176$.

So $x = \dfrac{D_x}{D} = \dfrac{182}{3}$ and $y = \dfrac{D_y}{D} = \dfrac{176°}{3}$.

Angles are $\dfrac{182°}{3}, \dfrac{182°}{3}$, and $\dfrac{176°}{3}$.

45. Yes, $|MN| = |M||N|$ since $|M| = 2$, $|N| = 3$, and $|MN| = \begin{vmatrix} 8 & 31 \\ 14 & 55 \end{vmatrix} = 6$.

47. $\begin{vmatrix} a & b \\ c & d \end{vmatrix} \begin{vmatrix} e & f \\ g & h \end{vmatrix} =$

$\begin{vmatrix} ae + bg & af + bh \\ ce + dg & cf + dh \end{vmatrix} =$

$(ae + bg)(cf + dh) - (ce + dg)(af + bh) =$
$aecf + bgcf + aedh + bgdh - ceaf - dgaf - cebh - dgbh = bgcf + aedh - dgaf - cebh =$
$ad(eh - gf) - bc(eh - gf) = (ad - bc)(eh - gf) =$

$= \begin{vmatrix} a & b \\ c & d \end{vmatrix} \begin{vmatrix} e & f \\ g & h \end{vmatrix}$

49. No, $|-2M| = \begin{vmatrix} -6 & -4 \\ -10 & -8 \end{vmatrix} = 8$ and $-2|M| = -4$.

For Thought

1. False, the sign array of A is used in evaluating $|A|$.
2. False, the last term should be $1 \cdot \begin{vmatrix} 3 & 4 \\ 0 & 0 \end{vmatrix}$.
3. True
4. True, $|A|$ was expanded about the third row.
5. False, $|A|$ can be expanded only about a row or column.
6. False, a minor is a 2×2 matrix only if it comes from a 3×3 matrix.
7. False, $x = \dfrac{D_x}{D}$.
8. True
9. False, it can happen that $D = 0$ and there are infinitely many solutions
10. False, Cramer's Rule applies only to a system of linear equations.

6.6 Exercises

1. $\begin{vmatrix} 5 & -6 \\ 9 & -8 \end{vmatrix} = -40 + 54 = 14.$

3. $\begin{vmatrix} 4 & 5 \\ 7 & 9 \end{vmatrix} = 36 - 35 = 1$

5. $\begin{vmatrix} 2 & 1 \\ 7 & -8 \end{vmatrix} = -16 - 7 = -23$

7. $\begin{vmatrix} 2 & 1 \\ 4 & -6 \end{vmatrix} = -12 - 4 = -16$

9. $1\begin{vmatrix} 1 & -2 \\ -1 & 5 \end{vmatrix} - (-3)\begin{vmatrix} -4 & 0 \\ -1 & 5 \end{vmatrix} + 3\begin{vmatrix} -4 & 0 \\ 1 & -2 \end{vmatrix}$
$= 1(3) - (-3)(-20) + 3(8) = -33$

11. $3\begin{vmatrix} 4 & -1 \\ 1 & -2 \end{vmatrix} - 0\begin{vmatrix} -1 & 2 \\ 1 & -2 \end{vmatrix} + 5\begin{vmatrix} -1 & 2 \\ 4 & -1 \end{vmatrix}$
$= 3(-7) - 0 + 5(-7) = -56$

13. $-2\begin{vmatrix} 0 & -1 \\ 2 & -7 \end{vmatrix} - (-3)\begin{vmatrix} 5 & 1 \\ 2 & -7 \end{vmatrix} + 0\begin{vmatrix} 5 & 1 \\ 0 & -1 \end{vmatrix}$
$= -2(2) - (-3)(-37) + 0 = -115$

15. $0.1\begin{vmatrix} 20 & 6 \\ 90 & 8 \end{vmatrix} - 0.4\begin{vmatrix} 30 & 1 \\ 90 & 8 \end{vmatrix} + 0.7\begin{vmatrix} 30 & 1 \\ 20 & 6 \end{vmatrix}$
$= 0.1(-380) - 0.4(150) + 0.7(160) = 14$

17. Expanding about the second row,
$D = -(-2)\begin{vmatrix} 3 & 5 \\ 3 & -4 \end{vmatrix} = -(-2)(-27) = -54$

19. Expanding about the first row,
$D = 1\begin{vmatrix} 2 & 2 \\ 4 & 4 \end{vmatrix} - 1\begin{vmatrix} 2 & 2 \\ 4 & 4 \end{vmatrix} + 1\begin{vmatrix} 2 & 2 \\ 4 & 4 \end{vmatrix}$
$= 1(0) - 1(0) + 1(0) = 0$

21. Expanding about the first row,
$D = -(-1)\begin{vmatrix} 3 & 6 \\ -2 & -5 \end{vmatrix} = -(-1)(-3) = -3$

23. Expanding about the second column,
$D = -9\begin{vmatrix} 2 & 1 \\ 4 & 6 \end{vmatrix} = -9(8) = -72$

25. Expanding about the first row,
$3\begin{vmatrix} -3 & 2 & 0 \\ 3 & 1 & 2 \\ -4 & 1 & 3 \end{vmatrix} + 0 + 1\begin{vmatrix} 2 & -3 & 0 \\ -2 & 3 & 2 \\ 2 & -4 & 3 \end{vmatrix} -$
$5\begin{vmatrix} 2 & -3 & 2 \\ -2 & 3 & 1 \\ 2 & -4 & 1 \end{vmatrix} =$
$= 3(-37) + 1(4) - 5(6) = -137$

27. Expand determinant about second row.
$-(1)\begin{vmatrix} -3 & 4 & 6 \\ 3 & 1 & -3 \\ 0 & 2 & 1 \end{vmatrix} + (-5)\begin{vmatrix} 2 & 4 & 6 \\ 1 & 1 & -3 \\ -2 & 2 & 1 \end{vmatrix} =$
$= -(1)(3) + (-5)(58) = -293$

29. One finds $D = \begin{vmatrix} 1 & 1 & 1 \\ 1 & -1 & 1 \\ 2 & 1 & 1 \end{vmatrix} = 2,$

$D_x = \begin{vmatrix} 6 & 1 & 1 \\ 2 & -1 & 1 \\ 7 & 1 & 1 \end{vmatrix} = 2, \; D_y = \begin{vmatrix} 1 & 6 & 1 \\ 1 & 2 & 1 \\ 2 & 7 & 1 \end{vmatrix} = 4,$

and $D_z = \begin{vmatrix} 1 & 1 & 6 \\ 1 & -1 & 2 \\ 2 & 1 & 7 \end{vmatrix} = 6.$

Since $x = \dfrac{D_x}{D} = 2/2 = 1$, $y = \dfrac{D_y}{D} = 4/2$,

and $z = \dfrac{D_z}{D} = 6/2 = 3$,

the solution is $(x,y,z) = (1,2,3)$.

31. One finds $D = \begin{vmatrix} 1 & 2 & 0 \\ 1 & -3 & 1 \\ 2 & -1 & 0 \end{vmatrix} = 5$,

$D_x = \begin{vmatrix} 8 & 2 & 0 \\ -2 & -3 & 1 \\ 1 & -1 & 0 \end{vmatrix} = 10$,

$D_y = \begin{vmatrix} 1 & 8 & 0 \\ 1 & -2 & 1 \\ 2 & 1 & 0 \end{vmatrix} = 15$,

and $D_z = \begin{vmatrix} 1 & 2 & 8 \\ 1 & -3 & -2 \\ 2 & -1 & 1 \end{vmatrix} = 25$.

Since $x = \dfrac{D_x}{D} = 10/5 = 2$,

$y = \dfrac{D_y}{D} = 15/5 = 3$,

and $z = \dfrac{D_z}{D} = 25/5 = 5$,

the solution is $(x,y,z) = (2,3,5)$.

33. One finds $D = \begin{vmatrix} 2 & -3 & 1 \\ 1 & 4 & -1 \\ 3 & -1 & 2 \end{vmatrix} = 16$,

$D_x = \begin{vmatrix} 1 & -3 & 1 \\ 0 & 4 & -1 \\ 0 & -1 & 2 \end{vmatrix} = 7$,

$D_y = \begin{vmatrix} 2 & 1 & 1 \\ 1 & 0 & -1 \\ 3 & 0 & 2 \end{vmatrix} = -5$,

and $D_z = \begin{vmatrix} 2 & -3 & 1 \\ 1 & 4 & 0 \\ 3 & -1 & 0 \end{vmatrix} = -13$.

Since $x = \dfrac{D_x}{D} = 7/16$,

$y = \dfrac{D_y}{D} = -5/16$, and $z = \dfrac{D_z}{D} = -13/16$,

$(x,y,z) = (7/16, -5/16, -13/16)$.

35. One finds $D = \begin{vmatrix} 1 & 1 & 1 \\ 2 & -1 & 3 \\ 3 & 1 & -1 \end{vmatrix} = 14$,

$D_x = \begin{vmatrix} 2 & 1 & 1 \\ 0 & -1 & 3 \\ 0 & 1 & -1 \end{vmatrix} = -4$,

$D_y = \begin{vmatrix} 1 & 2 & 1 \\ 2 & 0 & 3 \\ 3 & 0 & -1 \end{vmatrix} = 22$,

and $D_z = \begin{vmatrix} 1 & 1 & 2 \\ 2 & -1 & 0 \\ 3 & 1 & 0 \end{vmatrix} = 10$.

Since $x = \dfrac{D_x}{D} = -4/14$,

$y = \dfrac{D_y}{D} = 22/14$, and $z = \dfrac{D_z}{D} = 10/14$,

we have $(x,y,z) = (-2/7, 11/7, 5/7)$.

37. This system is dependent since the sum of the first two equations is the third equation. Adding the first and third equations, $3x - 3z = 4$ or $z = x - 4/3$. Substituting into the first equation,

$$x + y - 2\left(x - \dfrac{4}{3}\right) = 1$$

$$y = 1 - \dfrac{8}{3} + x$$

$$y = x - \dfrac{5}{3}$$

Solution set is

$$\left\{\left(x, x - \dfrac{5}{3}, x - \dfrac{4}{3}\right) : x \text{ is any real number}\right\}.$$

39. Multiply the first equation by -2 and add to the third equation.

$$-2x + 2y - 2z = -10$$
$$2x - 2y + 2z = 16$$
$$\overline{}$$
$$0 = 6$$

Inconsistent and the system has no solution.

6.6 THREE LINEAR SYSTEMS AND DETERMINANTS

41. Let $x, y,$ and z be the ages of Jackie, Rochelle, and Alisha, respectively. Since $\dfrac{x+y}{2} = 33$, $\dfrac{y+z}{2} = 25$, and $\dfrac{x+z}{2} = 19$ then

$$x + y = 66$$
$$y + z = 50$$
$$x + z = 38$$

One finds $D = \begin{vmatrix} 1 & 1 & 0 \\ 0 & 1 & 1 \\ 1 & 0 & 1 \end{vmatrix} = 2,$

$D_x = \begin{vmatrix} 66 & 1 & 0 \\ 50 & 1 & 1 \\ 38 & 0 & 1 \end{vmatrix} = 54,\ D_y = \begin{vmatrix} 1 & 66 & 0 \\ 0 & 50 & 1 \\ 1 & 38 & 1 \end{vmatrix} = 78,$ and $D_z = \begin{vmatrix} 1 & 1 & 66 \\ 0 & 1 & 50 \\ 1 & 0 & 38 \end{vmatrix} = 22.$

So Jackie is $x = \dfrac{D_x}{D} = 54/2 = 27$ years old, Rochelle is $y = \dfrac{D_y}{D} = 78/2 = 39$ years old, and Alisha is $z = \dfrac{D_z}{D} = 22/2 = 11$ years old.

43. Let $x, y,$ and z be the scores in the first test, second test, and final exam, respectively. So

$$x + y + z = 180$$
$$0.2x + 0.2y + 0.6z = 76$$
$$0.1x + 0.2y + 0.7z = 83$$

One finds $D = \begin{vmatrix} 1 & 1 & 1 \\ .2 & .2 & .6 \\ .1 & .2 & .7 \end{vmatrix} = -.04,$

$D_x = \begin{vmatrix} 180 & 1 & 1 \\ 76 & .2 & .6 \\ 83 & .2 & .7 \end{vmatrix} = -1.2,$

$D_y = \begin{vmatrix} 1 & 180 & 1 \\ .2 & 76 & .6 \\ .1 & 83 & .7 \end{vmatrix} = -2,$

and $D_z = \begin{vmatrix} 1 & 1 & 180 \\ .2 & .2 & 76 \\ .1 & .2 & 83 \end{vmatrix} = -4.$

Test scores are $x = \dfrac{D_x}{D} = 1.2/(.04) = 30$, $y = \dfrac{D_y}{D} = 2/(.04) = 50$, and the final exam is $z = \dfrac{D_z}{D} = 4/(.04) = 100.$

45. One finds $D = \begin{vmatrix} 0.2 & -0.3 & 1.2 \\ 0.25 & 0.35 & -0.9 \\ 2.4 & -1 & 1.25 \end{vmatrix} = -\dfrac{527}{800},$

$D_x = \begin{vmatrix} 13.11 & -0.3 & 1.2 \\ -1.575 & 0.35 & -0.9 \\ 42.02 & -1 & 1.25 \end{vmatrix} = -\dfrac{11,067}{1,000},$

$D_y = \begin{vmatrix} 0.2 & 13.11 & 1.2 \\ 0.25 & -1.575 & -0.9 \\ 2.4 & 42.02 & 1.25 \end{vmatrix} = -\dfrac{64,821}{8,000},$

and $D_z = \begin{vmatrix} 0.2 & -0.3 & 13.11 \\ 0.25 & 0.35 & -1.575 \\ 2.4 & -1 & 42.02 \end{vmatrix} = -\dfrac{3,689}{500}.$

Solution is
$$(x, y, z) = \left(\dfrac{D_x}{D}, \dfrac{D_y}{D}, \dfrac{D_z}{D}\right) = (16.8, 12.3, 11.2).$$

47. Let $x, y,$ and z be the prices per gallon of regular, plus, and supreme gasoline. One finds

$D = \begin{vmatrix} 1270 & 980 & 890 \\ 1450 & 1280 & 1050 \\ 1340 & 1190 & 1060 \end{vmatrix} = 18,038,000,$

$D_x = \begin{vmatrix} 3728.66 & 980 & 890 \\ 4496.82 & 1280 & 1050 \\ 4279.01 & 1190 & 1060 \end{vmatrix} = 19,823,762,$

$D_y = \begin{vmatrix} 1270 & 3728.66 & 890 \\ 1450 & 4496.82 & 1050 \\ 1340 & 4279.01 & 1060 \end{vmatrix} = 21,988,322,$

$D_z = \begin{vmatrix} 1270 & 980 & 3728.66 \\ 1450 & 1280 & 4496.82 \\ 1340 & 1190 & 4279.01 \end{vmatrix} = 23,070,602.$

Regular gas costs $\dfrac{D_x}{D} \approx \$1.099$,

Plus costs $\dfrac{D_y}{D} \approx \$1.219$, and

Supreme costs $\dfrac{D_z}{D} \approx \$1.279$.

49.

One finds $D = \begin{vmatrix} 1 & 1 & 1 & 1 \\ 2 & -1 & 1 & 3 \\ 1 & 2 & -1 & 2 \\ 1 & -1 & -1 & 4 \end{vmatrix} = 11,$

$D_w = \begin{vmatrix} 4 & 1 & 1 & 1 \\ 13 & -1 & 1 & 3 \\ -2 & 2 & -1 & 2 \\ 8 & -1 & -1 & 4 \end{vmatrix} = 11,$

$D_x = \begin{vmatrix} 1 & 4 & 1 & 1 \\ 2 & 13 & 1 & 3 \\ 1 & -2 & -1 & 2 \\ 1 & 8 & -1 & 4 \end{vmatrix} = -22,$

$D_y = \begin{vmatrix} 1 & 1 & 4 & 1 \\ 2 & -1 & 13 & 3 \\ 1 & 2 & -2 & 2 \\ 1 & -1 & 8 & 4 \end{vmatrix} = 33,$

and $D_z = \begin{vmatrix} 1 & 1 & 1 & 4 \\ 2 & -1 & 1 & 13 \\ 1 & 2 & -1 & -2 \\ 1 & -1 & -1 & 8 \end{vmatrix} = 22.$

Solution is
$\left(\dfrac{D_w}{D}, \dfrac{D_x}{D}, \dfrac{D_y}{D}, \dfrac{D_z}{D}\right) = (1, -2, 3, 2).$

51.

Note, $\begin{vmatrix} x & y & 1 \\ 3 & -5 & 1 \\ -2 & 6 & 1 \end{vmatrix} = 8 - 11x - 3y = 0.$

Since both points $(3, -5)$ and $(-2, 6)$ satisfies $8 - 11x - 3y = 0$, this is an equation of the line through the two points.

53. We will show it for one particular case and the other cases can proved similarly. Let us suppose $A = \begin{bmatrix} a & b & c \\ 0 & 0 & 0 \\ g & h & i \end{bmatrix}$. Then $|A| = a \cdot 0 \cdot i + b \cdot 0 \cdot g + c \cdot 0 \cdot h - g \cdot 0 \cdot c - h \cdot 0 \cdot a - i \cdot 0 \cdot b = 0$

55. We will prove it for one particular case and the other cases can be shown similarly. Let us suppose $A = \begin{bmatrix} a & b & c \\ d & e & f \\ kg & kh & ki \end{bmatrix}$. Then $|A| = aeki + bfkg + cdkh - kgec - khfa - kidb = k(aei + bfg + cdh - gec - hfa - idb) = k|A|.$

Review Exercises

1. $\begin{bmatrix} 2+3 & -3+7 \\ -2+1 & 4+2 \end{bmatrix} = \begin{bmatrix} 5 & 4 \\ -1 & 6 \end{bmatrix}$

3. $\begin{bmatrix} 4 & -6 \\ -4 & 8 \end{bmatrix} - \begin{bmatrix} 3 & 7 \\ 1 & 2 \end{bmatrix} = \begin{bmatrix} 1 & -13 \\ -5 & 6 \end{bmatrix}$

5. $AB = \begin{bmatrix} 6-3 & 14-6 \\ -6+4 & -14+8 \end{bmatrix} = \begin{bmatrix} 3 & 8 \\ -2 & -6 \end{bmatrix}$

7. $D + E$ is undefined

9. $AC = \begin{bmatrix} -2-9 \\ 2+12 \end{bmatrix} = \begin{bmatrix} -11 \\ 14 \end{bmatrix}$

11. $EF = \begin{bmatrix} 3 & 2 & -1 \\ -12 & -8 & 4 \\ 9 & 6 & -3 \end{bmatrix}$

13. $FG = [-3+2+2 \quad 2-3 \quad -1] = [1 \quad -1 \quad -1] =$

15. GF is undefined

17. On $\begin{bmatrix} 2 & -3 & | & 1 & 0 \\ -2 & 4 & | & 0 & 1 \end{bmatrix}$, use $R_1 + R_2 \to R_2$ to get

$\begin{bmatrix} 2 & -3 & | & 1 & 0 \\ 0 & 1 & | & 1 & 1 \end{bmatrix}$, use $-3R_2 + R_1 \to R_1$ to get

$\begin{bmatrix} 2 & 0 & | & 4 & 3 \\ 0 & 1 & | & 1 & 1 \end{bmatrix}$, use $\dfrac{1}{2}R_1 \to R_1$ to get

$\begin{bmatrix} 1 & 0 & | & 2 & 1.5 \\ 0 & 1 & | & 1 & 1 \end{bmatrix}$. So $A^{-1} = \begin{bmatrix} 2 & 1.5 \\ 1 & 1 \end{bmatrix}$.

19.
On $\begin{bmatrix} -1 & 0 & 0 & | & 1 & 0 & 0 \\ 1 & 1 & 0 & | & 0 & 1 & 0 \\ -2 & 3 & 1 & | & 0 & 0 & 1 \end{bmatrix}$,

use $R_1 + R_2 \to R_1$, $-2R_2 + R_3 \to R_3$, and $-R_1 \to R_1$ to get

$\begin{bmatrix} 1 & 0 & 0 & | & -1 & 0 & 0 \\ 0 & 1 & 0 & | & 1 & 1 & 0 \\ 0 & 3 & 1 & | & -2 & 0 & 1 \end{bmatrix}$,

use $-3R_2 + R_3 \to R_3$ to get

$\begin{bmatrix} 1 & 0 & 0 & | & -1 & 0 & 0 \\ 0 & 1 & 0 & | & 1 & 1 & 0 \\ 0 & 0 & 1 & | & -5 & -3 & 1 \end{bmatrix}$.

So $G^{-1} = \begin{bmatrix} -1 & 0 & 0 \\ 1 & 1 & 0 \\ -5 & -3 & 1 \end{bmatrix}$.

21.
From number 5, $(AB)^{-1} = \begin{bmatrix} 3 & 8 \\ -2 & -6 \end{bmatrix}^{-1}$.

On $\begin{bmatrix} 3 & 8 & | & 1 & 0 \\ -2 & -6 & | & 0 & 1 \end{bmatrix}$, use $-\frac{1}{2}R_2 \to R_2$

to get

$\begin{bmatrix} 3 & 8 & | & 1 & 0 \\ 1 & 3 & | & 0 & -0.5 \end{bmatrix}$, use $-3R_2 + R_1 \to R_1$

to get

$\begin{bmatrix} 0 & -1 & | & 1 & 1.5 \\ 1 & 3 & | & 0 & -0.5 \end{bmatrix}$, use $3R_1 + R_2 \to R_2$

to get

$\begin{bmatrix} 0 & -1 & | & 1 & 1.5 \\ 1 & 0 & | & 3 & 4 \end{bmatrix}$, use $R_1 \leftrightarrow R_2$ to get

$\begin{bmatrix} 1 & 0 & | & 3 & 4 \\ 0 & -1 & | & 1 & 1.5 \end{bmatrix}$, use $-R_2 \to R_2$ to get

$\begin{bmatrix} 1 & 0 & | & 3 & 4 \\ 0 & 1 & | & -1 & -1.5 \end{bmatrix}$.

So $(AB)^{-1} = \begin{bmatrix} 3 & 4 \\ -1 & -1.5 \end{bmatrix}$.

23.
$AA^{-1} = I = \begin{bmatrix} 1 & 0 \\ 0 & 1 \end{bmatrix}$

25. $|A| = 8 - 6 = 2$

27. Expanding about the third column,

$|G| = 1 \cdot \begin{vmatrix} -1 & 0 \\ 1 & 1 \end{vmatrix} = 1(-1) = -1$.

29. Solution is $(x, y) = (10/3, 17/3)$.

First solution is by Gauss-Jordan.

On $\begin{bmatrix} 1 & 1 & | & 9 \\ 2 & -1 & | & 1 \end{bmatrix}$, use $-2R_1 + R_2 \to R_2$ to get

$\begin{bmatrix} 1 & 1 & | & 9 \\ 0 & -3 & | & -17 \end{bmatrix}$, use $-\frac{1}{3}R_2 \to R_2$ to get

$\begin{bmatrix} 1 & 1 & | & 9 \\ 0 & 1 & | & 17/3 \end{bmatrix}$, use $-R_2 + R_1 \to R_1$ to get

$\begin{bmatrix} 1 & 0 & | & 10/3 \\ 0 & 1 & | & 17/3 \end{bmatrix}$.

Secondly, by matrix inversion note $A^{-1} = \begin{bmatrix} 1/3 & 1/3 \\ 2/3 & -1/3 \end{bmatrix}$ and $A^{-1} \begin{bmatrix} 9 \\ 1 \end{bmatrix} = \begin{bmatrix} 10/3 \\ 17/3 \end{bmatrix}$.

Thirdly, by Cramer's Rule note

$D = \begin{vmatrix} 1 & 1 \\ 2 & -1 \end{vmatrix} = -3$, $D_x = \begin{vmatrix} 9 & 1 \\ 1 & -1 \end{vmatrix} = -10$,

and $D_y = \begin{vmatrix} 1 & 9 \\ 2 & 1 \end{vmatrix} = -17$.

So $\frac{D_x}{D} = 10/3$, $\frac{D_y}{D} = 17/3$.

31. Solution is $(x, y) = (-2, 3)$.

First solution is by Gauss-Jordan. On

$\begin{bmatrix} 2 & 1 & | & -1 \\ 3 & 2 & | & 0 \end{bmatrix}$, use $-\frac{3}{2}R_1 + R_2 \to R_2$ to get

$\begin{bmatrix} 2 & 1 & | & -1 \\ 0 & 1/2 & | & 3/2 \end{bmatrix}$, use $-2R_2 + R_1 \to R_1$

and $2R_2 \to R_2$ to get

$\begin{bmatrix} 2 & 0 & | & -4 \\ 0 & 1 & | & 3 \end{bmatrix}$, use $\frac{1}{2}R_2 \to R_2$ to get

$\begin{bmatrix} 1 & 0 & | & -2 \\ 0 & 1 & | & 3 \end{bmatrix}$.

Secondly, by matrix inversion note $A^{-1} = \begin{bmatrix} 2 & -1 \\ -3 & 2 \end{bmatrix}$ and $A^{-1} \begin{bmatrix} -1 \\ 0 \end{bmatrix} = \begin{bmatrix} -2 \\ 3 \end{bmatrix}$.

Thirdly, by Cramer's Rule note

$$D = \begin{vmatrix} 2 & 1 \\ 3 & 2 \end{vmatrix} = 1, \quad D_x = \begin{vmatrix} -1 & 1 \\ 0 & 2 \end{vmatrix} = -2,$$

and $D_y = \begin{vmatrix} 2 & -1 \\ 3 & 0 \end{vmatrix} = 3.$

So $\dfrac{D_x}{D} = -2, \dfrac{D_y}{D} = 3.$

33. Solution set is $\{(x,y)|x - 5y = 9\}$.

Solution is by Gauss-Jordan.

On $\begin{bmatrix} 1 & -5 & | & 9 \\ -2 & 10 & | & -18 \end{bmatrix}$, use $2R_1 + R_2 \to R_2$

to get $\begin{bmatrix} 1 & -5 & | & 9 \\ 0 & 0 & | & 0 \end{bmatrix}$. Dependent system.

This cannot be solved by matrix inversion since

$A^{-1} = \begin{bmatrix} 1 & -5 \\ -2 & 10 \end{bmatrix}^{-1}$ does not exist.

Nor can it be solved by Cramer's rule

since $|D| = \begin{vmatrix} 1 & -5 \\ -2 & 10 \end{vmatrix} = 0.$

35. No solution as shown by Gauss-Jordan. On

$\begin{bmatrix} 0.05 & 0.1 & | & 1 \\ 10 & 20 & | & 20 \end{bmatrix}$, use $-200R_1 + R_2 \to R_2$

to get $\begin{bmatrix} 0.05 & 0.1 & | & 1 \\ 0 & 0 & | & -180 \end{bmatrix}$. Inconsistent.

This cannot be solved by matrix inversion since

$A^{-1} = \begin{bmatrix} 0.05 & 0.1 \\ 10 & 20 \end{bmatrix}^{-1}$ does not exist.

Nor can it be solved by Cramer's rule

since $|D| = \begin{vmatrix} 0.05 & 0.1 \\ 10 & 20 \end{vmatrix} = 0.$

37. Solution is $(x,y,z) = (1,2,3)$.

First solution is by Gauss-Jordan.

On $\begin{bmatrix} 1 & 1 & -2 & | & -3 \\ -1 & 2 & -1 & | & 0 \\ -1 & -1 & 3 & | & 6 \end{bmatrix}$, use

$R_1 + R_2 \to R_2$ to get

$\begin{bmatrix} 1 & 1 & -2 & | & -3 \\ 0 & 3 & -3 & | & -3 \\ 0 & 0 & 1 & | & 3 \end{bmatrix}$, use

$\dfrac{1}{3}R_2 \to R_2$ to get

$\begin{bmatrix} 1 & 1 & -2 & | & -3 \\ 0 & 1 & -1 & | & -1 \\ 0 & 0 & 1 & | & 3 \end{bmatrix}$, use

$R_3 + R_2 \to R_2$ and $2R_3 + R_1 \to R_1$ to get

$\begin{bmatrix} 1 & 1 & 0 & | & 3 \\ 0 & 1 & 0 & | & 2 \\ 0 & 0 & 1 & | & 3 \end{bmatrix}$, use

$-R_2 + R_1 \to R_1$ to get

$\begin{bmatrix} 1 & 0 & 0 & | & 1 \\ 0 & 1 & 0 & | & 2 \\ 0 & 0 & 1 & | & 3 \end{bmatrix}.$

Secondly, by matrix inversion note $A^{-1} =$

$\begin{bmatrix} 5/3 & -1/3 & 1 \\ 4/3 & 1/3 & 1 \\ 1 & 0 & 1 \end{bmatrix}$ and $A^{-1} \begin{bmatrix} -3 \\ 0 \\ 6 \end{bmatrix} = \begin{bmatrix} 1 \\ 2 \\ 3 \end{bmatrix}.$

Thirdly, by Cramer's Rule note

$$D = \begin{vmatrix} 1 & 1 & -2 \\ -1 & 2 & -1 \\ -1 & -1 & 3 \end{vmatrix} = 3,$$

$$D_x = \begin{vmatrix} -3 & 1 & -2 \\ 0 & 2 & -1 \\ 6 & -1 & 3 \end{vmatrix} = 3,$$

$$D_y = \begin{vmatrix} 1 & -3 & -2 \\ -1 & 0 & -1 \\ -1 & 6 & 3 \end{vmatrix} = 6,$$

and $D_z = \begin{vmatrix} 1 & 1 & -3 \\ -1 & 2 & 0 \\ -1 & -1 & 6 \end{vmatrix} = 9.$

So $\dfrac{D_x}{D} = 1, \dfrac{D_y}{D} = 2,$ and $\dfrac{D_z}{D} = 3.$

39. Solution is $(x,y,z) = (-3,4,1)$.

First solution is by Gauss-Jordan.

On $\begin{bmatrix} 0 & 1 & -3 & | & 1 \\ 1 & 2 & 0 & | & 5 \\ 1 & 0 & 4 & | & 1 \end{bmatrix}$, use

$-R_3 + R_2 \to R_2$ and $R_1 \leftrightarrow R_3$ to get

$\begin{bmatrix} 1 & 0 & 4 & | & 1 \\ 0 & 2 & -4 & | & 4 \\ 0 & 1 & -3 & | & 1 \end{bmatrix}$, use

$\dfrac{1}{2}R_2 \to R_2$ to get

$\begin{bmatrix} 1 & 0 & 4 & | & 1 \\ 0 & 1 & -2 & | & 2 \\ 0 & 1 & -3 & | & 1 \end{bmatrix}$, use

$-R_2 + R_3 \to R_3$ to get

$\begin{bmatrix} 1 & 0 & 4 & | & 1 \\ 0 & 1 & -2 & | & 2 \\ 0 & 0 & -1 & | & -1 \end{bmatrix}$, use

$-2R_3 + R_2 \to R_2$, $4R_3 + R_1 \to R_1$, and $-R_3 \to R_3$ to get

$\begin{bmatrix} 1 & 0 & 0 & | & -3 \\ 0 & 1 & 0 & | & 4 \\ 0 & 0 & 1 & | & 1 \end{bmatrix}$.

Secondly, by matrix inversion

$A^{-1} = \begin{bmatrix} 4 & -2 & 3 \\ -2 & 3/2 & -3/2 \\ -1 & 1/2 & -1/2 \end{bmatrix}$

and $A^{-1} \begin{bmatrix} 1 \\ 5 \\ 1 \end{bmatrix} = \begin{bmatrix} -3 \\ 4 \\ 1 \end{bmatrix}$.

Thirdly, by Cramer's Rule note

$D = \begin{vmatrix} 0 & 1 & -3 \\ 1 & 2 & 0 \\ 1 & 0 & 4 \end{vmatrix} = 2$,

$D_x = \begin{vmatrix} 1 & 1 & -3 \\ 5 & 2 & 0 \\ 1 & 0 & 4 \end{vmatrix} = -6$,

$D_y = \begin{vmatrix} 0 & 1 & -3 \\ 1 & 5 & 0 \\ 1 & 1 & 4 \end{vmatrix} = 8$,

and $D_z = \begin{vmatrix} 0 & 1 & 1 \\ 1 & 2 & 5 \\ 1 & 0 & 1 \end{vmatrix} = 2$.

So $\dfrac{D_x}{D} = -3$, $\dfrac{D_y}{D} = 4$, and $\dfrac{D_z}{D} = 1$.

41. Solution set is

$\left\{ \left(\dfrac{3y+3}{2}, y, \dfrac{1-y}{2} \right) \mid y \text{ is any real number} \right\}$.

Solution is by Gauss-Jordan.

On $\begin{bmatrix} 1 & -1 & 1 & | & 2 \\ 1 & -2 & -1 & | & 1 \\ 2 & -3 & 0 & | & 3 \end{bmatrix}$, use

$-R_1 + R_2 \to R_2$ and $-2R_1 + R_3 \to R_3$ to get

$\begin{bmatrix} 1 & -1 & 1 & | & 2 \\ 0 & -1 & -2 & | & -1 \\ 0 & -1 & -2 & | & -1 \end{bmatrix}$, use

$-R_2 + R_3 \to R_3$, $-R_2 + R_1 \to R_1$, and $-R_2 \to R_2$ to get

$\begin{bmatrix} 1 & 0 & 3 & | & 3 \\ 0 & 1 & 2 & | & 1 \\ 0 & 0 & 0 & | & 0 \end{bmatrix}$. Since $z = \dfrac{1-y}{2}$,

$x = 3 - 3z = 3 - 3\left(\dfrac{1-y}{2}\right) = \dfrac{3y+3}{2}$.

This cannot be solved by matrix inversion since

$A^{-1} = \begin{bmatrix} 1 & -1 & 1 \\ 1 & -2 & -1 \\ 2 & -3 & 0 \end{bmatrix}^{-1}$ does not exist.

Nor can it be solved by Cramer's rule

since $|D| = \begin{vmatrix} 1 & -1 & 1 \\ 1 & -2 & -1 \\ 2 & -3 & 0 \end{vmatrix} = 0$.

43. No solution as shown by Gauss-Jordan.

On $\begin{bmatrix} 1 & -3 & -1 & | & 2 \\ 1 & -3 & -1 & | & 1 \\ 1 & -3 & -1 & | & 0 \end{bmatrix}$, use

$-R_1 + R_2 \to R_2$ and $-R_1 + R_3 \to R_3$ to get

$\begin{bmatrix} 1 & -3 & -1 & | & 2 \\ 0 & 0 & 0 & | & -1 \\ 0 & 0 & 0 & | & -2 \end{bmatrix}$. Inconsistent.

This cannot be solved by matrix inversion since

$A^{-1} = \begin{bmatrix} 1 & -3 & -1 \\ 1 & -3 & -1 \\ 1 & -3 & -1 \end{bmatrix}^{-1}$ does not exist.

Nor can it be solved by Cramer's rule

since $|D| = \begin{vmatrix} 1 & -3 & -1 \\ 1 & 3 & 1 \\ 1 & -3 & -1 \end{vmatrix} = 0$.

45. Using $x = 9$ in $x + y = -3$, $9 + y = -3$ and $y = -12$. Solution set is $\{(9, -12)\}$.

47.

Note $\begin{bmatrix} x \\ y \end{bmatrix} = \begin{bmatrix} 1 & 1 \\ 2 & 1 \end{bmatrix}^{-1} \begin{bmatrix} 6 \\ 8 \end{bmatrix} = \begin{bmatrix} -1 & 1 \\ 2 & -1 \end{bmatrix} \begin{bmatrix} 6 \\ 8 \end{bmatrix} = \begin{bmatrix} 2 \\ 4 \end{bmatrix}$.

So $(x, y) = (2, 4)$.

49. System of equations can be written as
$$x + y = -3$$
$$-x = 0$$
Using $x = 0$ in $x + y = -3$, $y = -3$.
So $(x, y) = (0, -3)$.

51. System of equations can be written as
$$\begin{bmatrix} 1 & 1 & 0 \\ 0 & 1 & 1 \\ 1 & 0 & 1 \end{bmatrix} \begin{bmatrix} x \\ y \\ z \end{bmatrix} = \begin{bmatrix} 1 \\ 1 \\ 1 \end{bmatrix}.$$
By using matrix inversion,
$$\begin{bmatrix} x \\ y \\ z \end{bmatrix} = \begin{bmatrix} 1/2 & -1/2 & 1/2 \\ 1/2 & 1/2 & -1/2 \\ -1/2 & 1/2 & 1/2 \end{bmatrix} \begin{bmatrix} 1 \\ 1 \\ 1 \end{bmatrix} =$$
$$\begin{bmatrix} 1/2 \\ 1/2 \\ 1/2 \end{bmatrix}.$$ Solution is $\{(1/2, 1/2, 1/2)\}$.

53. By using the inverse of the coefficient matrix,
$$\begin{bmatrix} x \\ y \\ z \end{bmatrix} = \begin{bmatrix} 3/5 & -3/5 & 2/5 \\ 2/5 & 3/5 & -2/5 \\ -1/5 & 1/5 & 1/5 \end{bmatrix} \begin{bmatrix} -1 \\ 7 \\ 17 \end{bmatrix} =$$
$$\begin{bmatrix} 2 \\ -3 \\ 5 \end{bmatrix}.$$ Solution set is $\{(2, -3, 5)\}$.

55. Let x and y be the number of gallons of pollutant A and pollutant B, respectively. So
$$10x + 6y = 4060$$
$$\frac{3}{4} = \frac{x}{y}$$
Substitute $x = \frac{3}{4}y$ in $10x + 6y = 4060$.
Solving for x, one finds the quantities discharged are $x \approx 225.56$ gallons of pollutant A and $y \approx 300.74$ gallons of pollutant B.

57. Let $x, y,$ and z be the expenses including tax for water, gas, and electricity, respectively. So
$$x + y + z = 189.83$$
$$\frac{x}{1.04} + \frac{y}{1.05} + \frac{z}{1.06} = 180$$
$$z = 2y$$

Solving the system, one finds the expenses including taxes are $x = \$22.88$ for water, $y = \$55.65$ for gas, and $z = \$111.30$ for electricity.

Chapter 6 Test

1. On $\begin{bmatrix} 2 & -3 & | & 1 \\ 1 & 9 & | & 4 \end{bmatrix}$, use $-2R_2 + R_1 \to R_2$ to get
$\begin{bmatrix} 2 & -3 & | & 1 \\ 0 & -21 & | & -7 \end{bmatrix}$, use $\frac{1}{2}R_1 \to R_1$ and
$-\frac{1}{21}R_2 \to R_2$ to get
$\begin{bmatrix} 1 & -3/2 & | & 1/2 \\ 0 & 1 & | & 1/3 \end{bmatrix}$, use $\frac{3}{2}R_2 + R_1 \to R_1$ to get
$\begin{bmatrix} 1 & 0 & | & 1 \\ 0 & 1 & | & 1/3 \end{bmatrix}$. Solution set is $\{(1, 1/3)\}$.

2. On $\begin{bmatrix} 2 & -1 & 1 & | & 5 \\ 1 & -2 & -1 & | & -2 \\ 3 & -1 & -1 & | & 6 \end{bmatrix}$, use
$-\frac{1}{2}R_1 + R_2 \to R_2$ and
$-\frac{3}{2}R_1 + R_3 \to R_3$ to get
$\begin{bmatrix} 2 & -1 & 1 & | & 5 \\ 0 & -3/2 & -3/2 & | & -9/2 \\ 0 & 1/2 & -5/2 & | & -3/2 \end{bmatrix}$, use
$-\frac{2}{3}R_2 \to R_2$ and $3R_3 + R_2 \to R_2$ to get
$\begin{bmatrix} 2 & -1 & 1 & | & 5 \\ 0 & 1 & 1 & | & 3 \\ 0 & 0 & -9 & | & -9 \end{bmatrix}$,
use $-\frac{1}{9}R_3 \to R_3$ and $R_2 + R_1 \to R_1$ to get
$\begin{bmatrix} 2 & 0 & 2 & | & 8 \\ 0 & 1 & 1 & | & 3 \\ 0 & 0 & 1 & | & 1 \end{bmatrix}$, use
$-R_3 + R_2 \to R_2$ and $\frac{1}{2}R_1 \to R_1$ to get
$\begin{bmatrix} 1 & 0 & 1 & | & 4 \\ 0 & 1 & 0 & | & 2 \\ 0 & 0 & 1 & | & 1 \end{bmatrix}$, use $-R_3 + R_1 \to R_1$

CHAPTER 6 TEST

to get $\begin{bmatrix} 1 & 0 & 0 & | & 3 \\ 0 & 1 & 0 & | & 2 \\ 0 & 0 & 1 & | & 1 \end{bmatrix}$.

Solution set is $\{(3,2,1)\}$.

3.

On $\begin{bmatrix} 1 & -1 & -1 & | & 1 \\ 2 & 1 & -1 & | & 0 \\ 5 & -2 & -4 & | & 3 \end{bmatrix}$, use

$-2R_1 + R_2 \to R_2$ and
$-5R_1 + R_3 \to R_3$ to get
$\begin{bmatrix} 1 & -1 & -1 & | & 1 \\ 0 & 3 & 1 & | & -2 \\ 0 & 3 & 1 & | & -2 \end{bmatrix}$, use

$R_2 - R_3 \to R_3$ and $\frac{1}{3}R_2 \to R_2$ to get
$\begin{bmatrix} 1 & -1 & -1 & | & 1 \\ 0 & 1 & 1/3 & | & -2/3 \\ 0 & 0 & 0 & | & 0 \end{bmatrix}$, use

$R_2 + R_1 \to R_1$ to get
$\begin{bmatrix} 1 & 0 & -2/3 & | & 1/3 \\ 0 & 1 & 1/3 & | & -2/3 \\ 0 & 0 & 0 & | & 0 \end{bmatrix}$. Since

$x = \frac{2}{3}z + \frac{1}{3}$, $3x = 2z + 1$ and

$z = \frac{3x-1}{2}$. Since $y = -\frac{1}{3}z - \frac{2}{3}$,

$y = -\frac{1}{3}\left(\frac{3x-1}{2}\right) - \frac{2}{3}$

$y = \frac{-3x-3}{6}$

$y = \frac{-x-1}{2}$

Solution set is
$\left\{ \left(x, \frac{-x-1}{2}, \frac{3x-1}{2}\right) : x \text{ is any real number} \right\}$.

4.
$A + B = \begin{bmatrix} 3 & -4 \\ -6 & 10 \end{bmatrix}$

5.
$2A - B = \begin{bmatrix} 2 & -2 \\ -4 & 8 \end{bmatrix} - \begin{bmatrix} 2 & -3 \\ -4 & 6 \end{bmatrix} =$

$\begin{bmatrix} 0 & 1 \\ 0 & 2 \end{bmatrix} =$

6.
$AB = \begin{bmatrix} 2+4 & -3-6 \\ -4-16 & 6+24 \end{bmatrix} = \begin{bmatrix} 6 & -9 \\ -20 & 30 \end{bmatrix}$

7.
$AC = \begin{bmatrix} -2-1 \\ 4+4 \end{bmatrix} = \begin{bmatrix} -3 \\ 8 \end{bmatrix}$

8. CB is undefined

9. $FG = [-2 \quad 3-2 \quad 1+1] - [-2 \quad 1 \quad 2]$

10.
$EF = \begin{bmatrix} 2 & 0 & -2 \\ 3 & 0 & -3 \\ -1 & 0 & 1 \end{bmatrix}$

11.

On $\begin{bmatrix} 1 & -1 & | & 1 & 0 \\ -2 & 4 & | & 0 & 1 \end{bmatrix}$, use

$2R_1 + R_2 \to R_2$ to get
$\begin{bmatrix} 1 & -1 & | & 1 & 0 \\ 0 & 2 & | & 2 & 1 \end{bmatrix}$, use

$\frac{1}{2}R_2 + R_1 \to R_1$ and $\frac{1}{2}R_2 \to R_2$

to get $\begin{bmatrix} 1 & 0 & | & 2 & 1/2 \\ 0 & 1 & | & 1 & 1/2 \end{bmatrix}$.

So $A^{-1} = \begin{bmatrix} 2 & 1/2 \\ 1 & 1/2 \end{bmatrix}$

12.

On $\begin{bmatrix} -2 & 3 & 1 & | & 1 & 0 & 0 \\ -3 & 1 & 3 & | & 0 & 1 & 0 \\ 0 & 2 & -1 & | & 0 & 0 & 1 \end{bmatrix}$,

use $-\frac{1}{2}R_1 \to R_1$ and $-\frac{3}{2}R_1 + R_2 \to R_2$ to get

$\begin{bmatrix} 1 & -3/2 & -1/2 & | & -1/2 & 0 & 0 \\ 0 & -7/2 & 3/2 & | & -3/2 & 1 & 0 \\ 0 & 2 & -1 & | & 0 & 0 & 1 \end{bmatrix}$, use

$\frac{4}{7}R_2 + R_3 \to R_3$, $-\frac{3}{7}R_2 + R_1 \to R_1$,

and $-\frac{2}{7}R_2 \to R_2$ to get

$\begin{bmatrix} 1 & 0 & -8/7 & | & 1/7 & -3/7 & 0 \\ 0 & 1 & -3/7 & | & 3/7 & -2/7 & 0 \\ 0 & 0 & -1/7 & | & -6/7 & 4/7 & 1 \end{bmatrix}$, use

$-3R_3 + R_2 \to R_2$, $-8R_3 + R_1 \to R_1$,

and $-7R_3 \to R_3$ to get

201

$$\begin{bmatrix} 1 & 0 & 0 & 7 & -5 & -8 \\ 0 & 1 & 0 & 3 & -2 & -3 \\ 0 & 0 & 1 & 6 & -4 & -7 \end{bmatrix}.$$

So $G^{-1} = \begin{bmatrix} 7 & -5 & -8 \\ 3 & -2 & -3 \\ 6 & -4 & -7 \end{bmatrix}$

13. $|A| = 4 - 2 = 2$

14. $|B| = 12 - 12 = 0$

15. Expanding about the third row,
$$|G| = -2 \begin{vmatrix} -2 & 1 \\ -3 & 3 \end{vmatrix} + (-1) \begin{vmatrix} -2 & 3 \\ -3 & 1 \end{vmatrix} = -2(-3) + (-1)(7) = -1$$

16.
One finds $|D| = \begin{vmatrix} 1 & -1 \\ -2 & 4 \end{vmatrix} = 2,$

$|D_x| = \begin{vmatrix} 2 & -1 \\ 2 & 4 \end{vmatrix} = 10,\ |D_y| = \begin{vmatrix} 1 & 2 \\ -2 & 2 \end{vmatrix} = 6.$

Solution is $\left(\dfrac{D_x}{D}, \dfrac{D_y}{D} \right) = (5, 3).$

17. Cramer's Rule is not applicable since
$|D| = \begin{vmatrix} 2 & -3 \\ -4 & 6 \end{vmatrix} = 0.$ Rather, multiply first equation by 2 and add to the second one.

$$4x - 6y = 12$$
$$-4x + 6y = 1$$
$$\overline{}$$
$$0 = 13$$

Inconsistent and no solution.

18.
One finds $|D| = \begin{vmatrix} -2 & 3 & 1 \\ -3 & 1 & 3 \\ 0 & 2 & -1 \end{vmatrix} = -1,$

$|D_x| = \begin{vmatrix} -2 & 3 & 1 \\ -4 & 1 & 3 \\ 0 & 2 & -1 \end{vmatrix} = -6,$

$|D_y| = \begin{vmatrix} -2 & -2 & 1 \\ -3 & -4 & 3 \\ 0 & 0 & -1 \end{vmatrix} = -2,$ and

$|D_z| = \begin{vmatrix} -2 & 3 & -2 \\ -3 & 1 & -4 \\ 0 & 2 & 0 \end{vmatrix} = -4.$ Solution is

$\left(\dfrac{D_x}{D}, \dfrac{D_y}{D}, \dfrac{D_z}{D} \right) = (6, 2, 4).$

19. Inverse of coefficient matrix is given by
$A^{-1} = \begin{bmatrix} 2 & 1/2 \\ 1 & 1/2 \end{bmatrix}$. Since
$A^{-1} \begin{bmatrix} 1 \\ -8 \end{bmatrix} = \begin{bmatrix} -2 \\ -3 \end{bmatrix}$, the solution set is $\{(-2, -3)\}$.

20. Inverse of coefficient matrix is given by Number 12, i.e. $A^{-1} = \begin{bmatrix} 7 & -5 & -8 \\ 3 & -2 & -3 \\ 6 & -4 & -7 \end{bmatrix}.$

Since $A^{-1} \begin{bmatrix} 1 \\ 0 \\ -1 \end{bmatrix} = \begin{bmatrix} 15 \\ 6 \\ 13 \end{bmatrix},$
solution set is $\{(15, 6, 13)\}$.

21. Corresponding system of equations is
$$x - y = 12$$
$$35y - 10x = 730$$
Solving this system, one finds $x = 46$ copies were bought and $y = 34$ copies were sold.

22. Substitute $(0, 3)$, $(1, -1/2)$, and $(4, 3)$ into $y = ax^2 + b\sqrt{x} + c$. So
$$c = 3$$
$$a + b + c = -\dfrac{1}{2}$$
$$16a + 2b + c = 3$$
Solving this system, one finds $a = 0.5$, $b = -4$, $c = 3$, and the graph is given by $y = 0.5x^2 - 4\sqrt{x} + 3$.

Tying It All Together

1. Simplify the left-hand side.
$$2x + 6 - 5x = 7$$
$$-3x = 1$$
Solution set is $\{-1/3\}$.

2. Multiply equation by 30.
$$\frac{1}{2}x - \frac{1}{6} = \frac{4}{5}$$
$$15x - 5 = 24$$
$$15x = 29$$
Solution set is $\{29/15\}$.

3. Multiply $\frac{1}{2}$ to $(2x - 2)$.
$$(x - 1)(6x - 8) = 4$$
$$6x^2 - 14x + 8 = 4$$
$$6x^2 - 14x + 4 = 0$$
$$2(3x - 1)(x - 2) = 0$$
Solution set is $\{1/3, 2\}$.

4. Simplify left-hand side.
$$1 - (4x - 2) = 9$$
$$-4x + 3 = 9$$
$$-4x = 6$$
Solution set is $\{-3/2\}$.

5. Solution set is $\{(4, -2)\}$ as can be seen from the point of intersection.

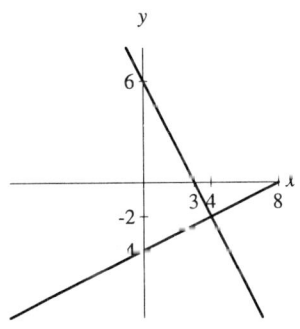

6. Using $y = 1 - 2x$ in $2x + 6y = 2$,
$$2x + 6(1 - 2x) = 2$$
$$2x + 6 - 12x = 2$$
$$-10x = -4$$
Using $x = \frac{2}{5}$ in $y = 1 - 2x$, we have
$y = 1 - \frac{4}{5} = \frac{1}{5}$. Solution set is $\left\{\left(\frac{2}{5}, \frac{1}{5}\right)\right\}$.

7. Multiply second equation by 6 and add to the first one.
$$2x - 0.06y = 20$$
$$18x + 0.06y = 120$$
$$\overline{}$$
$$20x = 140$$
Using $x = 7$ in $2x - 0.06y = 20$,
$$14 - 0.06y = 20$$
$$-0.06y = 6$$
$$y = -100$$
Solution set is $\{(7, -100)\}$.

8. On $\begin{bmatrix} 2 & -1 & | & -1 \\ 1 & 3 & | & -11 \end{bmatrix}$, use $-2R_2 + R_1 \to R_2$
to get
$\begin{bmatrix} 2 & -1 & | & -1 \\ 0 & -7 & | & 21 \end{bmatrix}$, use $-\frac{1}{7}R_2 \to R_2$
and $\frac{1}{2}R_1 \to R_1$ to get
$\begin{bmatrix} 1 & -1/2 & | & -1/2 \\ 0 & 1 & | & -3 \end{bmatrix}$, use $\frac{1}{2}R_2 + R_1 \to R_1$
to get $\begin{bmatrix} 1 & 0 & | & -2 \\ 0 & 1 & | & -3 \end{bmatrix}$.
Solution set is $\{(-2, -3)\}$.

9. Inverse of the coefficient matrix A is
$$A^{-1} = \begin{bmatrix} -1/2 & -5/2 \\ -1/2 & -3/2 \end{bmatrix}.$$
Since $A^{-1}\begin{bmatrix} -7 \\ 1 \end{bmatrix} = \begin{bmatrix} 1 \\ 2 \end{bmatrix}$,
the solution set is $\{(1, 2)\}$.

10.

One finds $|D| = \begin{vmatrix} 4 & -3 \\ 3 & -5 \end{vmatrix} = -11$,

$|D_x| = \begin{vmatrix} 5 & -3 \\ 1 & -5 \end{vmatrix} = -22$, and

$|D_y| = \begin{vmatrix} 4 & 5 \\ 3 & 1 \end{vmatrix} = -11$.

Solution is $\left(\dfrac{D_x}{D}, \dfrac{D_y}{D}\right) = (2, 1)$.

11. Using $x = y - 1$ in $x^2 + y^2 = 25$,
$$(y^2 - 2y + 1) + y^2 = 25$$
$$2y^2 - 2y - 24 = 0$$
$$2(y - 4)(y + 3) = 0$$
If $y = 4$ then $x = y - 1 = 4 - 1 = 3$.
If $y = -3$ then $x = (-3) - 1 = -4$.
Solution set is $\{(3, 4), (-4, -3)\}$.

12. Using $y = x^2 - 1$ in $x + y = 1$,
$$x + (x^2 - 1) = 1$$
$$x^2 + x - 2 = 0$$
$$(x + 2)(x - 1) = 0$$
If $x = -2$ then $y = x^2 - 1 = (-2)^2 - 1 = 3$.
If $x = 1$ then $y = 1^2 - 1 = 0$.
Solution set is $\{(-2, 3), (1, 0)\}$.

7.1 THE PARABOLA

For Thought

1. False, vertex is $(0, -1/2)$.   2. True

3. True, since $p = 3/2$ and the focus is $(4, 5)$, vertex is $(4 - 3/2, 5) = (5/2, 5)$.

4. False. Focus is at $(0, 1/4)$ since parabola opens upward.   5. True, $p = 1/4$.

6. False. Since $p = 4$ and the vertex is $(2, -1)$, equation of parabola is $y = \frac{1}{16}(x-2)^2 - 1$ and x-intercepts are $(6, 0), (-2, 0)$.

7. False, if $x = 0$ then $y = 0$ and y-intercept is $(0, 0)$.   8. True

9. False. Since $p = 1/4$ and the vertex is $(5, 4)$, the focus is $(5, 4 + 1/4) = (-5, 17/4)$.

10. False, it opens to the left.

7.1 Exercises

1. $y = \frac{1}{4}x^2$ 3. $y = -x^2$

5. One finds $p = \frac{3}{2}$. So $a = \frac{1}{4p} = \frac{1}{6}$ and vertex is $\left(3, 5 - \frac{3}{2}\right) = \left(3, \frac{7}{2}\right)$. Parabola is given by $y = \frac{1}{6}(x-3)^2 + \frac{7}{2}$.

7. One finds $p = -\frac{5}{2}$. So $a = \frac{1}{4p} = -\frac{1}{10}$ and vertex is $\left(1, -3 + \frac{5}{2}\right) = \left(1, -\frac{1}{2}\right)$. Parabola is given by $y = -\frac{1}{10}(x-1)^2 - \frac{1}{2}$.

9. One finds $p = 0.2$. So $a = \frac{1}{4p} = 1.25$ and vertex is $(-2, 1.2 - 0.2) = (-2, 1)$. Parabola is given by $y = 1.25(x+2)^2 + 1$.

11. Completing the square,
$$y = (x^2 - 8x + 16) - 16 + 3$$
$$= (x-4)^2 - 13.$$
Since $\frac{1}{4p} = 1$, $p = 0.25$.
Since vertex is $(4, -13)$, focus is $(4, -13 + 0.25) = (4, -12.75)$, and directrix is $y = -13 - p = -13.25$.

13. Completing the square,
$$y = 2(x^2 + 6x + 9) + 5 - 18$$
$$= 2(x+3)^2 - 13.$$
Since $\frac{1}{4p} = 2$, $p = 1/8$.
Since vertex is $(-3, -13)$, the focus is $(-3, -13 + 1/8) = (-3, -103/8)$, and directrix is $y = -13 - 1/8 = -105/8$.

15. Completing the square,
$$y = -2(x^2 - 3x + 9/4) + 1 + 9/2$$
$$= -2(x - 3/2)^2 + 11/2.$$
Since $\frac{1}{4p} = -2$, $p = -0.125$.
Since vertex is $(1.5, 5.5)$, the focus is $(1.5, 5.5 - 0.125) = (1.5, 5.375)$, and directrix is $y = 5.5 + 0.125 = 5.625$.

17. Completing the square,
$$y = 5(x^2 + 6x + 9) - 45$$
$$= 5(x+3)^2 - 45.$$
Since $\frac{1}{4p} = 5$, $p = 0.05$.
Since vertex is $(-3, -45)$, the focus is $(-3, -45 + 0.05) = (-3, -44.95)$, and directrix is $y = -45 - .05 = -45.05$.

19. Completing the square,
$$y = \frac{1}{8}(x^2 - 4x + 4) + \frac{9}{2} - \frac{1}{2}$$
$$= \frac{1}{8}(x-2)^2 + 4.$$
Since $\frac{1}{4p} = 1/8$, $p = 2$.
Since vertex is $(2, 4)$, the focus is $(2, 4+2) = (2, 6)$, and directrix is $y = 4 - 2 = 2$.

21. Note, $a = 1$, $b = -4$. So $x = \frac{-b}{2a} = \frac{4}{2} = 2$ and since $\frac{1}{4p} = a = 1$, $p = 1/4$.
Substituting $x = 2$, $y = 2^2 - 4(2) + 3 = -1$.
Vertex is $(2, -1)$ and focus is $(2, -1 + p) = (2, -3/4)$,
directrix is $y = -1 - p = -5/4$, and parabola opens up since $a > 0$.

23. Note, $a = -1$, $b = 2$. So $x = \frac{-b}{2a} = \frac{-2}{-2} = 1$ and since $\frac{1}{4p} = a = -1$, $p = -1/4$.
Substituting $x = 1$, $y = -(1)^2 + 2(1) - 5 = -4$.
Vertex is $(1, -4)$ and focus is $(1, -4 + p) = (1, -17/4)$,
directrix is $y = -4 - p = -15/4$, and parabola opens down since $a < 0$.

25. Note, $a = 3$, $b = -6$. So $x = \frac{-b}{2a} = \frac{6}{6} = 1$ and since $\frac{1}{4p} = a = 3$, $p = 1/12$. Substituting $x = 1$, $y = 3(1)^2 - 6(1) + 1 = -2$.
Vertex is $(1, -2)$ and focus is $(1, -2 + p) = (1, -23/12)$,
directrix is $y = -2 - p = -25/12$, and parabola opens up since $a > 0$.

27. Note, $a = -1/2$, $b = -3$.
So $x = \frac{-b}{2a} = \frac{3}{-1} = -3$ and since $\frac{1}{4p} = a = -1/2$, $p = -1/2$.

Substituting $x = -3$, we have
$$y = -\frac{1}{2}(-3)^2 - 3(-3) + 2 = \frac{13}{2}.$$
Vertex is $\left(-3, \frac{13}{2}\right)$ and
focus is $\left(-3, \frac{13}{2} + p\right) = (-3, 6)$,
directrix is $y = \frac{13}{2} - p = 7$, and parabola opens down since $a < 0$.

29. Note, $y = \frac{1}{4}x^2 + 5$ is of the form $y = a(x-h)^2 + k$.
So $h = 0$, $k = 5$, and $\frac{1}{4p} = a = \frac{1}{4}$ from which
we have $p = 1$. The vertex is $(h, k) = (0, 5)$,
focus is $(0, 5 + p) = (0, 6)$,
directrix is $y = 5 - p = 4$ and parabola opens up since $a > 0$.

31. From the given focus and directrix one finds $p = 1/4$. So $a = \frac{1}{4p} = 1$, vertex is
$(h, k) = (1/2, -2 - p) = (1/2, -9/4)$,
axis of symmetry is $x = 1/2$, and parabola
is given by $y = a(x-h)^2 + k = \left(x - \frac{1}{2}\right)^2 - \frac{9}{4}$.
If $y = 0$ then $x - \frac{1}{2} = \pm\frac{3}{2}$ or $x = 2, -1$.
x-intercepts are $(2, 0), (-1, 0)$.
If $x = 0$ then $y = \left(0 - \frac{1}{2}\right)^2 - \frac{9}{4} = -2$ and y-intercept is $(0, -2)$.

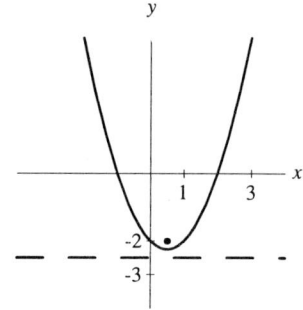

33. From the given focus and directrix one finds $p = -1/4$. So $a = \dfrac{1}{4p} = -1$, vertex is $(h,k) = (-1/2, 6-p) = (-1/2, 25/4)$, axis of symmetry is $x = -1/2$, and parabola is given by $y = a(x-h)^2 + k = -\left(x+\dfrac{1}{2}\right)^2 + \dfrac{25}{4}$.

If $y = 0$ then $x + \dfrac{1}{2} = \pm\dfrac{5}{2}$ or $x = -3, 2$. x-intercepts are $(-3, 0), (2, 0)$.

If $x = 0$ then $y = -\left(0+\dfrac{1}{2}\right)^2 + \dfrac{25}{4} = 6$ and y-intercept is $(0, 6)$.

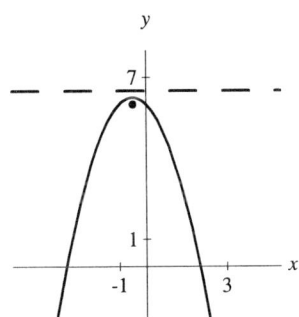

35. Since $\dfrac{1}{2}(x+2)^2 + 2$ is of the form $a(x-h)^2 + k$, vertex is $(h,k) = (-2, 2)$ and axis of symmetry is $x = -2$.

If $y = 0$ then $0 = \dfrac{1}{2}(x+2)^2 + 2$; this has no solution since left-hand side is always positive. No x-intercept.

If $x = 0$ then $y = \dfrac{1}{2}(0+2)^2 + 2 = 4$. y-intercept is $(0, 4)$.

Since $\dfrac{1}{4p} = a = \dfrac{1}{2}$, $p = \dfrac{1}{2}$,

focus is $(h, k+p) = (-2, 5/2)$, and directrix is $y = k - p = 3/2$.

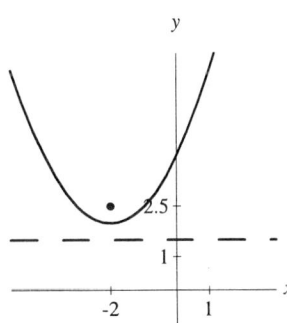

37. Since $-\dfrac{1}{4}(x+4)^2 + 2$ is of the form $a(x-h)^2 + k$, vertex is $(h,k) = (-4, 2)$ and axis of symmetry is $x = -4$. If $y = 0$ then

$$\dfrac{1}{4}(x+4)^2 = 2$$
$$x + 4 = \pm\sqrt{8}$$

x-intercepts are $(-4 \pm 2\sqrt{2}, 0)$.

If $x = 0$ then $y = -\dfrac{1}{4}(0+4)^2 + 2 = -2$. y-intercept is $(0, -2)$.

Since $\dfrac{1}{4p} = a = -\dfrac{1}{4}$, $p = -1$,

focus is $(h, k+p) = (-4, 1)$, and directrix is $y = k - p = 3$.

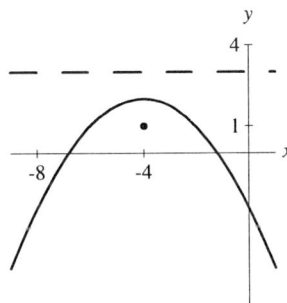

39. Since $\dfrac{1}{2}x^2 - 2$ is of the form $a(x-h)^2 + k$, vertex is $(h,k) = (0, -2)$ and axis of symmetry is $x = 0$. If $y = 0$ then

$$\dfrac{1}{2}x^2 = 2$$
$$x^2 = 4$$

x-intercepts are $(\pm 2, 0)$.

If $x = 0$ then $y = \dfrac{1}{2}(0)^2 - 2 = -2$.

y-intercept is $(0, -2)$.

Since $\dfrac{1}{4p} = a = \dfrac{1}{2}$, $p = 1/2$,

focus is $(h, k+p) = (0, -3/2)$, and directrix is $y = k - p = -5/2$.

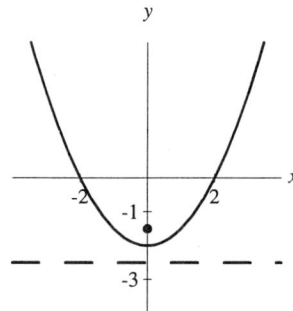

41. Since $y = (x-2)^2$ is of the form $a(x-h)^2 + k$, vertex is $(h, k) = (2, 0)$ and axis of symmetry is $x = 2$.

If $y = 0$ then $(x-2)^2 = 0$ and x-intercept is $(2, 0)$.

If $x = 0$ then $y = (0-2)^2 = 4$ and y-intercept is $(0, 4)$.

Since $\dfrac{1}{4p} = a = 1$, $p = 1/4$,

focus is $(h, k+p) = (2, 1/4)$, and directrix is $y = k - p = -1/4$.

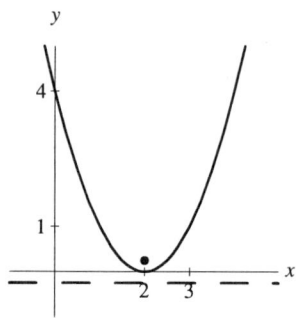

43. By completing the square,
$$y = \dfrac{1}{3}(x - 3/2)^2 - 3/4.$$
Vertex is $(h, k) = (3/2, -3/4)$ and axis of symmetry is $x = 3/2$.

If $y = 0$ then
$$\dfrac{1}{3}\left(x - \dfrac{3}{2}\right)^2 = \dfrac{3}{4}$$
$$\left(x - \dfrac{3}{2}\right)^2 = \dfrac{9}{4}$$
$$x = \dfrac{3}{2} \pm \dfrac{3}{2}$$

x-intercepts are $(3, 0), (0, 0)$.

If $x = 0$ then $y = \dfrac{1}{3}\left(0 - \dfrac{3}{2}\right)^2 - \dfrac{3}{4} = 0$ and y-intercept is $(0, 0)$.

Since $\dfrac{1}{4p} = a = 1/3$, $p = 3/4$,

focus is $(h, k+p) = (3/2, 0)$, and directrix is $y = k - p = -3/2$.

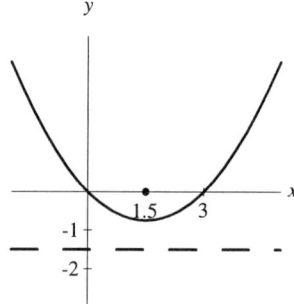

45. Since $x = -y^2$ is of the form $a(y-h)^2 + k$, vertex is $(k, h) = (0, 0)$ and axis of symmetry is $y = 0$.

If $y = 0$ then $x = -0^2 = 0$ and x-intercept is $(0, 0)$.

If $x = 0$ then $0 = -y^2$ and y-intercept is $(0, 0)$.

Since $\dfrac{1}{4p} = a = -1$, $p = -1/4$,

focus is $(k+p, h) = (-1/4, 0)$, and

directrix is $x = k - p = 1/4$.

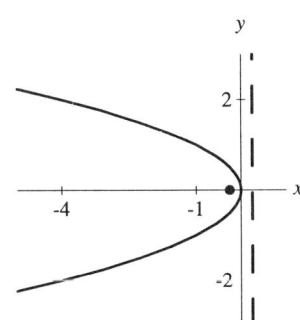

47. Since $x = -\dfrac{1}{4}y^2 + 1$ is of the form $a(y-h)^2 + k$, vertex is $(k,h) = (1,0)$ and axis of symmetry is $y = 0$.

If $y = 0$ then $x = 1$ and x-intercept is $(1,0)$.

If $x = 0$ then $\dfrac{1}{4}y^2 = 1$, $y^2 = 4$, and y-intercepts are $(0, \pm 2)$.

Since $\dfrac{1}{4p} = a = -\dfrac{1}{4}$, $p = -1$,

focus is $(k+p, h) = (0, 0)$, and directrix is $x = k - p = 2$.

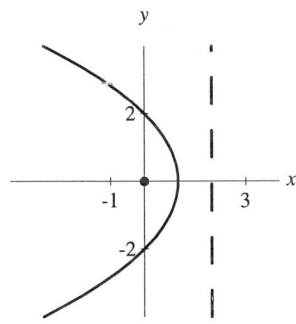

49. By completing the square,
$$x = (y + 1/2)^2 - 25/4.$$
Vertex is $(k, h) = (-25/4, -1/2)$ and axis of symmetry is $y = -1/2$.

If $y = 0$ then $x = 1/4 - 25/4 = -6$. x-intercept is $(-6, 0)$.

If $x = 0$ then
$$\left(y + \dfrac{1}{2}\right)^2 = \dfrac{25}{4}$$
$$y = -\dfrac{1}{2} \pm \dfrac{5}{2}$$
$$y = 2, -3$$

y-intercepts are $(0, 2), (0, -3)$.

Since $\dfrac{1}{4p} = a = 1$, $p = 1/4$,

focus is $(k+p, h) = (-6, -1/2)$, and directrix is $x = k - p = -13/2$.

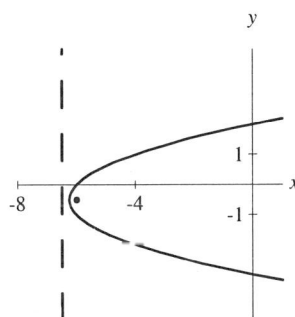

51. By completing the square,
$$x = -\dfrac{1}{2}(y+1)^2 - 7/2.$$
Vertex is $(k, h) = (-7/2, -1)$ and axis of symmetry is $y = -1$.

If $y = 0$ then $x = -1/2 - 7/2 = -4$. x-intercept is $(-4, 0)$.

If $x = 0$ then $0 = -\dfrac{1}{2}(y+1)^2 - 7/2 < 0$.

Inconsistent and so there is no y-intercept.

Since $\dfrac{1}{4p} = a = -1/2$, $p = -1/2$,

focus is $(k+p, h) = (-4, -1)$, and directrix is $x = k - p = -3$.

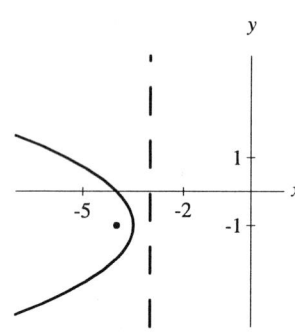

53. Since $x = 2(y-1)^2 + 3$ is of the form $a(y-h)^2 + k$, vertex is $(k, h) = (3, 1)$ and axis of symmetry is $y = 1$.
If $y = 0$ then $x = 2(-1)^2 + 3 = 5$ and x-intercept is $(5, 0)$.
If $x = 0$ then $2(y-1)^2 + 3 = 0$. Inconsistent, since the left-hand side is always positive. No y-intercept.
Since $\frac{1}{4p} = a = 2$, $p = 1/8$,
focus is $(k+p, h) = (25/8, 1)$, and directrix is $x = k - p = 23/8$.

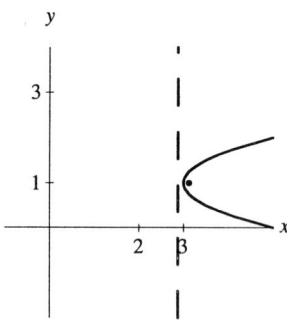

55. Since $x = -\frac{1}{2}(y+2)^2 + 1$ is of the form $a(y-h)^2 + k$, vertex is $(k, h) = (1, -2)$ and axis of symmetry is $y = -2$.
If $y = 0$ then $x = -\frac{1}{2}(2)^2 + 1 = -1$ and x-intercept is $(-1, 0)$.
If $x = 0$ then
$$\frac{1}{2}(y+2)^2 = 1$$
$$(y+2)^2 = 2$$
$$y = -2 \pm \sqrt{2}$$
y-intercepts are $(0, -2 \pm \sqrt{2})$.
Since $\frac{1}{4p} = a = -\frac{1}{2}$, $p = -\frac{1}{2}$,
focus is $(k+p, h) = (1/2, -2)$, and directrix is $x = k - p = 3/2$.

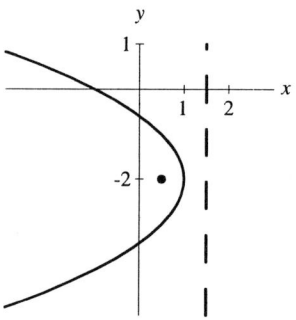

57. Since focus is 1 unit above the vertex $(1, 4)$, $p = 1$.
So $a = \frac{1}{4p} = \frac{1}{4}$ and parabola is given by $y = \frac{1}{4}(x-1)^2 + 4$.

59. Since vertex $(0, 0)$ is 2 units to the right of directrix, $p = 2$ and parabola opens to the right. So $a = \frac{1}{4p} = \frac{1}{8}$.
Parabola is given by $x = \frac{1}{8}y^2$.

61. Since the parabola opens up, $p = 55(12)$ inches, and the vertex is $(0, 0)$ then parabola is given by $y = \frac{1}{4p} = \frac{1}{2640}x^2$.
Thickness at the outside edge is

$23 + \dfrac{1}{2640}(100)^2 \approx 26.8$ inches.

63. $y = x^2$ has vertex $(0,0)$ and opens up. The second parabola can be written as $y = 2(x-1)^2 + 3$ and its vertex is $(1,3)$. In the given viewing window these graphs look alike.

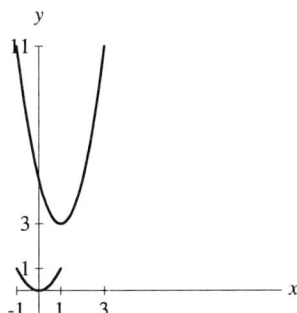

65. Two functions are $f_1(x) = \sqrt{-x}$ and $f_2(x) = -\sqrt{-x}$ where $x \le 0$.

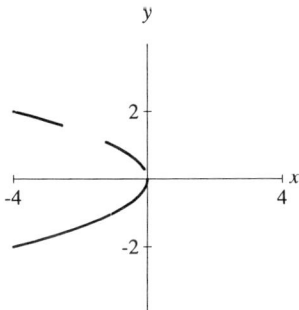

For Thought

1. False, y-intercepts are $(\pm 3, 0)$.

2. True, since it can be written as $\dfrac{x^2}{1/2} + y^2 = 1$.

3. True, length of major axis is $2a = 2(5) = 10$.

4. True, if $y = 0$ then $x^2 - \dfrac{1}{0.5} = 2$ and $x = \pm\sqrt{2}$.

5. True, if $x = 0$ then $y^2 = 3$ and $y = \pm\sqrt{3}$.

6. False, the center is not a point on the circle.

7. True   8. False, $(3, -1)$ satisfies equation.

9. False. No point satisfies equation since left-hand side is always positive.

10. False. Circle can be written as $(x-2)^2 + (y+1/2)^2 = 53/4$. So radius is $\sqrt{53}/2$.

7.2 Exercises

1. Since $c = 2$ and $b = 3$,
$a^2 = b^2 + c^2 = 9 + 4 = 13$ and $a = \sqrt{13}$.
Ellipse is given by $\dfrac{x^2}{13} + \dfrac{y^2}{9} = 1$.

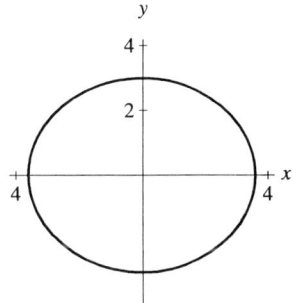

3. Since $c = 4$ and $a = 5$,
$b^2 = a^2 - c^2 = 25 - 16 = 9$ and $b = 3$.
Ellipse is given by $\dfrac{x^2}{25} + \dfrac{y^2}{9} = 1$.

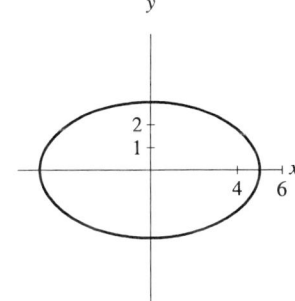

5. Since $c = 2$ and $b = 2$,
   $a^2 = b^2 + c^2 = 4 + 4 = 8$ and $a = \sqrt{8}$.

   Ellipse is given by $\dfrac{x^2}{4} + \dfrac{y^2}{8} = 1$.

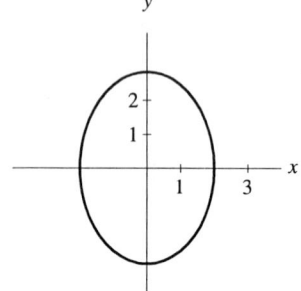

7. Since $c = 4$ and $a = 7$,
   $b^2 = a^2 - c^2 = 49 - 16 = 33$ and $b = \sqrt{33}$.

   Ellipse is given by $\dfrac{x^2}{33} + \dfrac{y^2}{49} = 1$.

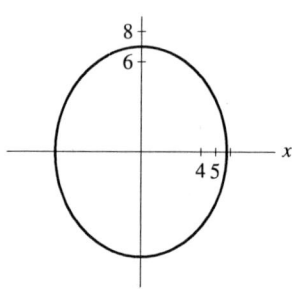

9. Since $c = \sqrt{a^2 - b^2} = \sqrt{16 - 4} = 2\sqrt{3}$,
   the foci are $(\pm 2\sqrt{3}, 0)$

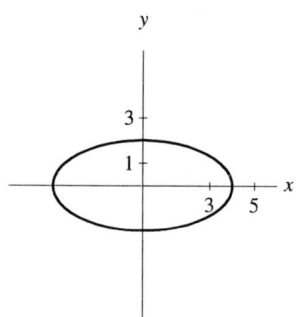

11. Since $c = \sqrt{a^2 - b^2} = \sqrt{36 - 9} = \sqrt{27}$,
    the foci are $(0, \pm 3\sqrt{3})$

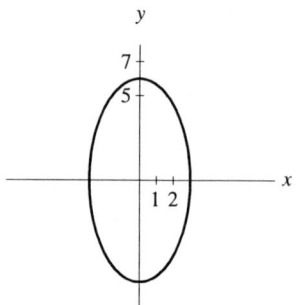

13. Since $c = \sqrt{a^2 - b^2} = \sqrt{25 - 1} = \sqrt{24}$,
    the foci are $(\pm 2\sqrt{6}, 0)$

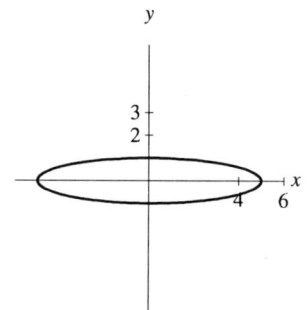

15. Since $c = \sqrt{a^2 - b^2} = \sqrt{25 - 9} = \sqrt{16}$,
    the foci are $(0, \pm 4)$

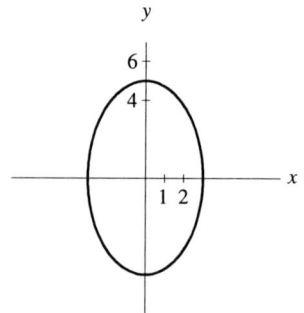

17. From $x^2 + \dfrac{y^2}{9} = 1$, one finds $c = \sqrt{a^2 - b^2} = \sqrt{9-1} = \sqrt{8}$ and foci are $(0, \pm 2\sqrt{2})$.

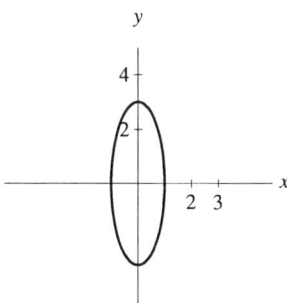

19. From $\dfrac{x^2}{9} + \dfrac{y^2}{4} = 1$, one finds $c = \sqrt{a^2 - b^2} = \sqrt{9-4} = \sqrt{5}$ and foci are $(\pm\sqrt{5}, 0)$.

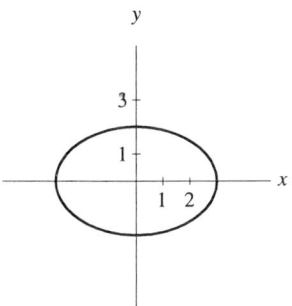

21. Since $c = \sqrt{a^2 - b^2} = \sqrt{16-9} = \sqrt{7}$, the foci are $(1 \pm c, -3) = (1 \pm \sqrt{7}, -3)$.

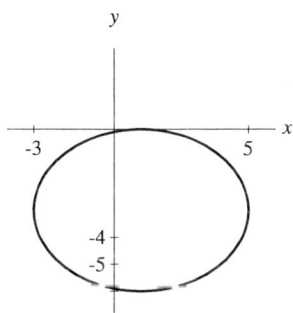

23. Since $c = \sqrt{a^2 - b^2} = \sqrt{25-9} = \sqrt{16} = 4$, the foci are $(3, -2 \pm c) = (3, 2), (3, -6)$.

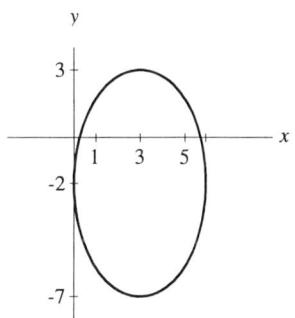

25. Since $\dfrac{(x+4)^2}{36} + (y+3)^2 = 1$, $c = \sqrt{a^2 - b^2} = \sqrt{36-1} = \sqrt{35}$ and foci are $(-4 \pm \sqrt{35}, -3)$.

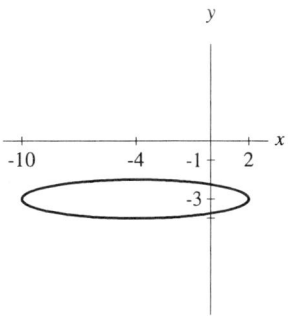

27. If one applies the method of completing the square, one obtains
$$9(x^2 - 2x + 1) + 4(y^2 + 4y + 4) = 11 + 9 + 16$$
$$9(x-1)^2 + 4(y+2)^2 = 36$$
$$\dfrac{(x-1)^2}{4} + \dfrac{(y+2)^2}{9} = 1.$$
From $a^2 = b^2 + c^2$ with $a = 3$ and $b = 2$, one finds $c = \sqrt{5}$. The foci are at the points $(1, -2 \pm \sqrt{5})$ and a sketch of the ellipse is given.

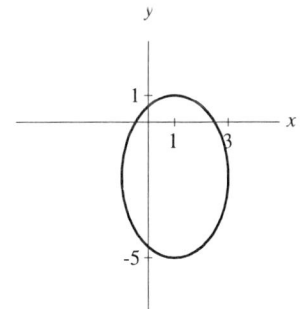

29. Since $\dfrac{x^2}{16} + \dfrac{y^2}{4} = 1$, $c = \sqrt{a^2 - b^2}$
    $= \sqrt{16 - 4} = \sqrt{12}$ and foci are $(\pm 2\sqrt{3}, 0)$.

31. Since $\dfrac{(x+1)^2}{4} + \dfrac{(y+2)^2}{16} = 1$, $c = \sqrt{a^2 - b^2}$
    $= \sqrt{16 - 4} = \sqrt{12}$ and foci are $(-1, -2 \pm 2\sqrt{3})$.

33. $x^2 + y^2 = 4$

35. Since $r = \sqrt{(4-0)^2 + (5-0)^2} = \sqrt{41}$,
    circle is given by $x^2 + y^2 = 41$.

37. Since $r = \sqrt{(4-2)^2 + (1+3)^2} = \sqrt{20}$,
    circle is given by $(x-2)^2 + (y+3)^2 = 20$.

39. Since center is $\left(\dfrac{3-1}{2}, \dfrac{4+2}{2}\right) = (1, 3)$ and
    $r = \sqrt{(3-1)^2 + (4-3)^2} = \sqrt{5}$,
    circle is given by $(x-1)^2 + (y-3)^2 = 5$.

41. center $(0,0)$, radius 10

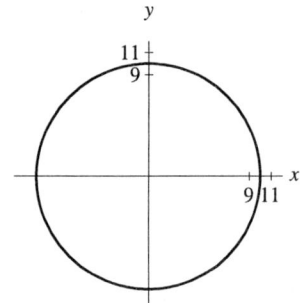

43. center $(1,2)$, radius 2

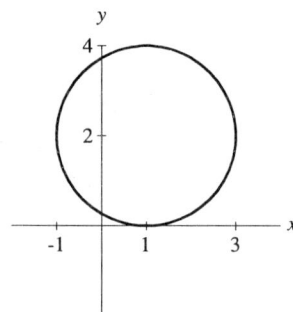

45. center $(-2,-2)$, radius $\sqrt{8}$ or $2\sqrt{2}$

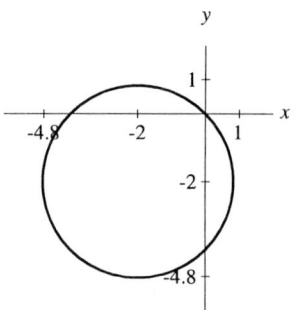

47. Completing the square,
$$x^2 + (y^2 + 2y + 1) = 8 + 1$$
$$x^2 + (y+1)^2 = 9$$
so center is $(0, -1)$ with radius 3.

49. Completing the square,
$$x^2 + 8x + 16 + y^2 - 10y + 25 = 16 + 25$$
$$(x+4)^2 + (y-5)^2 = 41$$
so center is $(-4, 5)$ with radius $\sqrt{41}$.

51. Completing the square,
$$(x^2 + 4x + 4) + y^2 = 5 + 4$$
$$(x+2)^2 + y^2 = 9$$
so center is $(-2, 0)$ with radius 3.

53. Completing the square,
$$x^2 - x + \dfrac{1}{4} + y^2 + y + \dfrac{1}{4} = \dfrac{1}{2} + \dfrac{1}{4} + \dfrac{1}{4}$$
$$(x - 1/2)^2 + (y + 1/2)^2 = 1$$
so center is $(1/2, -1/2)$ with radius 1.

55. Completing the square,
$$x^2 + \dfrac{2}{3}x + \dfrac{1}{9} + y^2 + \dfrac{1}{3}y + \dfrac{1}{36} = \dfrac{1}{9} + \dfrac{1}{9} + \dfrac{1}{36}$$
$$(x + 1/3)^2 + (y + 1/6)^2 = 1/4$$
so center is $(-1/3, -1/6)$ with radius $1/2$.

57. Divide equation by 2 and complete the square,
$$x^2 + 2x + y^2 = 1/2$$
$$(x^2 + 2x + 1) + y^2 = 1/2 + 1$$
$$(x+1)^2 + y^2 = 3/2$$
so center $(-1, 0)$ with radius $\sqrt{\dfrac{3}{2}}$ or $\dfrac{\sqrt{6}}{2}$.

59. Completing the square,
$$y^2 - y + x^2 = 0$$
$$y^2 - y + \frac{1}{4} + x^2 = \frac{1}{4}$$
$$(y - 1/2)^2 + x^2 = \frac{1}{4},$$
one identifies a circle.

61. Divide equation by 4 and complete the square.
$$x^2 + 3y^2 = 1$$
$$x^2 + \frac{y^2}{1/3} = 1$$
One identifies an ellipse.

63. Solve for y and complete the square.
$$y = -2x^2 - 4x - 4$$
$$y = -2(x^2 + 2x) - 4$$
$$y = -2(x^2 + 2x + 1) - 4 + 2$$
$$y = -2(x + 1)^2 - 2$$
One identifies a parabola.

65. Solve for y. Note, $(y - 2)^2 = (2 - y)^2$.
$$2 - x = (y - 2)^2$$
$$x = -(y - 2)^2 + 2$$
One identifies a parabola.

67. Simplify and note $(x - 4)^2 = (4 - x)^2$.
$$2(x - 4)^2 = 4 - y^2$$
$$2(x - 4)^2 + y^2 = 4$$
$$\frac{(x - 4)^2}{2} + \frac{y^2}{4} = 1$$
One identifies an ellipse.

69. Divide equation by 9 to get the circle given by $x^2 + y^2 = \frac{1}{9}$.

71. From the foci, $c = 2$ and $a^2 = b^2 + c^2 = b^2 + 4$. Equation of ellipse is of the form $\frac{x^2}{b^2 + 4} + \frac{y^2}{b^2} = 1$.

Substitute $x = 2$ and $y = 3$.
$$\frac{4}{b^2 + 4} + \frac{9}{b^2} = 1$$
$$4b^2 + 9(b^2 + 4) = b^2(b^2 + 4)$$
$$0 = b^4 - 9b^2 - 36$$
$$0 = (b^2 - 12)(b^2 + 3)$$
So $b^2 = 12$ and $a^2 = 12 + 4 = 16$.
Ellipse is given by $\frac{x^2}{16} + \frac{y^2}{12} = 1$.

73. If c is the distance between the center and focus $(0,0)$ then the other focus is $(2c, 0)$. Since the distance between the x-intercepts is $6 + 2c$, which is also the length of the major axis, then $6 + 2c = 2a$. Since $2a$ is the sum of the distances of $(5, 0)$ from the foci,
$$5 + \sqrt{25 + 4c^2} = 2a$$
$$5 + \sqrt{25 + 4c^2} = 6 + 2c$$
$$\sqrt{25 + 4c^2} = 1 + 2c$$
$$25 + 4c^2 = 1 + 4c + 4c^2$$
$$24 = 4c$$
$$6 = c$$
The other focus is $(2c, 0) = (12, 0)$.

75. Since the sun is a focus of the elliptical orbit, the length of the major axis is $2a = 521$ (the sum of the shortest distance, $P = 1$ AU, and longest distance, $A = 520$ AU, between the orbit and the sun, respectively). In addition, $c = 259.5$ AU (which is the distance from the center to a focus). The eccentricity is given by $e = \frac{c}{a} = \frac{259.5}{260.5} \approx .996$.
An equation of the orbit is $\frac{x^2}{260.5^2} + \frac{y^2}{520} = 1$

77. If $2a$ is the sum of the distances from Haley's comet to the two foci and c is the distance from the sun to the center of the ellipse then $c = a - 8 \times 10^7$.

Since $0.97 = c/a$, $0.97 = \frac{a - 8 \times 10^7}{a}$ and the solution of this equation is $a \approx 2.667 \times 10^9$. So $c = a(0.97) \approx 2.587 \times 10^9$. The maximum distance from the sun is $c + a \approx 5.25 \times 10^9$ km.

79. Solving for y, one finds
$$y^2 = 6360^2 - x^2$$
$$y = \pm\sqrt{6360^2 - x^2}.$$

A sketch of the circle is given.

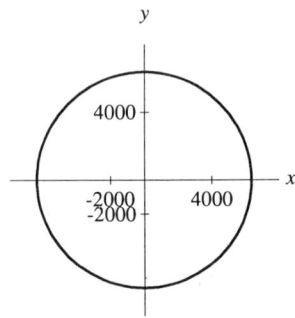

81.

(a) As derived below, an equation of the tangent line is given by
$$\frac{x_1 x}{a^2} + \frac{y_1 y}{b^2} = 1$$
$$\frac{-4x}{25} + \frac{\frac{9}{5}y}{9} = 1$$
$$\frac{1}{5}y = 1 + \frac{4x}{25}$$
$$y = \frac{4x}{5} + 5.$$

(b) The tangent line and the ellipse intersect at the point $\left(-4, \frac{9}{5}\right)$ as shown.

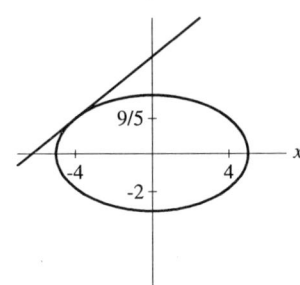

83. A parabolic reflector is preferable since otherwise one would have to place the moving quarterback on one focus of the ellipse.

85. If (x, y) is any point on the ellipse and the sum of the distances from the two foci $(0, \pm c)$ is $2a$ then
$$\sqrt{x^2 + (y-c)^2} + \sqrt{x^2 + (y+c)^2} = 2a$$

Simplify the inside of the radical and transpose a term to the right-hand side.

$$\sqrt{x^2 + y^2 - 2yc + c^2} = 2a - \sqrt{x^2 + y^2 + 2cy + c^2}$$

Squaring both sides,
$x^2 + y^2 - 2yc + c^2 =$
$4a^2 - 4a\sqrt{x^2 + y^2 + 2cy + c^2} + x^2 + y^2 + 2cy + c^2$
Cancel like terms and simplify.

$$-4yc - 4a^2 = -4a\sqrt{x^2 + y^2 + 2cy + c^2}$$
$$yc + a^2 = a\sqrt{x^2 + y^2 + 2cy + c^2}$$
$$y^2 c^2 + 2yca^2 + a^4 = a^2(x^2 + y^2 + 2cy + c^2)$$
$$y^2 c^2 + a^4 = a^2(x^2 + y^2 + c^2)$$
$$a^4 - a^2 c^2 = a^2 x^2 + y^2(a^2 - c^2)$$

Let $b^2 = a^2 - c^2$. So $a^2 b^2 = a^2 x^2 + b^2 y^2$ and $\frac{x^2}{b^2} + \frac{y^2}{a^2} = 1$ is an ellipse with foci $(0, \pm c)$ and one finds that the x-intercepts are $(\pm b, 0)$.

For Thought

1. False, it is a parabola.

2. False, it has no y-intercept.

3. True   4. True

5. False, $y = \frac{b}{a}x$ is an asymptote.

6. True   7. True, $c = \sqrt{16 + 9} = 5$.

8. True, $c = \sqrt{3 + 5} = \sqrt{8}$.

9. False, $y = \frac{2}{3}x$ is an asymptote.

10. False, it is a circle centered at $(0, 0)$.

7.3 Exercises

1. Note, $c = \sqrt{a^2 + b^2} = \sqrt{2^2 + 3^2} = \sqrt{13}$.

Foci $(\pm\sqrt{13}, 0)$, asymptotes $y = \pm\dfrac{b}{a}x = \pm\dfrac{3}{2}x$

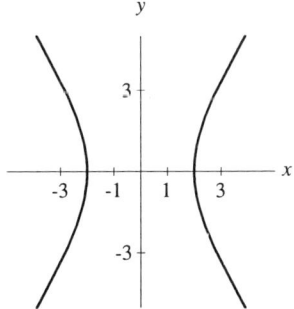

3. Note, $c = \sqrt{a^2 + b^2} = \sqrt{2^2 + 5^2} = \sqrt{29}$.

Foci $(0, \pm\sqrt{29})$, asymptotes $y = \pm\dfrac{a}{b}x = \pm\dfrac{2}{5}x$

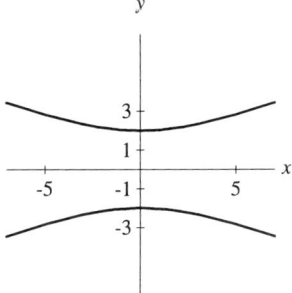

5. Note, $c = \sqrt{a^2 + b^2} = \sqrt{2^2 + 1^2} = \sqrt{5}$.

Foci $(\pm\sqrt{5}, 0)$, asymptotes $y = \pm\dfrac{b}{a}x = \pm\dfrac{1}{2}x$

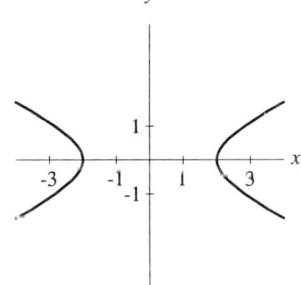

7. Note, $c = \sqrt{a^2 + b^2} = \sqrt{1^2 + 3^2} = \sqrt{10}$.

Foci $(\pm\sqrt{10}, 0)$, asymptotes $y = \pm\dfrac{b}{a}x = \pm 3x$

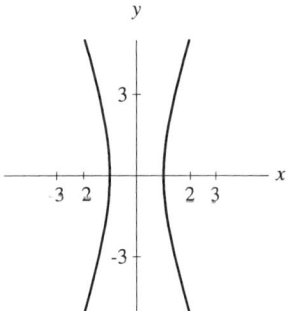

9. Dividing by 144, $\dfrac{x^2}{9} - \dfrac{y^2}{16} = 1$.

Note, $c = \sqrt{a^2 + b^2} = \sqrt{3^2 + 4^2} = 5$.

Foci $(\pm 5, 0)$, asymptotes $y = \pm\dfrac{b}{a}x = \pm\dfrac{4}{3}x$

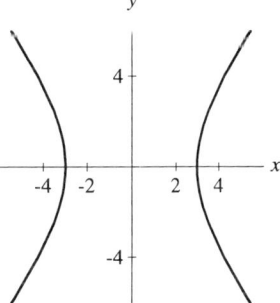

11. Note, $c = \sqrt{a^2 + b^2} = \sqrt{1^2 + 1^2} = \sqrt{2}$.

Foci $(\pm\sqrt{2}, 0)$, asymptotes $y = \pm\dfrac{b}{a}x = \pm x$

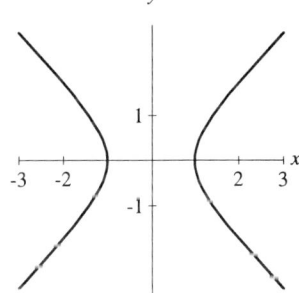

13. Note, $c = \sqrt{a^2 + b^2} = \sqrt{2^2 + 3^2} = \sqrt{13}$.
    Since the center is $(-1, 2)$,
    the foci are $(-1 \pm \sqrt{13}, 2)$.
    Since $y - 2 = \pm\dfrac{3}{2}(x + 1)$, solving
    for y the asymptotes are
    $y = \dfrac{3}{2}x + \dfrac{7}{2}$ and $y = -\dfrac{3}{2}x + \dfrac{1}{2}$.

    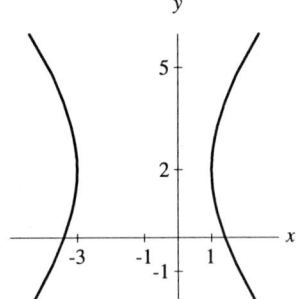

15. Note, $c = \sqrt{a^2 + b^2} = \sqrt{2^2 + 1^2} = \sqrt{5}$.
    Since the center is $(-2, 1)$,
    the foci are $(-2, 1 \pm \sqrt{5})$.
    Since $y - 1 = \pm 2(x + 2)$, solving
    for y the asymptotes are
    $y = 2x + 5$ and $y = -2x - 3$.

    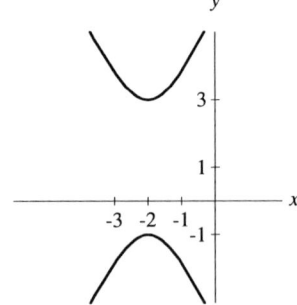

17. Note, $c = \sqrt{a^2 + b^2} = \sqrt{4^2 + 3^2} = 5$.
    Since the center is $(-2, 3)$,
    the foci are $(3, 3)$ and $(-7, 3)$.
    Since $y - 3 = \pm\dfrac{3}{4}(x + 2)$, solving
    for y the asymptotes are
    $y = \dfrac{3}{4}x + \dfrac{9}{2}$ and $y = -\dfrac{3}{4}x + \dfrac{3}{2}$.

    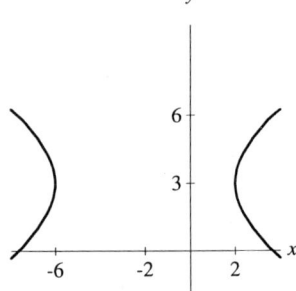

19. Note, $c = \sqrt{a^2 + b^2} = \sqrt{1^2 + 1^2} = \sqrt{2}$.
    Since the center is $(3, 3)$,
    the foci are $(3, 3 \pm \sqrt{2})$.
    Since $y - 3 = \pm(x - 3)$, solving
    for y the asymptotes are
    $y = x$ and $y = -x + 6$.

    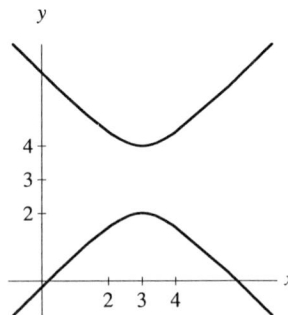

21. From the x-intercepts $(\pm 6, 0)$, the hyperbola is given by $\dfrac{x^2}{6^2} - \dfrac{y^2}{b^2} = 1$. From the asymptotes one gets $\dfrac{1}{2} = \dfrac{b}{6}$ and $b = 3$. Equation is $\dfrac{x^2}{36} - \dfrac{y^2}{9} = 1$.

23. From the x-intercepts $(\pm 3, 0)$, the hyperbola is given by $\dfrac{x^2}{3^2} - \dfrac{y^2}{b^2} = 1$. From the foci, $c = 5$ and $b^2 = c^2 - a^2 = 5^2 - 3^2 = 16$. Equation is $\dfrac{x^2}{9} - \dfrac{y^2}{16} = 1$.

25. From the vertices of the fundamental rectangle and since it opens sideways, one gets $a = 3$, $b = 5$, and the center is at the origin. Equation is $\dfrac{x^2}{9} - \dfrac{y^2}{25} = 1$.

27. $\dfrac{x^2}{9} - \dfrac{y^2}{16} = 1$

29. $\dfrac{y^2}{9} - \dfrac{(x-1)^2}{9} = 1$

31. Completing the square,

$$y^2 - (x^2 - 2x) = 2$$
$$y^2 - (x^2 - 2x + 1) = 2 - 1$$
$$y^2 - (x-1)^2 = 1$$

one gets a hyperbola.

33. $y = x^2 + 2x$ is a parabola.

35. Simplifying,

$$25x^2 + 25y^2 = 2500$$
$$x^2 + y^2 = 100$$

one gets a circle.

37. Simplifying,

$$25x = -100y^2 + 2500$$
$$x = -4y^2 + 100$$

one gets a parabola.

39. Completing the square,

$$2(x^2 - 2x) + 2(y^2 - 4y) = -9$$
$$2(x^2 - 2x + 1) + 2(y^2 - 4y + 4) = -9 + 2 + 8$$
$$2(x-1)^2 + 2(y-2)^2 = 1$$
$$(x-1)^2 + (y-2)^2 = \dfrac{1}{2}$$

one gets a circle.

41. Completing the square,

$$2(x^2 + 2x) + y^2 + 6y = -7$$
$$2(x^2 + 2x + 1) + y^2 + 6y + 9 = -7 + 2 + 9$$
$$2(x+1)^2 + (y+3)^2 = 4$$
$$\dfrac{(x+1)^2}{2} + \dfrac{(y+3)^2}{4} = 1$$

one gets an ellipse.

43. From the center $(0,0)$ and vertex $(0,8)$ one gets $a = 8$. From the foci $(0, \pm 10)$, $c = 10$. So $b^2 = c^2 - a^2 = 10^2 - 8^2 = 36$. Hyperbola is given by $\dfrac{y^2}{64} - \dfrac{x^2}{36} = 1$.

45. Multiply $16y^2 - x^2 = 16$ by 9 and add to $9x^2 - 4y^2 = 36$.

$$-9x^2 + 144y^2 = 144$$
$$9x^2 - 4y^2 = 36$$

$$140y^2 = 180$$
$$y^2 = \dfrac{9}{7}$$
$$y = \dfrac{3\sqrt{7}}{7}$$

Using $y^2 = \dfrac{9}{7}$ in $x^2 = 16(y^2 - 1)$, one obtains
$$x^2 = 16\left(\dfrac{2}{7}\right) \text{ or } x = \dfrac{4\sqrt{14}}{7}.$$
Exact location is $\left(\dfrac{4\sqrt{14}}{7}, \dfrac{3\sqrt{7}}{7}\right)$.

47. Since $c^2 = a^2 + b^2 = 1^2 + 1^2 = 2$, the foci of $x^2 - y^2 = 1$ are $A(\sqrt{2}, 0)$ and $B(-\sqrt{2}, 0)$. Note, $y^2 = x^2 - 1$.

    Suppose (x, y) is a point on the hyperbola whose distance from B is twice the distance between (x, y) and A. Then
    $$2\sqrt{(x-\sqrt{2})^2 + y^2} = \sqrt{(x+\sqrt{2})^2 + y^2}$$
    $$4((x-\sqrt{2})^2 + y^2) = (x+\sqrt{2})^2 + y^2$$
    $$4(x-\sqrt{2})^2 - (x+\sqrt{2})^2 + 3y^2 = 0$$
    $$3x^2 - 10\sqrt{2}x + 6 + 3y^2 = 0$$
    $$3x^2 - 10\sqrt{2}x + 6 + 3(x^2 - 1) = 0$$
    $$6x^2 - 10\sqrt{2}x + 6 = 0.$$

    Solving for x, one finds $x = \dfrac{3\sqrt{2}}{2}$ and $x = \dfrac{\sqrt{2}}{6}$; the second value must be excluded since it is out of the domain. Substituting $x = \dfrac{3\sqrt{2}}{2}$ into $y^2 = x^2 - 1$, one obtains $y = \pm\dfrac{\sqrt{14}}{2}$.

    By the symmetry of the hyperbola, there are four points that are twice as far from one focus as they are from the other focus. Namely, the points $\left(\pm\dfrac{3\sqrt{2}}{2}, \pm\dfrac{\sqrt{14}}{2}\right)$.

49. Note, the asymptotes are $y = \pm x$. The difference is
    $$50 - \sqrt{50^2 + 1} \approx 50 - 49.99 = 0.01$$

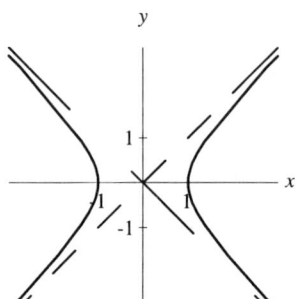

Review Exercises

1. If $y = 0$ then by factoring $0 = (x+6)(x-2)$ and x-intercepts are $(2, 0)$ & $(-6, 0)$.

   If $x = 0$, $y = -12$ and y-intercept is $(0, -12)$.

   By completing the square, $y = (x+2)^2 - 16$, vertex $(h, k) = (-2, -16)$, and axis of symmetry is $x = -2$.

   Since $p = \dfrac{1}{4a} = \dfrac{1}{4}$, the focus is $(h, k+p) = (-2, -63/4)$ and directrix is $y = k - p = -65/4$.

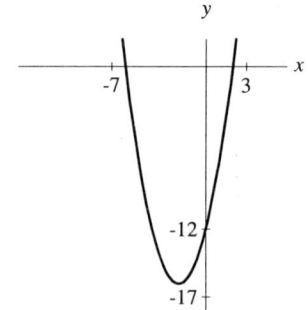

REVIEW EXERCISES

3. If $y = 0$ then by factoring $0 = x(6 - 2x)$ and x-intercepts are $(0,0)$ & $(3,0)$.

If $x = 0$, $y = 0$ and y-intercept is $(0,0)$.

By completing the square, one gets $y = -2(x - 3/2)^2 + 9/2$, vertex $(h,k) = (3/2, 9/2)$, and axis of symmetry is $x = 3/2$.

Since $p = \dfrac{1}{4a} = -\dfrac{1}{8}$, the focus is $(h, k+p) = (3/2, 35/8)$ and directrix is $y = k - p = 37/8$.

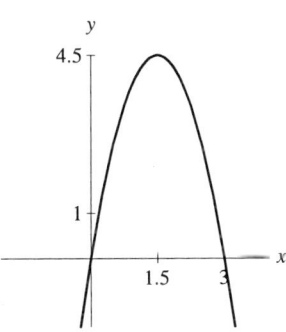

5. By completing the square, we have $x = (y+2)^2 - 10$. If $x = 0$ then $y + 2 = \pm\sqrt{10}$ and y-intercepts are $(0, -2 \pm \sqrt{10})$.

If $y = 0$, $x = -6$ and x-intercept is $(-6, 0)$.

Since $x = (y+2)^2 - 10$ is of the form $x = a(y - h)^2 + k$, the vertex is $(k, h) = (-10, -2)$, axis of symmetry is $y = -2$.

Since $p = \dfrac{1}{4a} = \dfrac{1}{4}$, the focus is $(k+p, h) = (-39/4, -2)$ and directrix is $x = k - p = -41/4$.

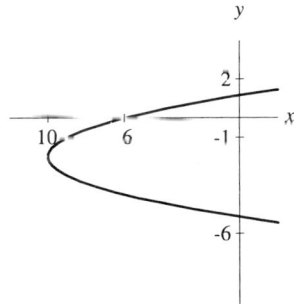

7. Since $c = \sqrt{a^2 - b^2} = \sqrt{36 - 16} = 2\sqrt{5}$, the foci are $(0, \pm 2\sqrt{5})$.

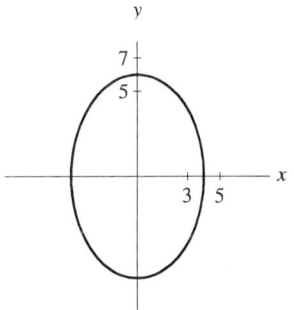

9. Since $c = \sqrt{a^2 - b^2} = \sqrt{24 - 8} = 4$, the foci are $(1, 1 \pm 4)$ or $(1, 5)$ & $(1, -3)$.

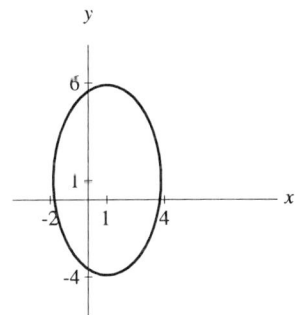

11. Since $c = \sqrt{a^2 - b^2} = \sqrt{10 - 8} = \sqrt{2}$, the foci are $(1, -3 \pm \sqrt{2})$.

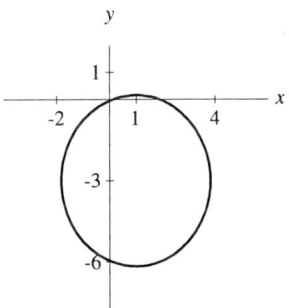

13. center $(0,0)$, radius 9

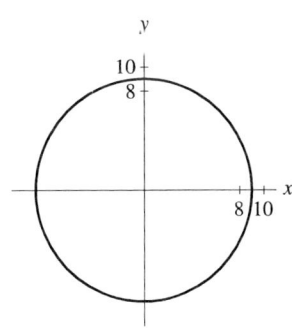

15. center $(-1, 0)$, radius 2

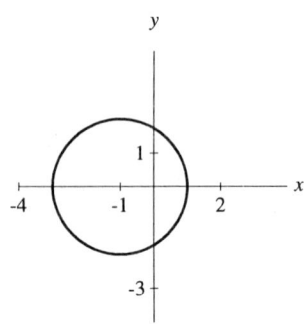

17. Completing the square,
$$x^2 + 5x + \frac{25}{4} + y^2 = -\frac{1}{4} + \frac{25}{4}$$
$$\left(x + \frac{5}{2}\right)^2 + y^2 = 6,$$
so center is $(-5/2, 0)$ and radius is $\sqrt{6}$.

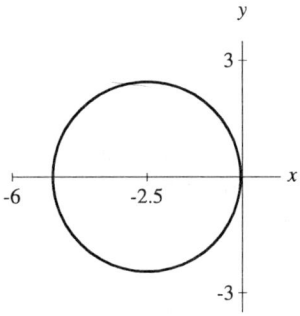

19. $x^2 + (y+4)^2 = 9$

21. $(x+2)^2 + (y+7)^2 = 6$

23. Since $c = \sqrt{a^2 + b^2} = \sqrt{8^2 + 6^2} = 10$,
foci $(\pm 10, 0)$, asymptotes $y = \pm \frac{b}{a} = \pm \frac{3}{4}x$

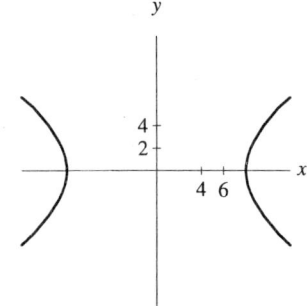

25. Since $c = \sqrt{a^2 + b^2} = \sqrt{8^2 + 4^2} = 4\sqrt{5}$,
the foci are $(4, 2 \pm 4\sqrt{5})$. Solving
for y in $y - 2 = \pm\frac{8}{4}(x - 4)$,
asymptotes are $y = 2x - 6$ and $y = -2x + 10$.

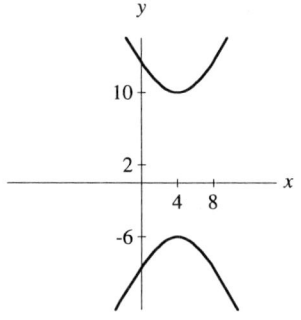

27. Completing the square,
$$x^2 - 4x - 4(y^2 - 8y) = 64$$
$$x^2 - 4x + 4 - 4(y^2 - 8y + 16) = 64 + 4 - 64$$
$$(x-2)^2 - 4(y-4)^2 = 4$$
$$\frac{(x-2)^2}{4} - (y-4)^2 = 1,$$
so $c = \sqrt{a^2 + b^2} = \sqrt{2^2 + 1^2} = \sqrt{5}$,
foci are $(2 \pm \sqrt{5}, 4)$, and solving for y
in $y - 4 = \pm\frac{1}{2}(x - 2)$, asymptotes
are $y = \frac{1}{2}x + 3$ and $y = -\frac{1}{2}x + 5$.

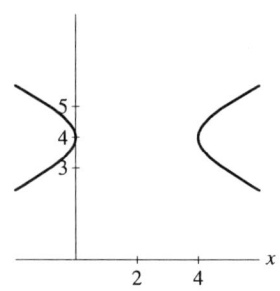

29. Hyperbola **31.** Ellipse

33. Parabola **35.** Hyperbola

37. $x^2 + y^2 = 4$ is a circle

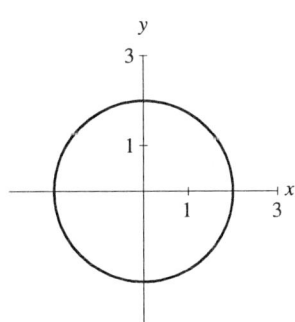

39. Since $4y = x^2 - 4$, $y = \frac{1}{4}x^2 - 1$.

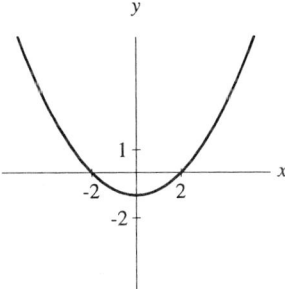

41. Since $x^2 + 4y^2 = 4$, $\frac{x^2}{4} + y^2 = 1$.

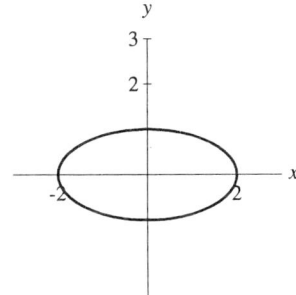

43. Since $x^2 - 4x + 4 + 4y^2 = 4$, $(x-2)^2 + 4y^2 = 4$ and $\frac{(x-2)^2}{4} + y^2 = 1$.

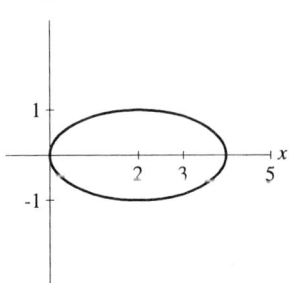

45. Since the vertex is midway between the focus $(1,3)$ and directrix $x = \frac{1}{2}$, the vertex is $\left(\frac{3}{4}, 3\right)$ and $p = \frac{1}{4}$. Since $a = \frac{1}{4p} = 1$, parabola is given by $x = (y-3)^2 + \frac{3}{4}$.

47. From the foci and vertices one gets $c = 4$ and $a = 6$, respectively. Since $b^2 = a^2 - c^2 = 36 - 16 = 20$, the ellipse is given by $\frac{x^2}{36} + \frac{y^2}{20} = 1$.

49. Radius is $\sqrt{(-1-1)^2 + (-1-3)^2} = \sqrt{20}$. Equation is $(x-1)^2 + (y-3)^2 = 20$.

51. From the foci and x-intercepts one gets $c = 3$ and $a = 2$, respectively. Since $b^2 = c^2 - a^2 = 9 - 4 = 5$, hyperbola is given by $\frac{x^2}{4} - \frac{y^2}{5} = 1$.

53. $(x+2)^2 + (y-3)^2 = 9$

55. $\frac{(x+2)^2}{9} + (y-1)^2 = 1$

57. $\frac{(y-1)^2}{9} - \frac{(x-2)^2}{4} = 1$

59. Equation is of the form $\dfrac{x^2}{100^2} - \dfrac{y^2}{b^2} = 1$. Since the hyperbola passes through $(120, 24\sqrt{11})$,

$$\dfrac{120^2}{100^2} - \dfrac{(24\sqrt{11})^2}{b^2} = 1$$

$$1.44 - \dfrac{6336}{b^2} = 1$$

$$b^2 = \dfrac{6336}{0.44}$$

$$b^2 = 120^2$$

Equation is $\dfrac{x^2}{100^2} - \dfrac{y^2}{120^2} = 1$.

61. Note $c = 30$, $a = 34$, and equation is of the form $\dfrac{x^2}{34^2} + \dfrac{y^2}{b^2} = 1$. Since $b^2 = a^2 - c^2 = 34^2 - 30^2 = 16^2$, equation is $\dfrac{x^2}{34^2} + \dfrac{y^2}{16^2} = 1$.

To find h, let $x = 32$. Substituting, one finds

$$\dfrac{32^2}{34^2} + \dfrac{y^2}{16^2} = 1$$

$$y^2 = \left(1 - \dfrac{32^2}{34^2}\right) 16^2$$

$$y \approx 5.406$$

Thus, $h = 2y \approx 10.81$ feet.

Chapter 7 Test

1. Circle $x^2 + y^2 = 8$

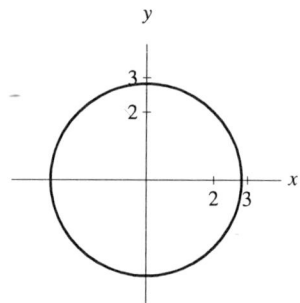

2. Ellipse $\dfrac{x^2}{9} + \dfrac{y^2}{100} = 1$

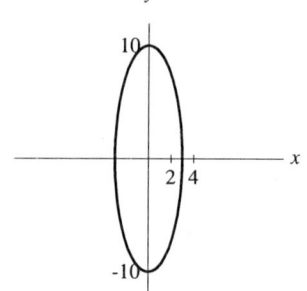

3. Parabola $y = x^2 + 6x + 8$

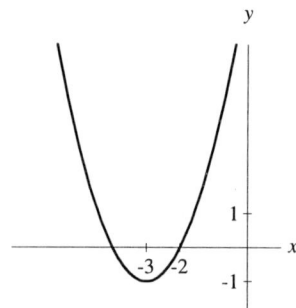

4. Hyperbola $\dfrac{y^2}{25} - \dfrac{x^2}{9} = 1$

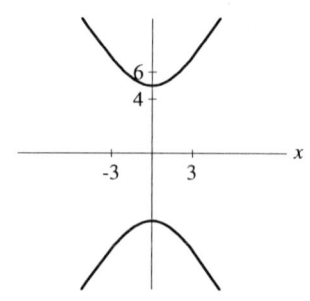

5. Circle $(x+3)^2 + (y-1)^2 = 10$

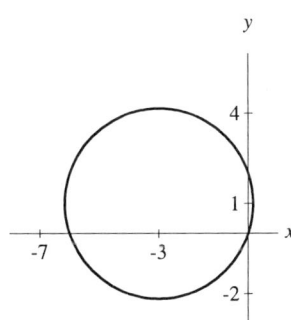

6. Hyperbola $\dfrac{(x-2)^2}{9} - \dfrac{(y+3)^2}{4} = 1$

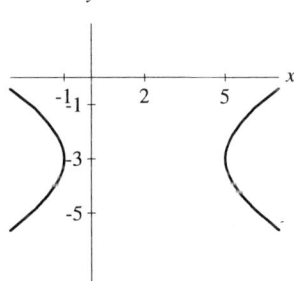

7. Hyperbola   8. Parabola

9. Circle   10. Ellipse

11. Since $(2\sqrt{3})^2 = 12$, equation is $(x+3)^2 + (y-4)^2 = 12$.

12. Midway between the focus $(2,0)$ and directrix $x = -2$ is the vertex $(0,0)$. Since $p = 2$, $a = \dfrac{1}{4p} = \dfrac{1}{8}$. Equation is $x = \dfrac{1}{8}y^2$.

13. Equation is of the form
$\dfrac{x^2}{2^2} + \dfrac{y^2}{a^2} = 1$. Since
$a^2 = b^2 + c^2 = 2^2 + \sqrt{6}^2 = 10$,
equation is $\dfrac{x^2}{4} + \dfrac{y^2}{10} = 1$.

14. From the foci and vertices one gets $c = 8$ and $a = 6$. Since $b^2 = c^2 - a^2 = 8^2 - 6^2 = 28$,
equation is $\dfrac{x^2}{36} - \dfrac{y^2}{28} = 1$.

15. Complete the square to get $y = (x-2)^2 - 4$.
So vertex is $(h,k) = (2,-4)$, axis of symmetry $x = 2$, $a = 1$, and $p = \dfrac{1}{4a} = \dfrac{1}{4}$.
Focus is $(h, k+p) = (2, -15/4)$ and directrix is $y = k - p = -17/4$.

16. Since $\dfrac{x^2}{16} + \dfrac{y^2}{4} = 1$,
$c = \sqrt{a^2 - b^2} = \sqrt{4^2 - 2^2} = 2\sqrt{3}$.
Foci are $(\pm 2\sqrt{3}, 0)$, length of major axis is $2a = 8$, length of minor axis is $2b = 4$.

17. Since $y^2 - \dfrac{x^2}{16} = 1$,
$c = \sqrt{a^2 + b^2} = \sqrt{1^2 + 4^2} = \sqrt{17}$.
Foci are $(0, \pm\sqrt{17})$, vertices $(0, \pm 1)$,
asymptotes $y = \pm \dfrac{1}{4}x$, length of transverse axis is $2a = 2$, length of conjugate axis is $2b = 8$.

18. Completing the square,
$$x^2 + x + \dfrac{1}{4} + y^2 - 3y + \dfrac{9}{4} = -\dfrac{1}{4} + \dfrac{1}{4} + \dfrac{9}{4}$$
$$\left(x + \dfrac{1}{2}\right)^2 + \left(y - \dfrac{3}{2}\right)^2 = \dfrac{9}{4}$$
so the center is $\left(-\dfrac{1}{2}, \dfrac{3}{2}\right)$ and the radius is $\dfrac{3}{2}$.

19. Since $\dfrac{x^2}{225} + \dfrac{y^2}{81} = 1$, $a = 15$ and $b = 9$.
Note, $c = \sqrt{a^2 - c^2} = \sqrt{15^2 - 9^2} = 12$.
Since the foci are $(\pm 12, 0)$,
the distance from the point of generation of the waves to the kidney stones is $2c = 24$ cm.

Tying It All Together

1. Parabola $y = 6x - x^2$

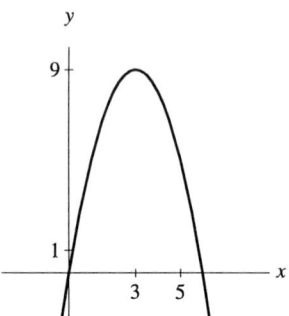

2. Line $y = 6x$

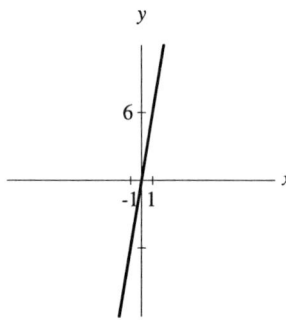

3. Parabola $y = 6 - x^2$

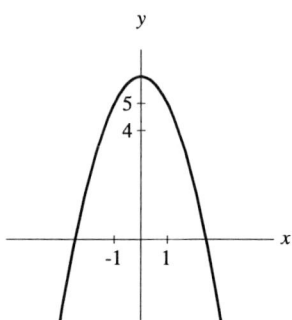

4. Circle $x^2 + y^2 = 6$

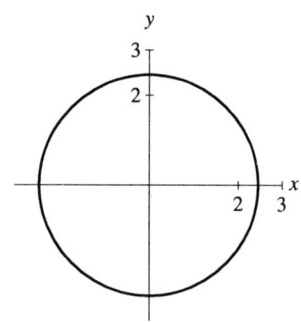

5. Line $y = 6 + x$

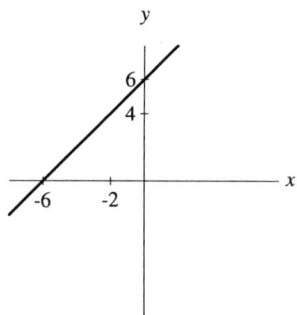

6. Parabola $x = 6 - y^2$

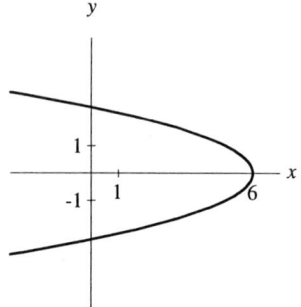

7. Parabola $y = (6 - x)^2$

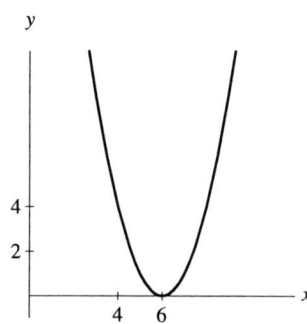

8. $y = |x + 6|$ goes through $(-6, 0)$, $(-5, 1)$

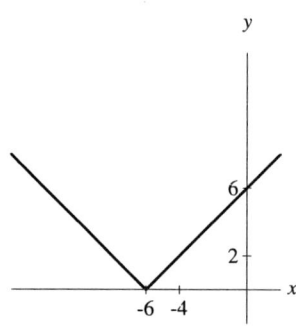

TYING IT ALL TOGETHER

9. $y = 6^x$ goes through $(0,1), (1,6)$

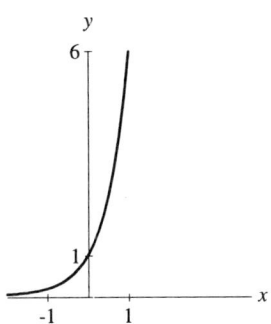

10. $y = \log_6(x)$ goes through $(6,1), (1/6, -1)$

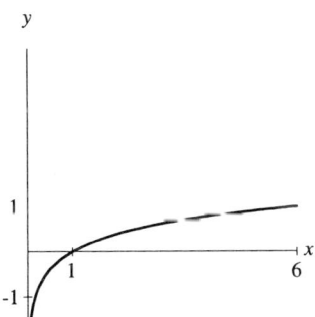

11. $y = \dfrac{1}{x^2 - 6}$ has asymptotes $x = \pm\sqrt{6}$

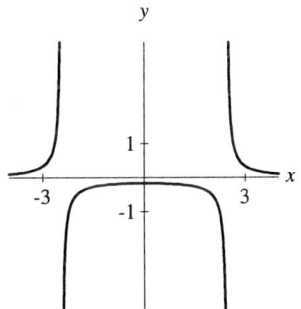

12. Ellipse $\dfrac{x^2}{9} + \dfrac{y^2}{4} = 1$

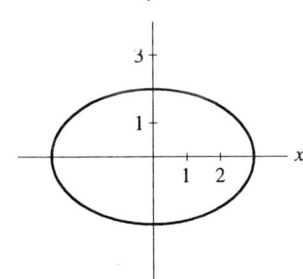

13. Hyperbola $\dfrac{x^2}{9} - \dfrac{y^2}{4} = 1$

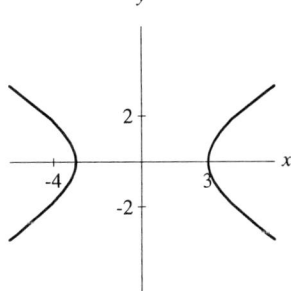

14. $y = x(6 - x^2)$ goes through $(0,0), (\pm\sqrt{6}, 0)$

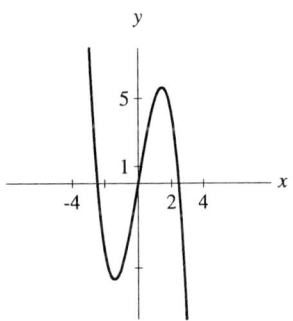

15. Line $2x + 3y = 6$

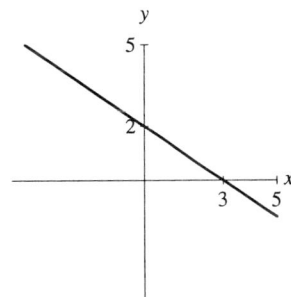

16. Hyperbola $\dfrac{(x-3)^2}{3} - \dfrac{y^2}{3} = 1$

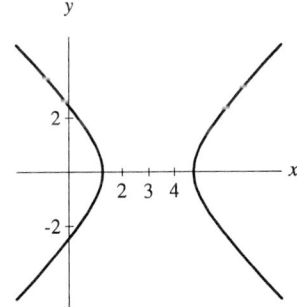

17. Solving for x,
$$3x - 9 + 5 = -9$$
$$3x = -5$$
solution set is $\{-5/3\}$.

18. Solving for x,
$$5x - 8x + 12 = 17$$
$$-3x = 5$$
solution set is $\{-5/3\}$.

19. Solving for x,
$$2x - \frac{2}{3} - \frac{3}{2} + 3x = \frac{3}{2}$$
$$5x - \frac{13}{6} = \frac{3}{2}$$
$$5x = \frac{11}{3}$$
solution set is $\{11/15\}$.

20. Solving for x,
$$-4x + 2 = 3x + 2$$
$$-7x = 0$$
solution set is $\{0\}$.

21. Solving for x,
$$\frac{1}{2}x - \frac{1}{4}x = \frac{3}{2} + \frac{1}{3}$$
$$\frac{1}{4}x = \frac{11}{6}$$
$$x = \frac{44}{6}$$
solution set is $\{22/3\}$.

22. Multiplying both sides by 100,
$$5(x - 20) + 2(x + 10) = 270$$
$$5x - 100 + 2x + 20 = 270$$
$$7x = 350$$
The solution set is $\{50\}$.

23. By using the quadratic formula to solve $2x^2 + 31x - 51 = 0$, one obtains
$$x = \frac{-31 \pm \sqrt{31^2 - 4(2)(-51)}}{4}$$
$$= \frac{-31 \pm \sqrt{1369}}{4}$$
$$= \frac{-31 \pm 37}{4}$$
$$= \frac{3}{2}, -17$$
The solution set is $\left\{\frac{3}{2}, -17\right\}$.

24. Since $x(2x + 31) = 0$, the solution set is $\left\{0, -\frac{31}{2}\right\}$.

25. By using the quadratic formula to solve $x^2 - 34x + 286 = 0$, one obtains
$$x = \frac{34 \pm \sqrt{34^2 - 4(1)(286)}}{2}$$
$$x = \frac{34 \pm \sqrt{12}}{2}$$
$$x = \frac{34 \pm 2\sqrt{3}}{2}$$
$$x = 17 \pm \sqrt{3}.$$
The solution set is $\left\{17 \pm \sqrt{3}\right\}$.

26. In solving $x^2 - 34x + 290 = 0$, one can use the quadratic formula. Then
$$x = \frac{34 \pm \sqrt{34^2 - 4(1)(290)}}{2}$$
$$x = \frac{34 \pm \sqrt{-4}}{2}$$
$$x = \frac{34 \pm 2i}{2}$$
$$x = 17 \pm i.$$
The solution set is $\{17 \pm i\}$.

For Thought

1. True

2. False, the domain of a finite sequence is $\{1, 2, ..., n\}$ for some positive integer n.

3. True

4. False, n is the independent variable.

5. False. First four terms are $1, -8, 27, -81$.

6. False, $a_5 = -3 + 4 \cdot 6 = 21$.

7. False. Common difference is $d = 4 - 7 = -3$.

8. False, there is no common difference.

9. False, since $14 = 4 + 2d$ then $d = 5$ and $a_4 = 4 + 3 \cdot 5 = 19$. 10. True

8.1 Exercises

1. $a_1 = 1^2 = 1, a_2 = 2^2 = 4,$
$a_3 = 3^2 = 9, a_4 = 4^2 = 16, a_5 = 5^2 = 25,$
$a_6 = 6^2 = 36, a_7 = 7^2 = 49.$

   First seven terms are $1, 4, 9, 16, 25, 36, 49$.

3. $b_1 = \dfrac{(-1)^2}{2} = \dfrac{1}{2}, b_2 = \dfrac{(-1)^3}{3} = -\dfrac{1}{3},$
$b_3 = \dfrac{(-1)^4}{4} = \dfrac{1}{4}, b_4 = \dfrac{(-1)^5}{5} = -\dfrac{1}{5},$
$b_5 = \dfrac{(-1)^6}{6} = \dfrac{1}{6}, b_6 = \dfrac{(-1)^7}{7} = -\dfrac{1}{7},$
$b_7 = \dfrac{(-1)^8}{8} = \dfrac{1}{8}, b_8 = \dfrac{(-1)^9}{9} = -\dfrac{1}{9}.$

   First eight terms are
$\dfrac{1}{2}, -\dfrac{1}{3}, \dfrac{1}{4}, -\dfrac{1}{5}, \dfrac{1}{6}, -\dfrac{1}{7}, \dfrac{1}{8}, -\dfrac{1}{9}.$

5. $c_1 = (-2)^0 = 1, c_2 = (-2)^1 = -2,$
$c_3 = (-2)^2 = 4, c_4 = (-2)^3 = -8,$
$c_5 = (-2)^4 = 16, c_6 = (-2)^5 = -32.$
First six terms are $1, -2, 4, -8, 16, -32$

7. $a_1 = 2^1 = 2, a_2 = 2^0 = 1,$
$a_3 = 2^{-1} = \dfrac{1}{2}, a_4 = 2^{-2} = \dfrac{1}{4}, a_5 = 2^{-3} = \dfrac{1}{8}.$

   First five terms are $2, 1, \dfrac{1}{2}, \dfrac{1}{4}, \dfrac{1}{8}.$

9. $a_1 = -6 + 0 = -6, a_2 = -6 - 4 = -10,$
$a_3 = -6 - 8 = -14, a_4 = -6 - 12 = -18,$
$a_5 = -6 - 16 = -22.$ First five
terms are $-6, -10, -14, -18, -22$.

11. $b_1 = 5 + 0 = 5, b_2 = 5 + 0.5 = 5.5,$
$b_3 = 5 + 1 = 6, b_4 = 5 + 1.5 = 6.5,$
$b_5 = 5 + 2 = 7, b_6 = 5 + 2.5 = 7.5,$
$b_7 = 5 + 3 = 8.$ First seven
terms are $5, 5.5, 6, 6.5, 7, 7.5, 8$.

13. $c_1 = \dfrac{1^2}{1!} = 1, c_2 = \dfrac{2^2}{2!} = 2,$
$c_3 = \dfrac{3^2}{3!} = \dfrac{3}{2}, c_4 = \dfrac{4^2}{4!} = \dfrac{2}{3},$
$c_5 = \dfrac{5^2}{5!} = \dfrac{5}{24}.$ First five
terms are $1, 2, \dfrac{3}{2}, \dfrac{2}{3}, \dfrac{5}{24}.$

15. $a_1 = 0! = 1, a_2 = 1! = 1,$
$a_3 = 2! = 2, a_4 = 3! = 6, a_5 = 4! = 24,$
$a_6 = 5! = 120, a_7 = 6! = 720.$ First seven
terms are $1, 1, 2, 6, 24, 120, 720$.

17. $a_1 = 8 + 0 = 8, a_2 = 8 - 3 = 5,$
$a_3 = 5 - 3 = 2, a_4 = 2 - 3 = -1,$ and
$a_{10} = 8 - 27 = -19.$ First four terms
are $8, 5, 2, -1$ and 10th term is -19.

19. One gets $a_1 = \dfrac{4}{2+1} = \dfrac{4}{3},$
$a_2 = \dfrac{4}{4+1} = \dfrac{4}{5}, a_3 = \dfrac{4}{6+1} = \dfrac{4}{7},$
$a_4 = \dfrac{4}{8+1} = \dfrac{4}{9}, a_{10} = \dfrac{4}{20+1} = \dfrac{4}{21}$
The first four terms are
$\dfrac{4}{3}, \dfrac{4}{5}, \dfrac{4}{7}, \dfrac{4}{9}$ and 10th term is $\dfrac{4}{21}$.

21. One gets $a_1 = \dfrac{(-1)^1}{2 \cdot 3} = -\dfrac{1}{6},$
$a_2 = \dfrac{(-1)^2}{3 \cdot 4} = \dfrac{1}{12}, a_3 = \dfrac{(-1)^3}{4 \cdot 5} = -\dfrac{1}{20},$
$a_4 = \dfrac{(-1)^4}{5 \cdot 6} = \dfrac{1}{30}, a_{10} = \dfrac{(-1)^{10}}{11 \cdot 12} = \dfrac{1}{132}.$

The first four terms are
$-\frac{1}{6}, \frac{1}{12}, -\frac{1}{20}, \frac{1}{30}$ and 10th term is $\frac{1}{132}$.

23. $a_1 = \frac{2^1}{1!} = 2$,

    $a_2 = \frac{2^2}{2!} = 2, a_3 = \frac{2^3}{3!} = \frac{4}{3}$,

    $a_4 = \frac{2^4}{4!} = \frac{2}{3}, a_{10} = \frac{2^{10}}{10!} = \frac{4}{14,175}$.

    First four terms are $2, 2, \frac{4}{3}, \frac{2}{3}$ and

    10th term is $\frac{4}{14,175}$.

25. $a_1 = \frac{(-2)^1}{0!} = -2$,

    $a_2 = \frac{(-2)^3}{1!} = -8, a_3 = \frac{(-2)^5}{2!} = -16$,

    $a_4 = \frac{(-2)^7}{3!} = -\frac{64}{3}$, and

    $a_{10} = \frac{(-2)^{19}}{9!} = -\frac{4096}{2835}$.

    First four terms are $-2, -8, -16, -\frac{64}{3}$

    and 10th term is $-\frac{4096}{2835}$.

27. $a_1 = -0.1 + 9 = 8.9$,

    $a_2 = -0.2 + 9 = 8.8, a_3 = -0.3 + 9 = 8.7$,

    $a_4 = -0.4 + 9 = 8.6, a_{10} = -1 + 9 = 8$.

    First four terms are $8.9, 8.8, 8.7, 8.6$ and

    and 10th term is 8.

29. $a_2 = 3a_1 + 2 = 3(-4) + 2 = -10$,
    $a_3 = 3a_2 + 2 = 3(-10) + 2 = -28$,
    $a_4 = 3a_3 + 2 = 3(-28) + 2 = -82$,
    $a_5 = 3a_4 + 2 = 3(-82) + 2 = -244$,
    $a_6 = 3a_5 + 2 = 3(-244) + 2 = -730$,
    $a_7 = 3a_6 + 2 = 3(-730) + 2 = -2188$,
    $a_8 = 3a_7 + 2 = 3(-2188) + 2 = -6562$.
    First four terms $-4, -10, -28, -82$ and
    8th term is -6562.

31. One finds $a_2 = a_1^2 - 3 = 2^2 - 3 = 1$,
    $a_3 = a_2^2 - 3 = 1^2 - 3 = -2$,
    $a_4 = a_3^2 - 3 = (-2)^2 - 3 = 1$.

    By a repeating pattern, $a_8 = 1$. First
    four terms are $2, 1, -2, 1$ and 8th term is 1.

33. $a_2 = a_1 + 7 = (-15) + 7 = -8$,
    $a_3 = a_2 + 7 = (-8) + 7 = -1$,
    $a_4 = a_3 + 7 = (-1) + 7 = 6$.

    There is a common difference of 7.
    So $a_8 = 34$. First four terms are $-15, -8, -1, 6$
    and 8th term is 34.

35. $a_n = 2n$

37. $a_n = 2n + 7$

39. $a_n = (-1)^{n+1}$

41. $a_n = n^3$

43. $a_n = e^n$

45. $a_n = \frac{1}{2^{n-1}}$

47. Yes, $d = 1$ is the common difference.

49. No, there is no common difference.

51. No, there is no common difference.

53. Yes, $d = \frac{\pi}{4}$ is the common difference.

55. Since $d = 5$ and $a_1 = 1$, $a_n = a_1 + (n-1)d = 1 + (n-1)5$. So $a_n = 5n - 4$.

57. Since $d = 2$ and $a_1 = 0$, $a_n = a_1 + (n-1)d = 0 + (n-1)2$. So $a_n = 2n - 2$.

59. Since $d = -4$ and $a_1 = 5$, $a_n = a_1 + (n-1)d = 5 + (n-1)(-4)$. So $a_n = -4n + 9$.

61. Since $d = 0.1$ and $a_1 = 1$, $a_n = a_1 + (n-1)d = 1 + (n-1)(0.1)$. So $a_n = 0.1n + 0.9$.

63. Since $d = \frac{\pi}{6}$ and $a_1 = \frac{\pi}{6}$, $a_n = a_1 + (n-1)d = \frac{\pi}{6} + (n-1)\frac{\pi}{6}$. So $a_n = \frac{\pi}{6}n$.

65. Since $d = 15$ and $a_1 = 20$,
    $a_n = a_1 + (n-1)d = 20 + (n-1)15$.
    So $a_n = 15n + 5$.

67. First four terms are $6, 3, 0, -3$ and
    10th term is $a_{10} = 6 + 9(-3) = -21$.

8.2 SERIES

69. First four terms are $1, 0.9, 0.8, 0.7$ and 10th term is $c_{10} = 1 + 9(-0.1) = 0.1$.

71. One finds $w_1 = -\frac{1}{3}1 + 5 = \frac{14}{3}$,
$w_2 = -\frac{1}{3}2 + 5 = \frac{13}{3}$,
$w_3 = -\frac{1}{3}3 + 5 = \frac{12}{3}$,
$w_4 = -\frac{1}{3}4 + 5 = \frac{11}{3}$, and
$w_{10} = -\frac{1}{3}10 + 5 = \frac{5}{3}$.

First four terms are $\frac{14}{3}, \frac{13}{3}, \frac{12}{3}, \frac{11}{3}$
and 10th term is $\frac{5}{3}$.

73. Since $a_n = a_1 + (n-1)d$,
$a_8 = (-3) + (7)5 = 32$.

75. Since $a_n = a_1 + (n-1)d$,
$$a_1 + 2d = 6$$
$$a_1 + 6d = 18$$

Subtracting the first equation from the second, $4d = 12$ and $d = 3$. From the first equation, $a_1 + 6 = 6$ and $a_1 = 0$. So $a_{10} = a_1 + 9d = 0 + 9 \cdot 3 = 27$. So $a_{10} = 27$.

77. Since $a_n = a_1 + (n-1)d$,
$a_{21} = 12 + 20d = 96$. So $20d = 84$ and the common difference is $d = 4.2$.

79. Since $a_n = a_1 + (n-1)d$,
$$a_1 + 2d = 10$$
$$a_1 + 6d = 20$$

Subtracting the first equation from the second, $4d = 10$ and $d = 2.5$. From the first equation, $a_1 + 5 = 10$ and $a_1 = 5$. Then $a_n = 5 + (n-1)(2.5) = 2.5n + 2.5$. So $a_n = 2.5n + 2.5$.

81. $a_1 = 3, a_n = a_{n-1} + 9$

83. $a_1 = \frac{1}{3}, a_n = 3a_{n-1}$

85. $a_1 = 16, a_n = \sqrt{a_{n-1}}$

87. The MSRP for the nth year (where $n \geq 1998$) is $P_n = (1.06)^{(n-1998)} 20,480$. For the next 5 years the prices are $P_{1999} = 21,709$, $P_{2000} = 23,011$, $P_{2001} = 24,392$, $P_{2002} = 25,856$, and $P_{2003} = 27,407$.

89. Number of pages read on the nth day of November is $a_n = 5 + 3(n-1)$. On November 30, they will read $a_{30} = 92$ pages.

91. On the nth year, annual cost is $a_n = 35,000 + 1800(n - 1996)$. In the year 2010, the cost is projected to be $a_{2010} = 35,000 + 1800(14) = \$60,200$.

93. Since there are four corners, $C_n = 4$. If the corner tiles are in place, the number of edge tiles needed for each side is $2(n-1)$. Since there are four sides, $E_n = 8(n-1)$.

If the corner tiles and edge tiles are in place, the interior tiles will occupy an $(n-1)$-by-$(n-1)$ square.

Note, the area of a tile is $\frac{1}{4}$ ft^2. Then the number of interior tiles is $I_n = \frac{(n-1)^2}{\frac{1}{4}}$ or equivalently $I_n = 4(n-1)^2$.

For Thought

1. True, $\sum_{i=1}^{3}(-2)^i = (-2)^1 + (-2)^2 + (-2)^3 = -6$

2. True, $\sum_{i=1}^{6}(0 \cdot i + 5) = (0 \cdot 1 + 5) + (0 \cdot 2 + 5) + (0 \cdot 3 + 5) + (0 \cdot 4 + 5) + (0 \cdot 5 + 5) + (0 \cdot 6 + 5) = 30$.

3. True, since $\sum_{i=1}^{k}(5i) = (5 \cdot 1) + ... + (5 \cdot k) = 5(1 + 2 + ... + k) = 5\sum_{i=1}^{k}(i)$.

4. True, $\sum_{i=1}^{k}(i^2 + 1) = (1^2 + 1) + ... + (k^2 + 1) = (1^2 + 2^2 + ... + k^2) + \underbrace{(1 + ... + 1)}_{k \text{ times}} = \sum_{i=1}^{k} i^2 + k$.

5. False, there are ten terms.

6. True, if $j = i - 1$ and $i = 2$ then $j = 1$.

7. True

8. True, $1 + 2 + ... + n = \dfrac{n}{2}(1 + n)$ represents an arithemetic series.

9. False, if $a_1 = 8$, $a_n = 68$, $d = 2$ then $68 = 8 + (n-1)2$ and $n = 31$. The sum of the arithmetic series is $S_{31} = \dfrac{31}{2}(8 + 68)$.

10. False, $\sum\limits_{i=1}^{10} i^2$ is not an arithmetic series.

8.2 Exercises

1. $1^2 + 2^2 + 3^2 + 4^2 + 5^2 = 55$

3. $(2-1)+(4-1)+(6-1)+(8-1)+(10-1)+(12-1) = 36$

5. $\dfrac{1}{4} + \dfrac{1}{8} + \dfrac{1}{16} + \dfrac{1}{32} = \dfrac{15}{32}$

7. $\sum\limits_{i=4}^{100} 5 = \underbrace{5 + ... + 5}_{97\ times} = 5(97) = 485$

9. $\underbrace{(-1)^4 + (-1)^5 + ... + (-1)^{47}}_{44\ terms} + (-1)^{48} = 0 + 1 = 1$

11. $[(-1)^7 + (-1)^8] + ... + [(-1)^{43} + (-1)^{44}] = 0 + ... + 0 = 0$

13. $\sum\limits_{i=1}^{6} i$

15. $\sum\limits_{i=1}^{5} (-1)^i (2i - 1)$

17. $\sum\limits_{i=1}^{5} i^2$

19. $\sum\limits_{i=1}^{5} \left(-\dfrac{1}{2}\right)^{i-1}$

21. $\sum\limits_{i=1}^{3} \ln(x_i)$

23. $\sum\limits_{i=1}^{11} ar^{i-1}$

25. Let $j = i - 1$. New series is $\sum\limits_{j=0}^{31} (-1)^{j+1}$

27. Let $j = i - 3$. New series is $\sum\limits_{j=1}^{10} (2j + 7)$

29. Let $j = x - 2$. New series is $\sum\limits_{j=0}^{8} \dfrac{10!}{(j+2)!(8-j)!}$

31. Let $j = n + 3$. New series is $\sum\limits_{j=5}^{9} \dfrac{5^{j-3} \cdot e^{-5}}{(j-3)!}$

33. $0.5 + 0.5r + 0.5r^2 + 0.5r^3 + 0.5r^4 + 0.5r^5$

35. $a^4 + a^3 b + a^2 b^2 + ab^3 + b^4$

37. $a^2 + 2ab + b^2$

39. $\dfrac{6 + 23 + 45}{3} = \dfrac{74}{3}$

41. $\dfrac{-6 + 0 + 3 + 4 + 3 + 92}{6} = 16$

43. $\dfrac{\sqrt{2} + \pi + 33.6 - 19.4 + 52}{5} \approx 14.151$

45. Note $a_1 = 1$, $a_n = 47$, $n = 47$. So the sum is $S_{47} = \dfrac{47}{2}(1 + 47) = 1128$.

47. Since $-16 = 8 + (n-1)(-3)$, $n = 9$. So the sum is $S_9 = \dfrac{9}{2}(8 + (-16)) = -36$.

49. Since $55 = 3 + (n-1)4$, $n = 14$. So the sum is $S_{14} = \dfrac{14}{2}(3 + 55) = 406$.

51. Since $5 = \dfrac{1}{2} + (n-1)\dfrac{1}{4}$, $n = 19$. So the sum is $S_{19} = \dfrac{19}{2}(1/2 + 5) = 52.25$.

53. Since $a_1 = -3$, $a_{12} = 6(12) - 9 = 63$, and $n = 12$, the sum is $S_{12} = \dfrac{12}{2}(-3 + 63) = 360$.

55. Since $a_3 = 0.7$, $a_{15} = -0.5$, and there are $n = 13$ terms in the series, the sum is $S_{13} = \dfrac{13}{2}(0.7 - 0.5) = 1.3$.

57. Since $a_1 = 30,000$, $d = 1,000$, and $a_{30} = 30,000 + 29(1,000) = 59,000$, so his total salary for 30 years of work is $S_{30} = \dfrac{30}{2}(30,000 + 59,000) = \$1,335,000$.

8.3 GEOMETRIC SEQUENCES AND SERIES

The mean annual salary for 30 years is $\frac{S_{30}}{30} = \$44,500$.

59. In the nth level the number of cans is $a_n = 12(10-n)$. In the mountain of cans there are $\sum_{i=1}^{9}(120-12n) = \sum_{j=1}^{9}(120-12(10-j)) = \sum_{j=1}^{9} 12j = \frac{9}{2}(12+108) = 540$ cans.

61. At 6% compounded annually, the future value of $1000 after i years is $\$1000(1.06)^i$. Then the amount in the account on January 1, 2000 is $\sum_{i=1}^{10} 1000(1.05)^i$.

63. With a calculator, one finds $200 + 200(.63) + 200(.63)^2 + 200(.63)^3 \approx 455.4$ mg.

65. Since $a_9 = 101$, $a_{60} = 356$, and there are $n = 52$ terms from a_9 to a_{60} then the mean of the 9th through the 60th term is $\frac{1}{52}\sum_{i=9}^{60} a_i = \frac{1}{52}\cdot\frac{52}{2}(101+356) = 228.5$.

67. One finds $a_2 = a_1^2 - 3 = (-2)^2 - 3 = 1$ and $a_3 = a_2^2 - 3 = 1^2 - 3 = -2$. By a repeating pattern, one derives $a_7 = a_9 = -2$ and $a_8 = a_{10} = 1$. The mean from the 7th term to the 10th term is $\frac{-2+1-2+1}{4} = -\frac{1}{2}$.

For Thought

1. False, the ratios $\frac{6}{2}$ and $\frac{24}{6}$ are not equal

2. True, $a_n = 3 \cdot 2^3 \cdot 2^{-n} = 24\left(\frac{1}{2}\right)^n$.

3. True, $a_1 = 5(0.3)^1 = 1.5$.

4. False, since $a_n = \left(\frac{1}{5}\right)^n$, the common ratio is $\frac{1}{5}$.

5. True **6.** False. Note, the ratio 2 does not satisfy $|r| < 1$.

7. False, $\sum_{i=1}^{9} 3(0.6)^i = \frac{1.8(1-0.6^9)}{1-0.6}$.

8. True, $\sum_{i=0}^{4} 2(10)^i = \frac{2(1-10^5)}{1-10} = 22,222$.

9. True, $\sum_{i=1}^{\infty} 3(0.1)^i = \frac{0.3}{1-0.1} = \frac{1}{3}$.

10. True, $\sum_{i=1}^{\infty} \left(\frac{1}{2}\right)^i = \frac{1/2}{1-1/2} = 1$.

8.3 Exercises

1. $a_n = \frac{1}{6} 2^{n-1}$

3. $a_n = (0.9)(0.1)^{n-1}$

5. $a_n = 4 \cdot (-3)^{n-1}$

7. arithmetic

9. geometric

11. neither

13. $a_1 = 2\cdot 1 = 2$, $a_2 = 2\cdot 2 = 4$, $a_3 = 2\cdot 3 = 6$, $a_4 = 2\cdot 4 = 8$. First four terms are $2,4,6,8$. Arithmetic sequence.

15. One finds $a_1 = 1^2 = 1$, $a_2 = 2^2 = 4$, $a_3 = 3^2 = 9$, $a_4 = 4^2 = 16$. First four terms are $1,4,9,16$. Neither sequence.

17. $a_1 = \left(\frac{1}{2}\right)^1 = \frac{1}{2}$, $a_2 = \left(\frac{1}{2}\right)^2 = \frac{1}{4}$, $a_3 = \left(\frac{1}{2}\right)^3 = \frac{1}{8}$, $a_4 = \left(\frac{1}{2}\right)^4 = \frac{1}{16}$. First four terms are $\frac{1}{2}, \frac{1}{4}, \frac{1}{8}, \frac{1}{16}$. Geometric sequence.

19. $b_1 = 2^3 = 8$, $b_2 = 2^5 = 32$, $b_3 = 2^7 = 128$, $b_4 = 2^9 = 512$. First four terms are $8,32,128,512$. Geometric sequence.

21. $c_2 = -3c_1 = -3\cdot 3 = -9$, $c_3 = -3c_2 = -3\cdot(-9) = 27$, $c_4 = -3c_3 = -3\cdot 27 = -81$. First four terms are $3,-9,27,-81$. Geometric sequence.

23. $a_1 = 3(-2)^0 = 3$, $a_2 = 3(-2)^1 = -6$, $a_3 = 3(-2)^2 = 12$, $a_{10} = 3(-2)^9 = -1536$. First three terms are $3,-6,12$ and $a_{10} = -1536$.

25. One finds $a_1 = 4(0.1)^0 = 4$, $a_2 = 4(0.1)^1 = 0.4$, $a_3 = 4(0.1)^2 = 0.04$, $a_{10} = 4(0.1)^9 = 4\times 10^{-9}$. First three terms are $4, 0.4, 0.04$ and $a_{10} = 4\times 10^{-9}$.

27. Since $a_n = a_1 r^{n-1}$,
$$\frac{3}{1024} = 3\left(\frac{1}{2}\right)^{n-1}$$
$$\frac{1}{1024} = \left(\frac{1}{2}\right)^{n-1}$$
$$\left(\frac{1}{2}\right)^{10} = \left(\frac{1}{2}\right)^{n-1}$$

So $n-1 = 10$ and the number of terms is $n = 11$.

29. Since $a_n = a_1 r^{n-1}$, $\frac{1}{81} = a_1 \left(\frac{1}{3}\right)^5$.

Solving, one finds the first term is $a_1 = 3$.

31. Since $a_n = a_1 r^{n-1}$, $6 = \frac{2}{3} r^2$.

So $9 = r^2$ and the common ratio is $r = 3$ or $r = -3$.

33. Since $a_6 = a_3 r^3$, $96 = -12 r^3$ and $r = -2$. Since $a_3 = a_1 r^2$, $-12 = a_1(-2)^2$. So $a_1 = -3$ and $a_n = (-3)(-2)^{n-1}$.

35. $\sum_{i=1}^{5} 6 \left(\frac{1}{3}\right)^{i-1} = \frac{6(1-(1/3)^5)}{1 - 1/3} = \frac{242}{27}$

37. $\sum_{i=1}^{8} 1.5 (-2)^{i-1} = \frac{1.5(1-(-2)^8)}{1-(-2)} = -127.5$

39. $\frac{2(1-1.05^{12})}{1-1.05} \approx 31.8343$

41. $\frac{200(1-1.01^8)}{1-1.01} \approx 1{,}657.13$

43. $\sum_{n=1}^{5} 3 \left(-\frac{1}{3}\right)^{n-1}$

45. $\sum_{n=1}^{\infty} 0.6(0.1)^{n-1}$

47. $\sum_{n=1}^{\infty} (-4.5) \left(-\frac{1}{3}\right)^{n-1}$

49. $S = \frac{a_1}{1-r} = \frac{3}{1-(-1/3)} = \frac{3}{4/3} = \frac{9}{4}$

51. $S = \frac{a_1}{1-r} = \frac{0.9}{1-0.1} = \frac{0.9}{0.9} = 1$

53. $S = \frac{a_1}{1-r} = \frac{-9.9}{1-(-1/3)} = \frac{-9.9}{4/3} =$
$-\frac{99}{10} \cdot \frac{3}{4} = -\frac{297}{40}$

55. $S = \frac{a_1}{1-r} = \frac{0.34}{1-0.01} = \frac{0.34}{0.99} = \frac{34}{99}$

57. No sum since $|r| = |-1.06| > 1$.

59. $S = \frac{a_1}{1-r} = \frac{0.6}{1-0.1} = \frac{0.6}{0.9} = \frac{2}{3}$

61. $S = \frac{a_1}{1-r} = \frac{34}{1-(-0.7)} = \frac{34}{1.7} = 20$

63. $\sum_{i=2}^{\infty} 4(0.1)^i = \frac{0.04}{1-0.1} = \frac{0.04}{0.9} = \frac{4}{90} = \frac{2}{45}$

65. $8.2 + 0.05454... = 8.2 + 0.1(0.5454...) =$
$8.2 + 0.1 \sum_{i=0}^{\infty} 0.54(0.01)^i =$
$8.2 + 0.1 \frac{0.54}{1-0.01} = 8.2 + \frac{0.054}{0.99} =$
$8.2 + \frac{54}{990} = \frac{8118 + 54}{990} = \frac{8172}{990}$

67. Using a formula for S_n, the sum is
$$\sum_{n=1}^{25} 100(.69)^{n-1} = \frac{100(1-(.69)^{25})}{1-.69}$$
$$\approx 322.55 \text{ mg}.$$

69. The total number of subscriptions in June is
$$\sum_{i=1}^{30} 2^{i-1} = \frac{1-2^{30}}{1-2} = 1{,}073{,}741{,}823.$$

71. At the end of the nth quarter the amount is $a_n = 4000(1.02)^n$. At the end of the 37th quarter it is $a_{37} = 4000(1.02)^{37} = \$8{,}322.74$.

73. At the end of the 12th month, the amount in the account is $\sum_{i=1}^{12} 200(1.01)^i =$
$$\frac{200(1.01)(1-1.01^{12})}{1-1.01} = \$2{,}561.87.$$

75. The value of this annuity immediately after the last payment is $\sum_{i=0}^{359} 100\left(1+\frac{0.09}{12}\right)^i =$
$$\frac{100(1-1.0075^{360})}{1-1.0075} = \$183{,}074.35.$$

77. Assume the ball has a small radius. Approximately, the distance it travels before it comes to a rest is

$$9 + \frac{2}{3} \cdot 9 + \frac{2}{3} \cdot 9 + \left(\frac{2}{3}\right)^2 \cdot 9 + \left(\frac{2}{3}\right)^2 \cdot 9 + \ldots =$$

$$9 + 18 \left[\frac{2}{3} + \left(\frac{2}{3}\right)^2 + \left(\frac{2}{3}\right)^3 + \ldots \right] =$$

$$9 + 18 \frac{2/3}{1 - 2/3} = 45 \text{ feet.}$$

79. The first amount spent in Hammond is $2 million, then 75% of $2 million, and so on. So the total economic impact is

$$\sum_{i=0}^{\infty} 2(10^6)(0.75)^i = \frac{2,000,000}{1 - 0.75} = \$8,000,000 \, .$$

81. Yes, since if $a = d = 0$ and r is any positive number then the arithmetic sequence is also a geometric sequence.

83. If $|r| < 1$ then $r^x \to 0$ as $x \to \infty$.
If $|r| > 1$ then $|r^x| \to \infty$ as $x \to \infty$.

Below are the graphs of $y = 2^x$ and $y = \left(\frac{1}{2}\right)^x$.

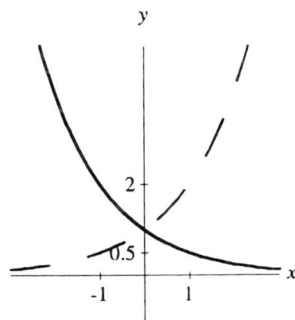

For Thought

1. False, there are $90 \cdot 26$ different codes.

2. False, there are 24^3 different possible fraternity names. **3.** False, there are $3 \cdot 5 \cdot 3 \cdot 1 = 30$ different outfits

4. True, $5! = 120$.

5. True, $P(10, 2) = \frac{10!}{8!} = 10 \cdot 9 = 90$.

6. False, the number of ways is 4^{20} **7.** True

8. True, $\frac{1000 \cdot 999 \cdot 998!}{998!} = 1000 \cdot 999 = 999,000$

9. False, $P(10, 1) = \frac{10!}{9!} = 10$ and

$$P(10, 9) = \frac{10!}{1!} = 10!.$$

10. False, $P(29, 1) = \frac{29!}{28!} = \frac{29 \cdot 28!}{28!} = 29$.

8.4 Exercises

1. Number of routes is $2 \cdot 4 = 8$.

3. There are $3! = 6$ different schedules. These schedules are given by:

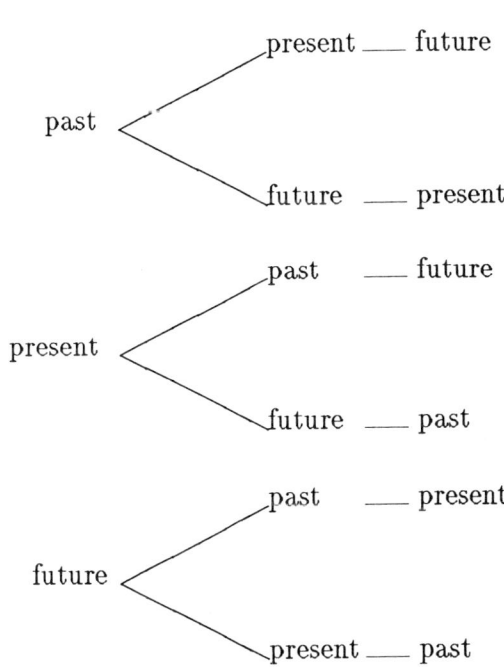

5. There are $4^5 = 1,024$ different possible hands.

7. There are 8 optional items since $2^8 = 256$.

9. $\frac{7!}{4!} = 210$

11. $\frac{99!}{99!} = 1$

13. $\frac{9!/4!}{5!} = \frac{9!}{4!5!} = 126$

15. $\frac{11!/8!}{3!} = \frac{11!}{8!3!} = 165$

17. 1820

19. $\dfrac{88\cdot 87\cdot 86\cdot 85!}{85!3!} = \dfrac{88\cdot 87\cdot 86}{3!} = 109,736$

21. Hercules can perform 12 tasks in $12! = 479,001,600$ ways.

23. The number of ways a first, second and a third restautant can be chosen is
$P(15,3) = \dfrac{15!}{12!} = 2730$ ways.

25. Since there are 8 half-hour shows from 8 : 00 p.m.
to 10 : 00 p.m., the number of schedules is $P(26,8) \approx 6.3 \times 10^{10}$.

27. Since each question has 4 possible answers, the number of ways to answer 6 questions is $4^6 = 4096$.

29. There are $4\cdot 3\cdot 6 = 72$ possible prizes.

31. There are $P(26,3) = 15,600$ possible passwords.

33. There are $3\cdot 10^4 = 30,000$ possible phone numbers.

35. Superman has $4! = 24$ ways to arrange these rescues.

37. Since the true-false questions can be answered in 2^5 ways and the multiple choice questions in 4^6 ways, the test can be answered in $2^5 \cdot 4^6 = 131,072$ ways.

39. On her list will be $P(7,3) = 210$ possible words.

For Thought

1. True, the number of ways is $C(5,3) = C(5,2) = 10$.

2. False, the number of ways is 3^5.

3. True, $P(5,2) = 20$.

4. False, it contains $n+1$ terms. 5. True

6. True, $C(n,r)$ is the number of subsets of size r and a set with n elements has 2^n subsets.

7. False, $P(8,3) = 336$ and $C(8,3) = 56$.

8. True 9. False, since $P(8,3) = 336$ and $P(8,5) = 6720$. 10. True

8.5 Exercises

1. $C(52,5) = 2,598,960$ possible poker hands

3. $C(49,6) = 13,983,816$ ways to choose six numbers

5. $C(5,3) = 10$ ways for an in-depth interview

7. There are $3^3 = 27$ ways to watch.

9. $C(8,3) = 56$ ways to make the selection

11. Since 3 men can be selected in $C(9,3)$ ways and 2 women in $C(6,2)$ ways, the team can be selected in $C(9,3)\cdot C(6,2) = 1260$ ways.

13. $12! = 479,001,600$ ways to return the papers

15. $6\cdot 6 = 36$ possible outcomes

17. Since there are 4! ways to arrange the bands in a line and 3! ways to line up the floats, the parade can be lead in $4!3! = 144$ ways.

19. Since A occurs 4 times, the number of permutations is $\dfrac{7!}{4!} = 210$.

21. These assignments can be done in $\dfrac{10!}{5!3!2!} = 2520$ ways.

23. There are 3, 6, 10, and $C(n,2)$ distinct chords, respectively,.

25. There are $\dfrac{10!}{3!2!5!} = 2520$ ways the assignments can be made.

27. Use Pascal's triangle. So $(a-2)^3 = 1(a^3) + 3(a^2)(-2) + 3(a)(-2)^2 + 1(-2)^3 = a^3 - 6a^2 + 12a - 8$.

29. Use Pascal's triangle. So $(2a+b^2)^3 = 1(2a)^3 + 3(2a)^2(b^2) + 3(2a)(b^2)^2 + 1(b^2)^3 = 8a^3 + 12a^2b^2 + 6ab^4 + b^6$.

31. Use Pascal's triangle. So $(x-2y)^4 = 1(x)^4 + 4(x)^3(-2y) + 6(x)^2(-2y)^2 + 4(x)(-2y)^3 + 1(-2y)^4 = x^4 - 8x^3y + 24x^2y^2 - 32xy^3 + 16y^4$.

8.6 PROBABILITY

33. Use Pascal's triangle. So $(x^2+1)^4 =$
$1(x^2)^4 + 4(x^2)^3(1) + 6(x^2)^2(1)^2 + 4(x^2)(1)^3 + 1(1)^4 =$
$x^8 + 4x^6 + 6x^4 + 4x^2 + 1$.

35. Use the Binomial Theorem. So $(x+y)^9 =$
$\binom{9}{0}x^9 + \binom{9}{1}x^8y + \binom{9}{2}x^7y^2 + ... =$
$x^9 + 9x^8y + 36x^7y^2 + ...$

37. Use the Binomial Theorem. So $(2x-y)^{12} =$
$\binom{12}{0}(2x)^{12} + \binom{12}{1}(2x)^{11}(-y) +$
$\binom{12}{2}(2x)^{10}(-y)^2 + ... =$
$4096x^{12} - 24{,}576x^{11}y + 67{,}584x^{10}y^2 + ...$

39. Use the Binomial Theorem. So $(2s-0.5t)^8 =$
$\binom{8}{0}(2s)^8 + \binom{8}{1}(2s)^7(-0.5t) +$
$\binom{8}{2}(2s)^6(-0.5t)^2 + ... =$
$256s^8 - 512s^7t + 448s^6t^2 + ...$

41. Use the Binomial Theorem. So $(m^2 - 2w^3)^9 =$
$\binom{9}{0}(m^2)^9 + \binom{9}{1}(m^2)^8(-2w^3) +$
$\binom{9}{2}(m^2)^7(-2w^3)^2 + ... =$
$m^{18} - 18m^{16}w^3 + 144m^{14}w^6 + ...$

43. $\binom{8}{5} = 56$

45. $\binom{13}{5}(-2)^5 = -41{,}184$

47. By looking at a particular term, one finds
$(a + [b+c])^{12} = ... + \binom{12}{10}a^2[b+c]^{10} + ... =$
$... + \binom{12}{10}a^2\left[... + \binom{10}{6}b^4c^6 + ...\right] + ...$
So the coefficient of $a^2b^4c^6$ is

$\binom{12}{10}\binom{10}{6} = \dfrac{12!}{2!4!6!} = 13{,}860$.

49. Coefficient of a^3b^7 in $(a+b+2c)^{10}$ is the number of rearrangements of aaabbbbbbb, which is $\dfrac{10!}{3!7!} = 120$.

For Thought

1. False, $P(E) = \dfrac{n(E)}{n(S)}$.

2. False, there are $2^4 = 16$ outcomes.

3. True, since possible outcomes are (H,H), (H,T), (T,H), and (T,T) then
$P(at\ least\ one\ tail) = \dfrac{3}{4} = 0.75$.

4. True, $\dfrac{5}{6}$ is the probability of not getting a four when a die is tossed then
$P(at\ least\ one\ four) = 1 - P(no\ four) =$
$1 - \dfrac{5}{6}\dfrac{5}{6} = \dfrac{11}{36}$.

5. False, since $\dfrac{1}{2}$ is the probability of getting a tail when a coin is tossed,
$P(at\ least\ one\ head) = 1 - P(four\ tails) =$
$1 - \left(\dfrac{1}{2}\right)^4 = \dfrac{15}{16}$.

6. False, the complement is getting either no head or exactly one head.

7. True, the complement of getting exactly 3 tails in a toss of three coins is getting at least one head.

8. False. Odds in favor of snow is
$\dfrac{P(snow)}{P(no\ snow)} = \dfrac{7/10}{3/10} = \dfrac{7}{3}$ i.e. 7 to 3.

9. False, $P(E) = 3/7$.

10. False, it is equivalent to 1 to 4.

8.6 Exercises

1. (a) 1/3, (b) 2/3, (c) 0

3. (a) $\dfrac{n(\{3,4,5,6\})}{6} = \dfrac{4}{6} = \dfrac{2}{3}$,

   (b) $\dfrac{n(\{1,2,3,4,5,6\})}{6} = \dfrac{6}{6} = 1$,

   (c) $\dfrac{n(\{1,2,3,5,6\})}{6} = \dfrac{5}{6}$,

   (d) 0, (e) $\dfrac{n(\{1\})}{6} = \dfrac{1}{6}$

5. (a) $\dfrac{n\{(T,T)\}}{4} = 1/4$,

   (b) $1 - P(2\ tails) = 1 - 1/4 = 3/4$,

   (c) $\dfrac{n\{(H,H)\}}{4} = 1/4$,

   (d) $\dfrac{n\{(T,T),(H,T),(T,H)\}}{4} = 3/4$

7. Note there are 36 possible outcomes.

   (a) $\dfrac{n\{(3,3)\}}{36} = 1/36$,

   (b) Since there are 11 possible outcomes with at least one 3, $P(at\ least\ one\ 3) = 11/36$,

   (c) Since there are 5 possible outcomes with a sum of 6, $P(sum\ is\ 6) = 5/36$,

   (d) $1 - P(sum\ is\ 1) = 1 - 1/36 = 35/36$,

   (e) $P(sum\ is\ 2) = 1/36$

9. (a) 3/13, (b) 9/13, (c) 9/13,

   (d) $P(marble\ is\ yellow) = 4/13$

11. There are 72 possible outcomes.

    (a) $\dfrac{n\{(1,9)\}}{72} = 1/72$,

    (b) $\dfrac{n\{(1,3),(3,1)\}}{72} = 2/72 = 1/36$,

    (c) $\dfrac{n\{(1,4),(4,1),(2,3),(3,2)\}}{72} = 4/72 = 1/18$

13. (a) $\dfrac{1}{C(52,5)} = \dfrac{1}{2,598,960}$

    (b) By using the counting principle, the probability of one 3, one 4, one 5, one 6, and one 7 is $\dfrac{4^5}{C(52,5)} = \dfrac{1024}{2,598,960}$.

15. $1 - P(surviving) = 1 - 0.001 = 0.999$

17. Note there are 36 possible outcomes.

    (a) $\dfrac{n\{(4,4)\}}{36} = 1/36$,

    (b) $1 - P(pair\ of\ 4's) = 1 - 1/36 = 35/36$,

    (c) 35/36, since this is the same event as (b)

19. (a) $e^{-3} \approx 0.05$

    (b) $1 - P(no\ hits\ in\ next\ 1\ million\ years) = 1 - 0.05 = 0.95$

21. odds in favor of eye coming ashore is $\dfrac{0.8}{0.2} = 4$ i.e. 4 to 1

23. (a) since $\dfrac{1/4}{3/4} = \dfrac{1}{3}$, the odds in favor of stock going up is 1 to 3

    (b) odds against stock going up is 3 to 1

25. If p is the probability of rain today, then
$$\dfrac{p}{1-p} = \dfrac{4}{1}$$
$$p = 4 - 4p$$
$$p = \dfrac{4}{5}.$$

   The probability of rain today is $\dfrac{4}{5}$.

27. (a) 1 to 9 (b) 9/10

29. Since $P(2\ heads) = \dfrac{C(4,2)}{2^4} = 3/8$ and $\dfrac{3/8}{5/8} = 3/5$, the odds in favor of getting 2 heads in four tosses is 3 to 5

31. Since $\dfrac{P(sum\ of\ 7)}{P(sum\ that\ is\ not\ 7)} = \dfrac{1/6}{5/6} = 1/5$, odds in favor of getting a sum of 7 is 1 to 5.

33. 1 to 1,999,999

35. $\dfrac{1}{1+31} = 1/32$

37. $P(high-risk\ or\ a\ woman) = P(high-risk) + P(woman) - P(high-risk\ or\ a\ woman) = 38\% + 64\% - 24\% = 78\%$.

39. One finds the probabilities
$P(3 \text{ boys}) = P(3 \text{ girls}) = 1/8$. By the addition rule for mutually exclusive events,
$P(3 \text{ boys or } 3 \text{ girls}) = 1/8 + 1/8 = 1/4$.

41. One can finds the probabilities
$P(heart) = 13/52$, $P(king) = 4/52$,
$P(heart \text{ and } king) = 1/52$. By the addition rule, $P(heart \text{ or } king) = 13/52 + 4/52 - 1/52 = 4/13$.

43. (a) 34%, (b) 22% + 18% = 40%,
(c) 34% + 22% + 18% = 74%

For Thought

1. True, $4 \cdot 1 - 2 = 2 \cdot 1^2$ when $n = 1$.

2. False, when $n = 3$ one gets $27 < 12 + 15$, which is inconsistent.

3. False, $\dfrac{100 - 1}{100 + 1} \approx 0.98$.

4. False. It can prove it for the positive integers.

5. True   6. False, rather $\sum_{i=1}^{k+1} \dfrac{1}{i(i+1)} = \dfrac{k+1}{k+2}$.

7. False   8. False, since $0 > 0$ is inconsistent.

9. True. To see this let S_n be the inequality given by $n^2 - n > 0$. Note S_2 is true since $2^2 - 2 > 0$ holds. Suppose S_k is true. One needs to show S_{k+1} is true. Note,
$$(k+1)^2 - (k+1) = (k^2 + 2k + 1) - k - 1$$
$$= (k^2 - k) + 2k$$
$$> 0 + 2k \text{ since } S_k \text{ is true}$$
$$= 2k$$
$$> 0$$
So S_{k+1} holds and S_n is true for integers $n > 1$.

10. False, when $n = 1$ one gets $1^2 - 1 > 0$, which is inconsistent.

8.7 Exercises

1. If $n = 1$, $3 - 1 = 2$ and $\dfrac{3 \cdot 1^2 + 1}{2} = \dfrac{4}{2}$.
If $n = 2$, $(3 - 1) + (6 - 1) = 7$ and
$\dfrac{3 \cdot 2^2 + 2}{2} = \dfrac{14}{2} = 7$.
If $n = 3$, $(3-1) + (6-1) + (9-1) = 15$
and $\dfrac{3 \cdot 3^2 + 3}{2} = \dfrac{30}{2} = 15$.
It is true for $n = 1, 2$, and 3.

3. If $n = 1$, $\dfrac{1}{1(1+1)} = \dfrac{1}{1+1}$.
If $n = 2$, $\dfrac{1}{2} + \dfrac{1}{6} = \dfrac{4}{6}$ and
$\dfrac{2}{2+1} = \dfrac{2}{3}$.
If $n = 3$, $\dfrac{1}{2} + \dfrac{1}{6} + \dfrac{1}{12} = \dfrac{9}{12}$
and $\dfrac{3}{3+1} = \dfrac{3}{4}$.
It is true for $n = 1, 2$, and 3.

5. If $n = 1$, one has $1 = 4 - 3$.
If $n = 2$, $1 + 4 = 5$ and $4(2) - 3 = 5$.
If $n = 3$, $1 + 4 + 9 = 14$ and $4(3) - 3 = 9$.
It is true for $n = 1, 2$ and false for $n = 3$.

7. If $n = 1$, one has $1 < 1$.
If $n = 2$, one has $4 < 8$. If $n = 3$, $9 < 27$.
It is true for $n = 2, 3$ and false for $n = 1$.

9. $S_1 : 2(1) = 1(1+1)$
$S_k : \sum_{i=1}^{k} 2i = k(k+1)$
$S_{k+1} : \sum_{i=1}^{k+1} 2i = (k+1)(k+2)$

11. $S_1 : 2 = 2 \cdot 1^2$
$S_k : 2 + 6 + \ldots + (4k - 2) = 2k^2$
$S_{k+1} : \begin{array}{l} 2 + 6 + \ldots + (4(k+1) - 2) = \\ 2 + 6 + \ldots + (4k + 2) = 2(k+1)^2 \end{array}$

13. $S_1 : 2 = 2^2 - 2$
$S_k : \sum_{i=1}^{k} 2^i = 2^{k+1} - 2$
$S_{k+1} : \sum_{i=1}^{k+1} 2^i = 2^{k+2} - 2$

15. $S_1 : (ab)^1 = a^1 b^1$
$S_k : (ab)^k = a^k b^k$
$S_{k+1} : (ab)^{k+1} = a^{k+1} b^{k+1}$

17. $S_1 :$ If $0 < a < 1$ then $0 < a^1 < 1$
$S_k :$ If $0 < a < 1$ then $0 < a^k < 1$
$S_{k+1} :$ If $0 < a < 1$ then $0 < a^{k+1} < 1$

19. Let $T_n : 1 + 2 + \ldots + n = \dfrac{n(n+1)}{2}$.

Step 1: If $n = 1$ then $T_1 : 1 = \dfrac{1(2)}{2}$.
So T_1 is true.

Step 2: Assume $T_k : 1 + 2 + \ldots + k = \dfrac{k(k+1)}{2}$
is true. Add $(k+1)$ to both sides and get
$$1 + 2 + \ldots + k + (k+1) = \dfrac{k(k+1)}{2} + (k+1)$$
$$= \dfrac{k(k+1) + 2(k+1)}{2}$$
$$= \dfrac{(k+1)(k+2)}{2}$$
So the truth of T_k implies the truth of T_{k+1}.
T_n is true for every positive integer n.

21. Let $T_n : 3 + 7 + \ldots + (4n-1) = n(2n+1)$.

Step 1: If $n = 1$ then $T_1 : 3 = 1(2+1)$.
So T_1 is true.

Step 2: Assume T_k is true i.e.
$3 + 7 + \ldots + (4k-1) = k(2k+1)$.
Since $4(k+1) - 1 = 4k+3$, add $4k+3$ to both sides. So
$3 + \ldots + (4k-1) + (4k+3) =$
$$= k(2k+1) + (4k+3)$$
$$= 2k^2 + k + 4k + 3$$
$$= 2k^2 + 5k + 3$$
$$= (k+1)(2k+3)$$
$$= (k+1)(2(k+1) + 1)$$
The truth of T_k implies the truth of T_{k+1}.
T_n is true for every positive integer n.

23. Let $T_n : \sum_{i=1}^{n} 2^i = 2^{n+1} - 2$.

Step 1: If $n = 1$ then $T_1 : 2 = 2^2 - 2$.
So T_1 is true.

Step 2: Assume $T_k : \sum_{i=1}^{k} 2^i = 2^{k+1} - 2$ is true
Add 2^{k+1} to both sides.
$$\sum_{i=1}^{k} 2^i + 2^{k+1} = 2^{k+1} - 2 + 2^{k+1}$$
$$= 2 \cdot 2^{k+1} - 2$$
$$= 2^{k+2} - 2$$
$$= 2^{(k+1)+1} - 2$$
So the truth of T_k implies the truth of T_{k+1}.
T_n is true for every positive integer n.

25. Let $T_n : \sum_{i=1}^{n} (3i - 1) = \dfrac{3n^2 + n}{2}$.

Step 1: If $n = 1$ then $T_1 : 3 - 1 = \dfrac{3+1}{2}$.
So T_1 is true.

Step 2: Assume $T_k : \sum_{i=1}^{k} (3i-1) = \dfrac{3k^2 + k}{2}$
is true. Since $3(k+1) - 1 = 3k + 2$, add
$3k + 2$ to both sides and get
$$\sum_{i=1}^{k} (3i-1) + 3k + 2 = \dfrac{3k^2 + k}{2} + (3k+2)$$
$$= \dfrac{3k^2 + k + 2(3k+2)}{2}$$
$$= \dfrac{3k^2 + 7k + 4}{2}$$
$$= \dfrac{3(k^2 + 2k + 1) + (k+1)}{2}$$
$$= \dfrac{3(k+1)^2 + (k+1)}{2}$$
So the truth of T_k implies the truth of T_{k+1}.
T_n is true for every positive integer n.

27. Let $T_n : 1^2 + 2^2 + \ldots + n^2 = \dfrac{n(n+1)(2n+1)}{6}$.

Step 1: If $n = 1$ then $T_1 : 1 = \dfrac{1(2)(3)}{6}$.
So T_1 is true.

Step 2: Assume T_k is true i.e. we
assume $1^2 + 2^2 + \ldots + k^2 = \dfrac{k(k+1)(2k+1)}{6}$
Add $(k+1)^2$ to both sides.
$1^2 + 2^2 + \ldots + k^2 + (k+1)^2 =$

$$= \frac{k(k+1)(2k+1)}{6} + (k+1)^2$$
$$= \frac{k(k+1)(2k+1) + 6(k+1)^2}{6}$$
$$= \frac{(k+1)[k(2k+1) + 6(k+1)]}{6}$$
$$= \frac{(k+1)[2k^2 + 7k + 6]}{6}$$
$$= \frac{(k+1)[k+2][2k+3]}{6}$$
$$= \frac{(k+1)[(k+1)+1][2(k+1)+1]}{6}$$

So the truth of T_k implies the truth of T_{k+1}. T_n is true for every positive integer n.

29. Let $T_n: 1 \cdot 3 + 2 \cdot 4 + \ldots + n \cdot (n+2) = \frac{n}{6}(n+1)(2n+7)$.

Step 1: If $n=1$ then $T_1: 1 \cdot 3 = \frac{1}{6}(2)(2+7)$. So T_1 is true.

Step 2: Assume T_k is true i.e. we assume $1 \cdot 3 + 2 \cdot 4 + \ldots + k \cdot (k+2) = \frac{k}{6}(k+1)(2k+7)$. Add $(k+1)(k+3)$ to both sides. Then
$1 \cdot 3 + 2 \cdot 4 + \ldots + k \cdot (k+2) + (k+1)(k+3) =$
$$= \frac{k}{6}(k+1)(2k+7) + (k+1)(k+3)$$
$$= \frac{k(k+1)(2k+7) + 6(k+1)(k+3)}{6}$$
$$= \frac{(k+1)(2k^2 + 13k + 18)}{6}$$
$$= \frac{(k+1)(2k+9)(k+2)}{6}$$
$$= \frac{(k+1)}{6}[k+2][2k+9]$$
$$= \frac{(k+1)}{6}[(k+1)+1][2(k+1)+7]$$

So the truth of T_k implies the truth of T_{k+1}. T_n is true for every positive integer n.

31. Let $T_n: \frac{1}{1 \cdot 3} + \frac{1}{3 \cdot 5} + \ldots + \frac{1}{(2n-1)(2n+1)} = \frac{n}{2n+1}$.

Step 1: If $n=1$ then $T_1: \frac{1}{1 \cdot 3} = \frac{1}{2+1}$. So T_1 is true.

Step 2: Assume T_k is true i.e. we assume
$$\frac{1}{1 \cdot 3} + \ldots + \frac{1}{(2k-1)(2k+1)} = \frac{k}{2k+1}.$$
Since $\frac{1}{[2(k+1)-1][2(k+1)+1]} = \frac{1}{(2k+1)(2k+3)}$, add $\frac{1}{(2k+1)(2k+3)}$ to both sides.
So $\frac{1}{1 \cdot 3} + \ldots$
$$\ldots + \frac{1}{(2k-1)(2k+1)} + \frac{1}{(2k+1)(2k+3)} =$$
$$= \frac{k}{2k+1} + \frac{1}{(2k+1)(2k+3)}$$
$$= \frac{k(2k+3) + 1}{(2k+1)(2k+3)}$$
$$= \frac{2k^2 + 3k + 1}{(2k+1)(2k+3)}$$
$$= \frac{(2k+1)(k+1)}{(2k+1)(2k+3)}$$
$$= \frac{k+1}{2(k+1)+1}$$

So the truth of T_k implies the truth of T_{k+1}. T_n is true for every positive integer n.

33. Let T_n: If $0 < a < 1$ then $0 < a^n < 1$.

Step 1: Clearly, T_1 is true for $n=1$.

Step 2: Assume $0 < a < 1$ implies $0 < a^k < 1$ i.e. assume T_k is true. Multiply $0 < a^k < 1$ by a to get $0 < a^{k+1} < a$. Since $a < 1$ and by transitivity, $0 < a^{k+1} < 1$.

So the truth of T_k implies the truth of T_{k+1}. T_n is true for every positive integer n.

35. Let $T_n: n < 2^n$.

Step 1: If $n = 1$ then $1 < 2^1$. So T_1 is true.

Step 2: Assume $k < 2^k$. Add 1 to both sides.
$$k+1 < 2^k + 1$$
$$< 2^k + 2^k$$
$$= 2 \cdot 2^k$$
$$= 2^{k+1}$$

So $k + 1 < 2^{k+1}$.
The truth of T_k implies the truth of T_{k+1}.
T_n is true for every positive integer n.

37. Let $T_n : 5^n - 1$ is divisible by 4.

    Step 1: If $n = 1$ then $5^1 - 1$ is divisible by 4.
    So T_1 is true.

    Step 2: Assume $5^k - 1$ is divisible by 4.
    Observe that $5^{k+1} - 1 = 5(5^k - 1) + 4$. Since sums of multiples of 4 are again multiples of 4, $5^{k+1} - 1$ is a multiple of 4.

    The truth of T_k implies the truth of T_{k+1}.
    T_n is true for every positive integer n.

39. Let $T_n : \sum_{i=0}^{n} x^i = \dfrac{x^{n+1} - 1}{x - 1}$, $x \neq 1$.

    Step 1: If $n = 1$ then $T_1 : 1 + x = \dfrac{x^2 - 1}{x - 1}$.
    So T_1 is true since $(x^2 - 1) = (x - 1)(x + 1)$.

    Step 2: Assume $T_k : \sum_{i=0}^{k} x^i = \dfrac{x^{k+1} - 1}{x - 1}$ is true.
    Add x^{k+1} to both sides.
    $$\sum_{i=0}^{k} x^i + x^{k+1} = \dfrac{x^{k+1} - 1}{x - 1} + x^{k+1}$$
    $$= \dfrac{x^{k+1} - 1 + (x - 1)x^{k+1}}{x - 1}$$
    $$= \dfrac{x^{k+1} - 1 + x^{k+2} - x^{k+1}}{x - 1}$$
    $$= \dfrac{x^{k+2} - 1}{x - 1}$$

    So the truth of T_k implies the truth of T_{k+1}.
    T_n is true for every positive integer n.

41. Let $T_n : (a^m)^n = a^{mn}$.

    Step 1: If $n = 1$ then $(a^m)^1 = a^m$.
    So T_1 is true.

    Step 2: Assume $(a^m)^k = a^{mk}$ i.e. assume T_k is true. So
    $$(a^m)^{k+1} = (a^m)^k a^m$$
    $$= a^{mk} a^m \text{ since } T_k \text{ is true}$$
    $$= a^{mk+m}$$
    $$= a^{m(k+1)}$$

    So the truth of T_k implies the truth of T_{k+1}.
    T_n is true for every positive integer n.

Review Exercises

1. One gets $a_1 = 2^0 = 1$, $a_2 = 2^1 = 2$, $a_3 = 2^2 = 4$, $a_4 = 2^3 = 8$, $a_5 = 2^4 = 16$

    First five terms are $1, 2, 4, 8, 16$

3. $a_1 = \dfrac{(-1)^1}{1!} = -1$, $a_2 = \dfrac{(-1)^2}{2!} = \dfrac{1}{2}$, $a_3 = \dfrac{(-1)^3}{3!} = \dfrac{-1}{6}$, $a_4 = \dfrac{(-1)^4}{4!} = \dfrac{1}{24}$.

    First five terms are $-1, \dfrac{1}{2}, -\dfrac{1}{6}, \dfrac{1}{24}$.

5. $a_1 = 3(0.5)^0 = 3$, $a_2 = 3(0.5)^1 = 1.5$, and $a_3 = 3(0.5)^2 = 0.75$.

    First three terms are $3, 1.5, 0.75$.

7. $a_1 = -3 + 6 = 3$, $a_2 = -6 + 6 = 0$, and $a_3 = -9 + 6 = -3$.

    First three terms are $3, 0, -3$.

9. Since $S = \dfrac{a_1(1 - r^n)}{1 - r}$ is the sum of a geometric series, $S = \dfrac{0.5(1 - 0.5^4)}{1 - 0.5} = 0.9375$.

11. Since $S = \dfrac{n}{2}(a_1 + a_n)$ is the sum of an arithmetic series, $S = \dfrac{50}{2}(11 + 207) = 5450$.

13. Since $S = \dfrac{a_1}{1 - r}$ is the sum of a geometric series, $S = \dfrac{0.3}{1 - 0.1} = \dfrac{0.3}{0.9} = \dfrac{1}{3}$.

15. Since $S = \dfrac{a_1(1 - r^n)}{1 - r}$ is the sum of a geometric series, $S = \dfrac{1000(1 - 1.05^{20})}{1 - 1.05} \approx 33,065.9541$

17. $a_n = \dfrac{(-1)^n}{n + 2}$

19. $a_n = 6\left(\dfrac{1}{6}\right)^{n-1}$

21. $\sum_{i=1}^{\infty} (-1)^{i+1} \dfrac{1}{i + 1}$

23. $\sum_{i=1}^{14} 2i$

REVIEW EXERCISES 243

25. Note, $a_n = a_1 r^{n-1}$. Since $256 = 4r^6$, $64 = r^6$ and common ratio is $r = \pm 2$.

27. $100 \left(1 + \dfrac{0.09}{4}\right)^{40} = \$243.52$

29. $\sum\limits_{i=1}^{10} 1000(1.06)^i = \dfrac{1000(1.06)(1 - 1.06^{10})}{1 - 1.06} = \$13,971.64$

31. If the pattern continues, the total number of Tummy Masters that could be sold is
$\sum\limits_{i=0}^{\infty} 100,000(0.9)^i = \dfrac{100,000}{1 - 0.9} = 1$ million.

33. $a^4 + \binom{4}{1}a^3(2b) + \binom{4}{2}a^2(2b)^2 + \binom{4}{3}a(2b)^3 + (2b)^4 =$
$a^4 + 8a^3b + 24a^2b^2 + 32ab^3 + 16b^4$

35. $(2a)^5 + \binom{5}{1}(2a)^4(-b) + \binom{5}{2}(2a)^3(-b)^2 + \binom{5}{3}(2a)^2(-b)^3 + \binom{5}{4}(2a)(-b)^4 + (-b)^5 =$
$32a^5 - 80a^4b + 80a^3b^2 - 40a^2b^3 + 10ab^4 - b^5$

37. $a^{10} + \binom{10}{1}a^9b + \binom{10}{2}a^8b^2 + \ldots =$
$a^{10} + 10a^9b + 45a^8b^2 + \ldots$

39. $(2x)^8 + \binom{8}{1}(2x)^7\left(\dfrac{y}{2}\right) + \binom{8}{2}(2x)^6\left(\dfrac{y}{2}\right)^2 + \ldots$
$= 256x^8 + 512x^7y + 448x^6y^2 + \ldots$

41. $\binom{13}{9} = 715$

43. $\dfrac{11!}{2!3!6!}2^3 = 36,960$

45. 24 terms

47. $5^9 = 1,953,125$ ways to mark the answers

49. $P(7,3) = 210$ possible three-letter 'words'

51. $3 \cdot 5 \cdot 4 = 60$ ways to place advertisements

53. $C(8,5) = 56$ possible vacations where order is not taken into account

55. (a) $C(8,4) = 70$ possible councils,
(b) $C(5,4) = 5$ possible councils from Democrats,
(c) $C(5,2) \cdot C(3,2) = 30$ possible councils with two Democrats and two Republicans

57. For KANSAS, $\dfrac{6!}{2!} = 180$ arrangements; for TEXAS, $5! = 120$ arrangements.

59. $2^7 = 128$ different families

61. Since there are 2^{10} ways to answer the test, the probabilities are
$P(\text{all 10 correct}) = \dfrac{1}{1024}$ and
$P(\text{all 10 wrong}) = \dfrac{1}{1024}$.

63. (a) 5/13, (b) 8/13, (c) 0, (d) 1

65. Note, $P(3 \text{ boys}) = \dfrac{1}{8}$. So odds in favor of 3 boys is $\dfrac{1/8}{7/8}$ i.e. 1 to 7.

67. Odds in favor of catching a perch is $\dfrac{0.9}{0.1}$ i.e. 9 to 1.

69. By the Addition Rule, $P(Math\ or English) = 0.7 + 0.6 - 0.4 = 0.9$ i.e. 90%.

71. $40,320$

73. 20

75. 1680

77. $\dfrac{8!}{2!6!} = 28$

79. $\dfrac{8!}{4!} = 1680$

81. $\dfrac{8!}{7!1!} = 8$

83. Let $T_n : 3 + 6 + 9 + \ldots + 3n = \dfrac{3}{2}(n^2 + n)$.

Step 1: If $n = 1$ then $T_1 : 3 = \dfrac{3}{2}(1 + 1)$.

So T_1 is true.

Step 2: Assume $T_k : 3 + 6 + ... + 3k = \dfrac{3}{2}(k^2 + k)$ is true. Add $3(k+1)$ to both sides and get
$3 + 6 + ... + 3k + 3(k+1) + 3(k+1) =$
$$= \dfrac{3}{2}(k^2+k) + 3(k+1)$$
$$= \dfrac{3(k^2+k) + 6(k+1)}{2}$$
$$= \dfrac{3k^2 + 9k + 6}{2}$$
$$= \dfrac{3}{2}(k^2 + 3k + 2)$$
$$= \dfrac{3}{2}((k+1)^2 + (k+1))$$

So the truth of T_k implies the truth of T_{k+1}. T_n is true for every positive integer n.

9. Since $a_n = a_1 + d(n-1)$, $9 = -3 + 8d$.
Solving for d, one gets $d = 1.5$.
So $a_n = -3 + 1.5(n-1)$ or $a_n = 1.5n - 4.5$.

10. $\dfrac{9!}{4!2!2!} = 3780$ possible nine-letter 'words'

11. Since $S = \dfrac{n}{2}(a_1 + a_n)$ is the sum of an arithmetic series, the mean daily sales is $\dfrac{1}{30} \sum_{i=0}^{29} 300 + 10i = \dfrac{1}{30} \dfrac{30}{2}(300 + 590) = \$445$

Chapter 8 Test

1. $a_1 = 2.3$, $a_2 = 2.3 + 0.5$,
$a_3 = 2.3 + 1$, and $a_4 = 2.3 + 1.5$.
First four terms are $2.3, 2.8, 3.3, 3.8$.

2. One finds $a_2 = \dfrac{1}{2} 20 = 10$,
$a_3 = \dfrac{1}{2} 10 = 5$, and $a_4 = \dfrac{1}{2} 5 = 2.5$.
First four terms are $20, 10, 5, 2.5$.

3. $a_n = (-1)^{n-1}(n-1)^2$

4. $a_n = 7 + 3(n-1) = 3n + 4$

5. $a_n = \dfrac{1}{3}\left(-\dfrac{1}{2}\right)^{n-1}$

6. Since $S = \dfrac{n}{2}(a_1 + a_n)$, the sum is $\dfrac{54}{2}(-5 + 154) = 4023$.

7. Since $S = \dfrac{a_1(1 - r^n)}{1 - r}$, the sum is $\dfrac{300(1.05)(1 - 1.05^{23})}{1 - 1.05} = 13,050.5997$.

8. Since $S = \dfrac{a_1}{1 - r}$, the sum is $\dfrac{0.98}{1 - 0.98} = 49$.

12. There are $10^4 = 10,000$ possible secret numbers. The probability of guessing the correct number in 3 tries is the probability of getting it in the 1st try plus the probability of getting it in the 2nd try plus the probability of getting it in the 3rd try. Assuming the person remembers what he tries and tries a different number after a failure, this probability is $\frac{1}{10,000} + \frac{9,999}{10,000} \cdot \frac{1}{9,999} + \frac{9,999}{10,000} \cdot \frac{9,998}{9,999} \cdot \frac{1}{9,998} = \frac{3}{10,000}$

13. At the end of the 300th month the value of the annuity is $\sum_{i=1}^{300} 700\left(1+\frac{0.06}{12}\right)^i =$
$\sum_{i=1}^{300} 700(1.005)^i = 700(1.005)\left(\frac{1-1.005^{300}}{1-1.005}\right) =$
$\$487,521.25$.

The cost of the house at the end of 25 years is $120,000(1.06)^{25} = \$515,024.49$.

No, they will not have enough money to buy the house.

14. $a^5 + \binom{5}{1}a^4(-2x) + \binom{5}{2}a^3(-2x)^2 +$
$\binom{5}{3}a^2(-2x)^3 + \binom{5}{4}a(-2x)^4 + (-2x)^5 =$
$a^5 - 10a^4x + 40a^3x^2 - 80a^2x^3 + 80ax^4 - 32x^5$

15. $x^{24} + \binom{24}{1}x^{23}(y^2) + \binom{24}{2}x^{22}(y^2)^2 + \ldots =$
$x^{24} + 24x^{23}y^2 + 276x^{22}y^4 + \ldots$

16. $\sum_{i=0}^{30} \binom{30}{i} m^{30-i}y^i$

17. Since $P(sum\ is\ 7) = \frac{6}{36}$, odds in favor of 7 is $\frac{6/36}{30/36} = \frac{1}{5}$ i.e. 1 to 5.

18. $C(12,3) = 220$ selections

19. $C(12,2) \cdot C(10,2) = 66 \cdot 45 = 2970$

20. $1/P(8,3) = \frac{1}{336}$ is the probability of getting the horses and the order of finish correctly.

21. Let $T_n : \sum_{i=1}^{n} \left(\frac{1}{2}\right)^i = 1 - 2^{-n}$.

Step 1: If $n = 1$ then $T_1 : \frac{1}{2} = 1 - 2^{-1}$.
So T_1 is true.

Step 2: Assume $T_k : \sum_{i=1}^{k} \left(\frac{1}{2}\right)^i = 1 - 2^{-k}$ is true.

Add $\left(\frac{1}{2}\right)^{k+1}$ to both sides.

$\sum_{i=1}^{k} \left(\frac{1}{2}\right)^i + \left(\frac{1}{2}\right)^{k+1} = 1 - 2^{-k} + \left(\frac{1}{2}\right)^{k+1}$

$= 1 - 2 \cdot 2^{-(k+1)} + 2^{-(k+1)}$

$= 1 - 2^{-(k+1)}$

So the truth of T_k implies the truth of T_{k+1}. T_n is true for every positive integer n.

Since $1 - 2^{-n} < 1$ for all positive integers n and by transitivity, $\sum_{i=1}^{n} \left(\frac{1}{2}\right)^i < 1$ for all positive integers n.